ELECTROPHYSIOLOGY

OF THE

CENTRAL NERVOUS SYSTEM

ELECTROPHYSIOLOGY

OF THE

CENTRAL NERVOUS SYSTEM

Edited by V. S. Rusinov

Institute of Higher Nervous Activity and Neurophysiology
Academy of Sciences of the USSR, Moscow

Translated from Russian by
Basil Haigh
Cambridge, England

Translation edited by
Robert W. Doty
Center for Brain Research
University of Rochester
Rochester, New York

℗ SPRINGER SCIENCE+BUSINESS MEDIA, LLC • 1970

ACKNOWLEDGMENTS

A number of papers in this collection were originally written in English; one was written in French. The authors of these papers were kind enough to provide us with a copy of their manuscript, thus allowing us to print them directly as written in English or as translated from French.

We extend our thanks to the following authors for having made their manuscripts available.

Papers submitted in English:

P. Andersen and T. Lømo, Mechanisms of Control of Pyramidal Cell Activity (p. 25)

M. A. B. Brazier, The Biologist and the Mathematician: A Necessary Symbiosis (p. 77)

P. Buser, G. Viala, L. Chertok, and G. Fontaine, Inhibition of the "Escape Reaction" in Rabbit: An Analysis of the Mechanism of Hypnotic Akinesia (p. 87)

R. W. Doty, On Butterflies in the Brain (p. 97)

E. Grastyán and L. Angyán, Electrical Correlates of the Direction of Movement Elicited by Subcortical Stimulation (p. 155)

Y. Hori, I. Toyohara, and N. Yoshii, Conditioning of Single Unit Activity by Intercerebral Stimulation (p. 167)

E. R. John, Current Problems in Electrophysiological Studies of Memory (p. 179)

D. P. Purpura, Functional Properties of Dendrites in the Mammalian Brain (p. 331)

Paper submitted in French:

A. Remond, Analysis of Information Concerning the Electrical Activity of the Brain (p. 349)

The original Russian text, first published by Nauka Press in Moscow in 1967, has been corrected by the editor for this edition. The present translation is published under an agreement with Mezhdunarodnaya Kniga, the Soviet book export agency. Since the authors of two papers that appeared in the Russian edition were planning independent publication of expanded versions of their material at the time the translation was being prepared, the following are not included in this volume: "Mechanisms for the Selection and Preservation of Acquired Stimulus—Response Patterns," by H. H. Jasper; "Alpha-Rhythms in Dogs," by W. Storm Van Leeuwen, A. Kamp, M. L. Kok, and A. M. Tielen.

В. С. РУСИНОВ

Современные проблемы электрофизиологии
центральной нервной системы

SOVREMENNYE PROBLEMY ELEKTROFIZIOLOGII
TSENTRAL'NOI NERVNOI SISTEMY

Library of Congress Catalog Card Number 69-12542
SBN 306-30381-7

ISBN 978-1-4684-1757-9   ISBN 978-1-4684-1755-5 (eBook)
DOI 10.1007/978-1-4684-1755-5

© Springer Science+Business Media New York 1970
Originally published by Plenum Press, New York in 1970
Softcover reprint of the hardcover 1st edition 1970

To M. N. Livanov

This collection commemorates the sixtieth birthday of Professor
M. N. Livanov, a Corresponding Member of the Academy of
Sciences of the USSR and a leading Soviet electrophysiologist, and
marks his completion of 35 years of scientific and teaching activities.

The name of Mikhail Nikolaevich Livanov is well known
among Soviet and Western specialists. He has developed Pavlov's
teaching along a completely new line: the study of higher nervous
activity in man and animals by electrophysiological methods;
his work has provided a deeper insight into the detailed mechanisms of conditioned reflex activity. Many Soviet and Western
workers have followed this path. Livanov's approach to his chosen
problems is characteristically analytico-synthetic, and he com-
bines this with the use of highly precise methods of investigation.
He was one of the first neurophysiologists in the Soviet Union to
investigate functional organization of the cerebral cortex at the
neuronal level, and as a result of this work he shed considerable
light on the mechanisms of higher nervous activity. He has put
forward an original hypothesis of the organization of functions in
systems of cortical neurons and has obtained facts confirming its
validity.

M. N. Livanov is unrelenting in his quest for the new. After
diligent and successful trials he invented an original method of
electroencephaloscopy capable of undertaking a detailed analysis
of potentials at many points of the brain simultaneously. By combining this method with computer techniques, fundamentally new
opportunities were presented for the detailed quantitative analysis
of the integrative activity of the brain. Livanov has used these
methods of investigation in clinical practice and in aerospace medicine. They have proved most fruitful in the field of psychiatry,
and have given a fuller and deeper understanding of the pathophys-
iological mechanisms underlying various nervous and mental
diseases.

Foreword

The most important yet the most difficult scientific task confronting man is how his brain produces his behavior and his subjective experience. The complexity of this problem is ineffably vast, exceeding by many orders of magnitude the theoretical and technical achievements concerning atomic energy or the exploration of space.

Unlike these areas of endeavor, neuroscience is fortunate in knowing no national rivalries, and its only secrecies are those of language. The latter, however, are often highly effective in concealing from workers in Los Angeles the discoveries of their colleagues in Moscow. A cogent example is provided in this volume by Roy John (p. 179) whose experiments proceeded for several years before he discovered the important body of data accumulated earlier by Prof. Livanov and his colleagues utilizing the same ingenious technique of the "tracer stimulus."

Reduction of such occurrences is certainly one of the goals of the present book, which now becomes a double translation, a dozen of the papers having originally been translated into Russian. One third of the 36 papers in this volume come from Prof. Livanov's immediate colleagues in the Laboratory of Higher Nervous Activity and Neurophysiology, Academy of Sciences of the USSR; one third from other colleagues in the USSR; and the remaining third from laboratories scattered throughout the world. It is particularly appropriate to honor Prof. Livanov by inclusion of six papers from his laboratory (by Aslanov; Gavrilova; Knipst; Korol'kova and Shvets; Rusinov and Grindel'; and Yanson et al.) concerning electroencephaloscopy as well as papers on this topic by the other two major exponents of this technique, Adey and Remond. The ability to make relatively simple, quantitative comparisons of activity concurrently in a great many areas of the brain will revolutionize both theoretical and clinical aspects of neuroscience, much as did the introduction of electroencephalography. Livanov has certainly led the way to this new era, and the exciting prospects for the future are unequivocally clear in these pages. We trust that Prof. Livanov will long continue such fertile and highly original contributions to knowledge of the brain.

R. Doty

Rochester, N.Y.
12 January 1970

Contents

Another promising direction of neurophysiological research has been his use for the first time in biology of experiments controlled by electronic computers, i.e., the provision of a feedback between computer and object. In this way an attack could be made on one of the fundamental problems in electrophysiology: elucidation of the functional significance of bioelectrical phenomena.

M. N. Livanov has proved a worthy representative of Soviet science abroad. He is permanent representative of the Soviet Union to the International Association of Electroencephalography and Neurophysiology, and member of the International Brain Research Organization (IBRO). In the Soviet Union, M. N. Livanov is a leading scientific administrator.

His creative power is at its height and he teems with new ideas and plans. We wish him good health and many years of life to the benefit of Soviet science.

The Organization of Cerebral Tissue in Transaction and Storage of Information*

W. R. Adey

Space Biology Laboratory
Brain Research Institute
University of California
Los Angeles, California

Studies of functions of the brain on a behavioristic basis, divorced in greater or lesser degree from associated activity in brain systems, has afforded the psychologist a broad field of endeavor in the past decade, with manipulation of schedules of reinforcement dependent on both peripheral and central stimuli. Although it has been claimed for these studies that they have yielded information about brain organization in perception and learning, with disclosure, for example, of specific brain regions that constitute "centers," this notion itself has little place in the physiologist's current scheme of cerebral functions. For whatever may be the apparent dominance of one brain region in the behavioral expression of a particular function, our enthusiasm must be tempered by knowledge of profoundly influential, wide-ranging and substantially reciprocal pathways involving such centers; of continuous activity in neuronal firing patterns related complexly and in stochastic ways to sensory stimuli; and of brain substance organized in a series of tissue compartments, each contributing functionally to the whole, and carrying with them the keys to that most unique function of brain tissue, the storage of information.

It may be argued that there persists a gulf that deepens and widens between physiological activity of any recordable kind, and mental activity in perception and learning. The physiologist discerns processes having varying degrees of correlation with men-

---

* Pages 324-340 in the Russian edition.

tal activity, and at this stage, he has barely begun evaluation of these correlates. He will concern himself with elucidation of their progressively finer patterns for some time to come. The hierarchical nature of these correlates, however, as they have already been disclosed from the level of the single cell to cerebral system interrelations in man, invites consideration of the possibility of causal significance in at least some of these physiological events (Adey, 1966).

We face the problem of what constitutes information at the input of cerebral systems, what are its transforms in transactional processes, and what are the bases of storage and recall. We have tended to assume that, at each stage of these presumably sequential events, the electrical activity of the brain might provide the sole, or at least an adequate measure of the state of tissue in which any or all of these processes might occur simultaneously. Yet, as will be discussed elsewhere, we do not yet know to which of a variety of electrical processes that occur simultaneously within cerebral nerve cells, and in certain respects are apparently unique to cerebral neurons, we may attach special significance in initiation of storage, as opposed to transaction, of information in brain tissue. It is also quite conceivable that the changing states of brain tissue deriving from storage of information may not be reflected in electrophysiological records, except during information retrieval. A gamut of new techniques quite unrelated to classical electrophysiological recording methods may thus be necessary to detect such changes (Adey et al., 1962b; Adey et al., 1963; Adey et al., 1965; Adey et al., 1966).

Our confrontation of such problems here will concern itself with the salient features of a structural model of cortical tissue, and with the genesis of electrical waves in cerebral tissue, and their relationship to cellular firing. We will consider briefly aspects of EEG wave activity accompanying conditional behavioral responses in animals, and the development of a "normative library" of EEG patterns accompanying sleep, wakeful states, and conditional behavior in man. We will examine the electrical impedance characteristics of cerebral tissue and the occurrence of regional differences in responses to alerting, orienting, and conditional stimuli. Finally, it is proposed to evaluate the basis for these impedance phenomena in the frame of altered conductance in perineuronal elements, including a recently disclosed macromolecular "glue" of mucopolysaccharides and mucoproteins that forms a series of well organized coats integral with neuronal and neuroglial membranes, and which may control the passage of ionic material across the neuronal membrane in a fashion determined by their ion-binding abilities, as discussed below. The discussion thus leads to consideration of a model of cerebral organization in learning, in which the baffling complexity of the tiny cosmos of the neuronal membrane, and the narrow clefts that mark off each neuron from adjacent neurons and neuroglial cells, may all play a part in the transaction and storage of information.

A TRICOMPARTMENTAL MODEL OF CELLULAR ORGANIZATION IN CEREBRAL STRUCTURES

Recent electron microscopy studies of brain tissue have revealed a series of intimate interrelations between cells of neuronal and non-neuronal systems, and have required a reappraisal of structural organization within organized populations of neurons in a domain of cerebral tissue. The laminar pattern said to characterize nerve cells in mammalian cerebral cortex (Sholl, 1956) is also the basis of a unique feature seen also in cerebral ganglia of many invertebrates, in the great overlap of dendritic branches, or "tree" of one neuron with those of adjacent neurons. Electron micrographs show that, in some instances, these contacts between dendrites of different neurons are as close as in synaptic junctions, or about 100 Å (van der Loos, 1962; Rall et al., 1966). Although functional interactions which might be initiated in this way are unknown, it is necessary to take account of such structural relations, since they may determine the degree of "coupling" between a neuron and others in its vicinity, in their mutual interaction in slow wave processes described below.

Neuroglial cells intervene between nerve cells and the vascular apparatus, and may thus exercise a regulatory function on neuronal metabolism. The neuroglial envelope has been described by some authors as essentially complete around individual neurons, while others have described a more restricted "packeting" of glial elements around synaptic terminals with relatively bare intervening areas (Peters and Palay, 1964). By cytochemical techniques combined with electron microscopy, it has been possible to display the distribution of a variety of enzymes in the interfaces between these cellular elements. These enzymes lie at adjacent membranes where neuronal and neuroglial cells are in contact, but not where neuronal elements adjoin each other (Barrnett, 1963).

In addition to neuronal and neuroglial compartments, there is a third zone, the extracellular compartment, traditionally given scant attention in most histological descriptions. The role of this compartment may be highly significant in the modulation and control of ionic movements between neuronal and neuroglial elements. It would be easy to dismiss this compartment as a mere bucket of saline, if our notions of cerebral organization required it merely to contribute sodium ions, in accordance with certain popular neuronal models. Attention, however, has recently been focused on its content of mucoprotein and mucopolysaccharide material (Pease, 1966). Considerable organization may be exhibited by these large molecules. Mucopolysaccharides have been shown to be chemically disordered in the types and location of their sugar molecules in mental disorders, such as infantile amaurotic idiocy (Barker et al., 1962). Susceptibility of these macromolecules to volume changes in the presence of divalent cations, such as calcium, has been emphasized by Katchalsky (1964), so that their ion-binding ability and capacity to retain water in a variably structured molecular lattice (Bennett, 1963; Brandt, 1962) may bear directly on the transfer of ions across the neuronal membrane, and also on the size of the extracellular space.

The volume of the extracellular compartment has been estimated at quite small values (between 1 and 4%) in material fixed with osmic acid, but van Harreveld, Crowell and Malhotra (1965) have shown that brain tissue rapidly cooled to -207°C shows as much as 24% extracellular space. Failure to cool rapidly, and particularly, allowing an interval of 90 seconds or more to pass after circulatory arrest and before fixation, leads to obliteration of the extracellular space, and transfer of fluid to an intraglial location. Chemical estimates of the extracellular space (Reed et al., 1964) have indicated about 14% of cerebral volume, so that the figure derived from osmium-fixed electron micrographs may be too small.

In summary, these recent studies have emphasized the unique nature of cerebral cellular organization, with emphasis on a dendritic tree of substantially greater volume than the cell body, and overlapping and perhaps physically contiguous with those of neighboring neurons; a neuroglial compartment intimately involved in neural metabolic exchanges; and an extracellular compartment disposed between neuronal and neuroglial elements and characterized by a macromolecular "glue."

NEUROELECTRIC ASPECTS OF CEREBRAL ACTIVITY: THE GENESIS OF WAVES AND THEIR RELATION TO UNIT FIRING

Early hopes that the phenomenon of alpha blocking in the EEG, or similarly obvious changes from sleep to wakefulness, might have their counterparts during finer perceptual and learned performances have not borne fruit. Unit firing patterns recorded simultaneously with wave activity in the same tissue have failed to disclose constant phase relations between the waves recorded grossly and the unit firing, even in the very regular waves of hippocampal theta trains, thus giving no support to the notion that the waves arise as the envelope of the firing of many neurons in a population (Li et al., 1957; Green et al., 1961; Jasper and Stefanis, 1965).

Since, therefore, the slow wave-like processes of the EEG may be generated in this form in individual neurons, Elul (1962) has examined the spontaneous EEG recorded between paired microelectrodes at progressively decreasing distances. With tip separation of the microelectrodes of only 30 mμ, he recorded a spontaneous EEG within the cortex of apparently normal form and amplitude. Moreover, no similarity could be found between monopolar and differential derivations of either spikes or slow waves recorded with such small tip separations, suggesting that the dimensions of cortical dipoles generating the EEG are limited to distances of cellular magnitude. More recently, Elul (1965a, 1965b) has found large rhythmic waves in unanesthetized cortical neurons by intracellular recording, with an amplitude of 5 to 15 mV, thus involving a major portion of the membrane potential of 70 to 90 mV, and hundreds of times larger than the EEG recorded in adjacent extracellular tissue, or on the cortical surface (Fig. 1).

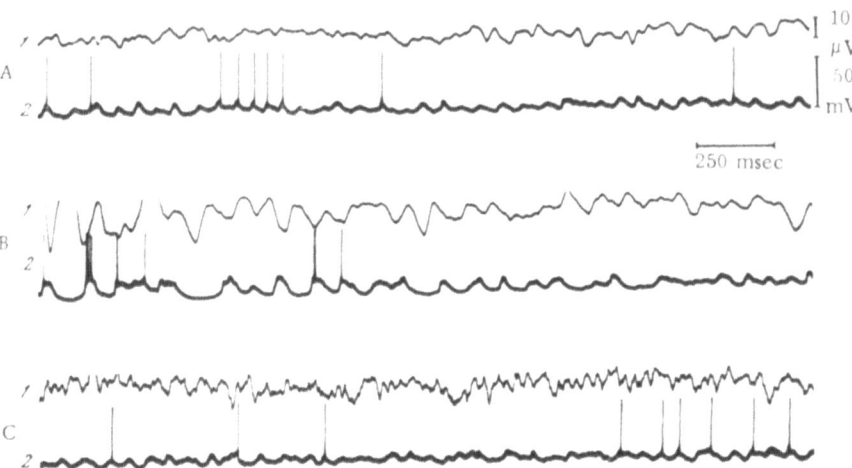

Fig. 1. Examples of large intracellular waves and simultaneous EEG records from cortical surface in sleeping cat (A and B) and faster intracellular and surface records when awake (C). (From Elul, 1965.)

Similar findings have been reported by Creutzfeldt et al. (1964), Fujita and Sato (1964), and Jasper and Stefanis (1965).

In our laboratory, Elul has found a strong general relationship between frequencies in these intracellular waves and the EEG recorded grossly from the same region. In sleep with large regular slow waves, similar patterns characterize intracellular records, and the rapid surface EEG in the aroused state is accompanied by similarly increased frequencies in intracellular waves (Elul, 1965a, 1965b). Spectral analysis techniques applied by Elul to the EEG and to the intracellular waves have confirmed essentially identical spectral density contours across the spectrum from 1 to 30 cps, and have also revealed a virtual absence of coherence, or linear predictability, between the two wave trains, when coherence is measured over successive ten-second epochs for many hundreds of seconds (Elul, 1966). In modeling the contribution of individual cellular generators to the configuration of the EEG, Elul has applied the central limit theorem of statistics (Cramer, 1962), and considers that cortical neuronal generators meet the theorem requirements of individual amplitude distributions, which are nonlinearly related, and possess a mean, and a finite standard deviation (Loève, 1955). He concludes that the EEG may be accounted for as the normal distribution ensuing from combination of activity of nonlinearly related neuronal generators.

Relationships of Intracellular Waves to Firing of the Neuron

It was noted in our studies that firing of the neuron occurs near, but not necessarily at, the peaks of the depolarizing phase of the intracellular wave. However, we have consistently observed that the level of depolarization reached on the depolarizing peaks of these waves is not a critical determinant of the initiation of firing. In many instances, depolarizing peaks exceeded those on which firing occurred without initiation of a spike, suggesting that the relationship between intracellular waves and the spike output may not always be a linear one (Elul and Adey, 1965). These findings suggest that either the spike-triggering zones may be multiple, or that, if the spike discharge occurs only from a single restricted zone in the vicinity of the axon hillock, then the internal organization of the neuron may involve selective internal current paths to the spike-triggering zone, with different abilities in various parts of the neuronal membrane to influence current flow through the membrane of the axon hillock.

Unit Response Patterns During Classical Conditioning

On the basis of the previous discussion of neuronal activity in both slow processes and simultaneous pulse coded firing, we may turn to conditional firing patterns in cerebral neurons. Medial thalamic units can be repeatedly extinguished and retrained during classical conditioning (Kamikawa et al., 1964). The paradigm involved pairing a light flash as a conditional stimulus with an unconditional shock train to the sciatic nerve (Fig. 2). These cells exhibited considerable plasticity in such a paradigm, with gradual development of firing patterns which might be the converse of those initially elicited. A series of "extinction" trials followed each group of 50 "training" trials, and showed progressive rebound phenomena, so that over a period of several hours, it was possible to see here also a series of gradual changes that increased in magnitude, as well as showing qualitative differences from those in the first extinction trial.

Two salient features emerged from this study which offer small clues to an understanding of the processes that might under-

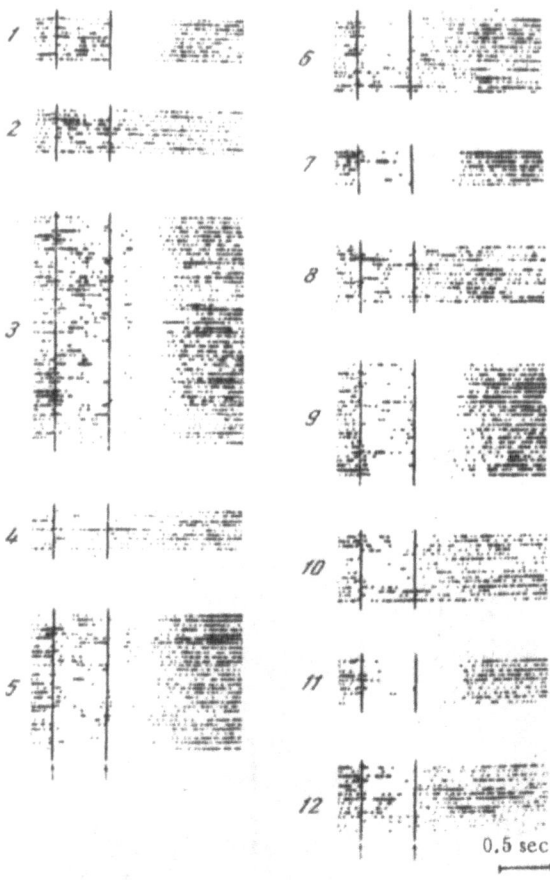

Fig. 2. Development of an inhibitory conditional response in an habenular unit. Each dot represents a unit
discharge, and each horizontal row of dots, a single trial.
Trials are grouped according to stimulus conditions, A:
sciatic nerve stimulation only (US) control; B: flash only
(CS) control; C: flash and sciatic stimulation (first sequence of training trials, etc.). Left vertical line (marked
flash) indicates time of CS presentation. Vertical line
(marked sciatic) indicates time of onset of US. (From
Kamikawa et al., 1964.)

lie the lasting structural changes associated with a memory trace. Tests of conditional-unconditional stimulus (CS-US) intervals in the range 200 to 800 msec indicated that intervals shorter than 300 msec failed to elicit a conditional response. With training trials given every 10 sec, a change in firing in the CS-US interval characteristic of a conditional response required a minimum of about 50 trials, presented over a period of about 20 minutes. The requirement for a minimum CS-US interval of the order of 300 msec suggests a time scale comparable with electrophysiological events at the neuronal membrane, such as the prolonged inhibitory postsynaptic potentials seen in cortical neurons (Andersen et al., 1963), lasting up to several hundred milliseconds, and without an equivalently persistent counterpart in spinal motoneurons; or the wave processes in cortical neurons described above (Elul and Adey, 1965). A simple origin for the latter in synaptic potentials remains uncertain at this time. On the other hand, the requirement that training trials have a time course of 20 min to generate a conditional response suggests a much slower process, perhaps the synthesis of a macromolecule, protein in nature, and located at the neuronal membrane, where it might influence directly the excitability of the cell by synaptic volleys. The possible role of a conditioned synaptic input, which is both excitatory and inhibitory, and is a dynamic process as proposed by Roitbak (1960), has been discussed elsewhere (Adey, 1966).

Relationships of EEG Wave Trains
to Conditional Responses

We may examine, as indices of learning, alterations in spontaneous rhythms, including induction of "activated" EEG patterns and those characterized by rhythmic slow waves; or alterations in evoked potentials following either single stimuli or trains of repetitive stimuli. It may be remarked at the outset that these procedures probably do not rank equivalently in our attempts to unravel the electrical signal patterns correlated with learning. Spontaneous rhythms and the changes induced in them in the course of conditioning may be presumed to relate in varying degree to the processing of information arriving in an essentially continuous fashion from environmental stimuli. On the other hand, environmental stimuli to which conditioning occurs do not typically possess rhythmic characters, nor is the environment normally perceived as a

series of tachistoscopic impressions. In the frame of a communication system, these brief, iterative signals constitute transients
in a system already occupied in varying degree with processing of
information on a continuous basis. As discussed in detail elsewhere (Adey, 1966), the arrival of these transients in a "signal
space" so occupied might reflect in evoked potential configurations
only residual availability of "space" not occupied in the primary
processing of information essential to the conditional process.

The characteristics of "spontaneous" EEG rhythm changes
occurring in cortical and subcortical structures in classical conditioning and in visual discrimination have been extensively re -
viewed elsewhere (Adey, 1966). Studies of the activated EEG in
classical conditioning have contributed to the view that the desyn-
chronizing response represents a stereotype that can scarcely be
pursued further in finer analysis of correlates with behavioral responsiveness and conditioning. Yet, it has long been recognized
(Iwama, 1950) that certain manipulations of a classical condition -
ing procedure leads to trains of synchronous slow waves, as when
the CS-US interval was prolonged, or in cortical areas surrounding a focal activated response. Recognition that a synchronous,
rather than activated, pattern of cortical activity accompanies
certain aspects of classical conditioning has led to a finer analysis of distribution of wave activity in cortical and subcortical
structures regularly occurring in certain operant performances,
and to extensive computer analyses of their patterns (Adey et al.,
1961; Adey et al., 1962c; Adey and Walter, 1963; Walter and
Adey, 1963, 1965; Walter, 1963).

Characteristics of EEG Activity in Orienting
and Discriminative Behavior in the Cat. Attention
will be focused here on the allocortical structures of the hippocampal system in the laying down of the memory trace (Baldwin and
Bailey, 1958; Adey, 1959; Gastaut and Lammers, 1961; Alajouanine,
1961). The dramatic memory defects resulting from interference
with temporal lobe structures (Klüver and Bucy, 1939) have been
comprehensively reviewed by Drachman and Ommaya (1964), who
conclude that medial temporal lobe damage is associated with loss
of retention and impairment of acquisition, rather than with impaired short-term memory. The memory trace may be laid down
outside the hippocampal system (Penfield, 1958), but integrity of

its interrelations with these seemingly unrelated cortical and subcortical regions may be vital to the appropriate recall of previously learned discriminative habits (Adey et al., 1962c). Moreover,
many classical conditioned reflexes in the cardiovascular and res-
piratory systems persist after disruption by large diencephalic
lesions of major connections between the cerebrum and more cau-
dal levels of the brainstem (Doty et al., 1959).

Attempts to find electrophysiological correlates of the
orienting reflex led Grastyàn and his colleagues (Grastyàn, 1959;
Grastyàn et al., 1959) to postulate a specific relationship between
hippocampal theta wave trains and orienting behavior. However,
the exquisite plasticity of hippocampal theta rhythms in changing
behavioral states, including the appearance of bursts of waves in a
narrow spectral range during performance of a visual discriminative task, have suggested more subtle and specific relations to
discriminative functions and judgment capability (Adey, 1964;
Adey et al., 1960; Adey et al., 1962a; Adey et al., 1962c).

Radulovacki and Adey (1965) found it possible to distinguish
hippocampal EEG activity in three basic states in the cat: in alert
but nonperforming animals, in the course of discriminative performance, and during orienting behavior. Alert but nonperforming animals exhibited a wide spectrum of "theta" waves in the
range 3 to 7 cps on first introduction into the test situation, without overt aspects of orienting behavior. This activity persisted
in EEG epochs between discriminative and orienting trials throughout many months of training. During T-box discriminative performance, theta waves regularized at 6 cps, and appeared as a regular waveform in the computed daily average. Computed averages
in orienting trials, given in the same numbers on each test day and
randomly interspersed with the discriminative trials, showed
slower and less regular averages at 4 to 5 cps.

Single doses of lysergic acid diethylamide (LSD-25) were
followed by prolonged disinhibition of inhibited orienting behavior,
and by gradual appearance of a regular EEG average during orien-
tations 5 to 10 days after the drug, and declining after 15 to 20
days, concurrently with the decline of orienting behavior. A similar but accelerated series of behavioral and EEG changes were induced by a psychotomimetic cyclohexamine, CL-400. Our studies indicate that in the cat, at least, it is necessary to take ac-

count of hippocampal wave trains with characteristic features that relate in clear and specifiable ways to the performance of a discriminative task, and in different, but equally recognizable patterns, to aspects of orienting behavior.

This exquisite sensitivity of neuroelectric processes in the hippocampus to subtle shifts in cerebral states, and indications that hippocampal theta activity during discrimination has the characteristics of a "pacemaker," with fragmentary and less regular rhythms in the midbrain reticular formation, subthalamus and primary sensory cortical areas, have suggested the deposition of a "memory trace" in extrahippocampal systems, and subsequent recall on the stochastic re-establishment of similar wave patterns (Adey and Walter, 1963).

These hypotheses based on animal studies have led us to the more challenging task of seeking comparable regularity of EEG patterns in man, as concomitants of task performances and also in general relation to behavioral states.

Analysis of a Baseline EEG Library and Development of Simple Pattern Recognition Techniques. It has long been a matter of concern that definition of EEG patterns has rested, not only on the subjective opinion of the investigator, but also on wide individual variations in apparently normal subjects. We have, therefore, sought to establish by computer analysis the presence of common EEG factors in a significant population of young adult males, both in relation to task performances and in assessment of sleep states.

In detailed studies to be reported elsewhere (Walter et al., 1966), a series of 200 astronaut candidates were tested in perceptual and learning tasks by means of a programming device developed in our laboratory, and using a magnetic tape command system to ensure accurate and identical timing of presentations for all subjects. Subject testing and EEG recording were performed by Dr. P. Kellaway and R. Maulsby at the Methodist Hospital, Houston. Physiological data were recorded on magnetic tape, together with the command signals, for subsequent computer analysis. These data constitute a "normative library," and include not only 18 EEG channels from all scalp areas, but also the electrooculogram, electrocardiogram, galvanic skin responses, and respiration.

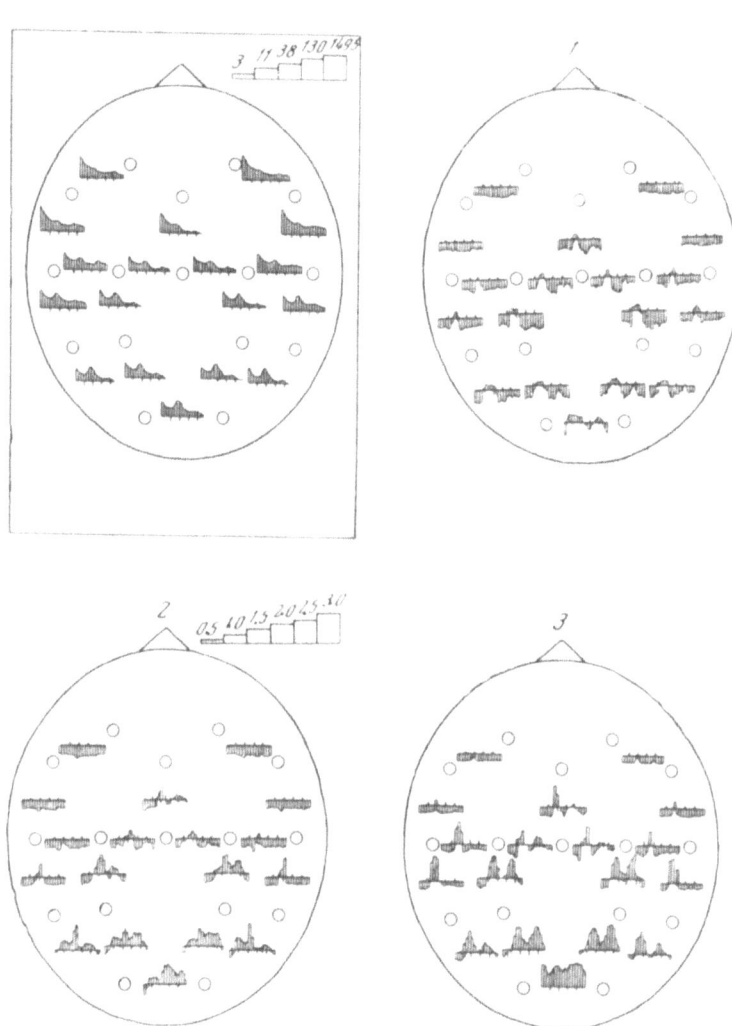

Fig. 3

Caption and continued figure on p. 14

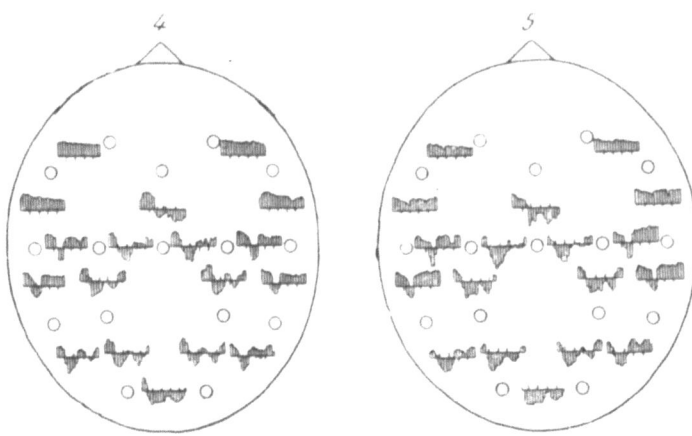

Fig. 3. Spectral analyses from 50 astronaut candidates, pooled into
averages for each scalp location (see text). Top left figure is average for all subjects over 12 situations. Bars cover spectrum from 1
to 25 cps at intervals of 1 cps. The average over 12 situations was
used as the mean for measurements of variance at each frequency
in the five situations shown for the population of 50 candidates.
(From Walter et al., 1966.)

Intensive spectral analyses were performed on 50 subjects
randomly chosen from the total of 200. The scope of the analysis
may be judged from the requirement for 25 hr of main computation
time, wherein multiplications were performed at approximately
500,000/sec. This comprehensive analysis appears well justified,
since it has allowed selection of variables for a possible on-line
system that would be far less demanding in computer requirements.

In spectral methods of analysis, functions relating intensity
to frequency in any one lead are classified as autospectra, whereas
crossspectra describe shared intensities across a band of frequencies (Walter, 1963). Problems in the electronic design of analog
filters with appropriately narrow skirt characteristics have led to
the development of digital filters, in which the digital computer
provides weighting factors by which the time function is multiplied.
The sum of these products is taken as the output of the digital fil-
ter. The application of a set of digital filters to a function of time
can be viewed as a discrete version of a Fourier transform (Wal-

ter, 1963; Adey, 1965), with substantial superiority over analog methods in our capacity to precisely specify the bandpass characteristics in the low frequency range between 0.5 and 10 cps.

Since its phase shift is zero, it has become possible to measure for the first time the phase relations between two EEG wave trains at each frequency across the spectrum, as well as shared amplitudes between them at each frequency. This has led us to the calculation of the coherence function as a measure of statistical variability in linear interrelations between brain regions. Its magnitude may be expressed as $coh(f) = MAGS(f)/ASX(f)\ ASY(f)$, where $MAGS(f)$ is the mean cross spectral magnitude at frequency f and $ASX(f)$ is the autospectrum of X and $ASY(f)$ the autospectrum of Y, at the respective frequencies. The coherence function is a measure of the linear predictability of activity in any area, on the basis of knowing the activity in any other area, or series of other areas. Its value in pattern recognition procedures will be discussed below.

To synthesize the data from the 50 subjects, an averaging procedure was adopted on the spectral outputs from all subjects in the various test situations. Epochs in different levels of sleep were separately averaged. These averages were made for each scalp region, and displayed as a series of bar graphs covering the spectrum from 0 to 25 cps. First an average was prepared of spectral densities at each recording site for all test epochs, including sitting with eyes closed at rest, eyes closed during 1/sec flash stimuli, during an auditory vigilance task, during visual discriminations at 3 second intervals, and a similar series of more difficult discriminations at 1 second intervals (Fig. 3, top left).

The contours of these "lumped," or collected spectra were then used as a mean for comparison with the spectra for the individual situations. The subsequent graphs in Fig. 3 thus show the variations about the mean established by the average over 12 situations shown in the top left figure. It will be seen that such a display clearly separates spectral density distributions for the 50 subjects in the five situations shown. In particular, the distributions for more difficult visual discriminations in one second (Fig. 3. 5) already characterize discriminations made in 3 seconds (Fig. 3. 4). Moreover, further analysis, taking account of coherence functions between different scalp leads, allows sharp separa-

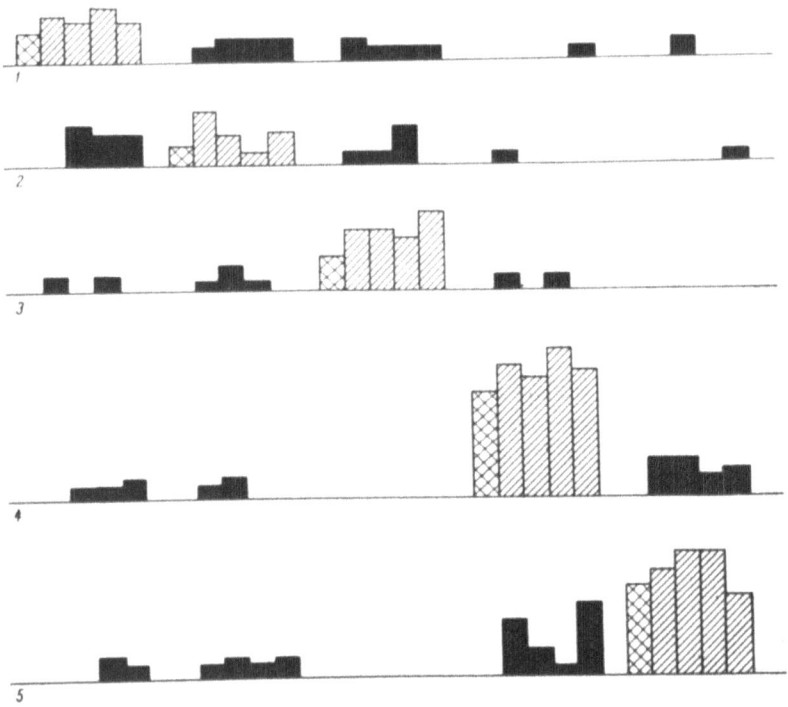

Fig. 4. Step-wise discriminant analysis as applied to the spectral outputs of EEG records in five situations: eyes closed, resting (EC-R), eyes open, resting (EO-R), during performance of auditory vigilance task with eyes closed (EC-T), performance of a visual discrimination in 3 sec (EO-T-3), and performance of a similar more difficult task in 1 sec (EO-T-1). The cross-hatched bars are for the simultaneous calculation of the most significant variables in 4 subjects, and the striped bars for each subject individually. (From Walter et al., 1965.)

tion of these two situations in an automated pattern analysis (see below). It is also possible to compare an individual with the mean for the group, or with his own mean, using a two-color display technique.

Discriminant analysis was applied to these spectral outputs in four subjects (Walter et al., 1965), covering five situations: eyes closed at rest, eyes open at rest, an auditory vigilance task, and the two visual discriminative tasks described above (Fig. 4). A computer program attempted to assign each EEG segment to the situation from which it came, using measurements derived from four EEG channels: left and right parieto-occipital, vertex, and bioccipital. Activity in each channel was analyzed into four frequency bands covering the spectrum from 1.5 to 25 cps. In each band, measurements were made of the strength of activity in each channel, mean frequency within the band (the dominant frequency when present), band width within the band (an expression of the regularity of the dominant frequency), and the coherence between pairs of channels.

The four variables which best distinguished among the five situations were: left parieto-occipital alpha intensity, the mean frequency of theta-band activity in the vertex, the coherence in the theta band between left parieto-occipital and vertex leads, and coherence in the beta band between vertex and bioccipital leads.

The separate analysis of each subject's records in the same way yielded a higher proportion of correct classifications than in the group analysis, with his own best four measurements, between 62 and 69% of a single subject's samples were correctly classified, as contrasted with 51% for the subjects simultaneously. An even greater disparity appeared after 15 measurements were selected. Individually, 95, 93, 96, and 90% were correct, while for the subjects together, only 65% were correct (Fig. 4). It would appear that each subject may have a spatially and numerically characterized individual EEG "signature," as to which measurements are most effective in distinguishing different situations (Walter et al., 1965).

FOCAL MEASUREMENTS OF ELECTRICAL
IMPEDANCE IN DIFFERENT CEREBRAL
STRUCTURES IN ALERTING, ORIENTING
AND DISCRIMINATIVE RESPONSES

Measurements were made in the hippocampus, amygdala, and midbrain reticular formation during alerting, orienting, and discriminative performances in the cat (Adey et al., 1966). Measurements were made in focal volumes of approximately 1.0 mm^3 at 1000 cps with coaxial electrodes, with current densities of $10^{-13}/\mu^2$ of electrode surface.

In the fully trained animal, computed daily averages of hippocampal impedance decreased by as much as 8% of baseline during

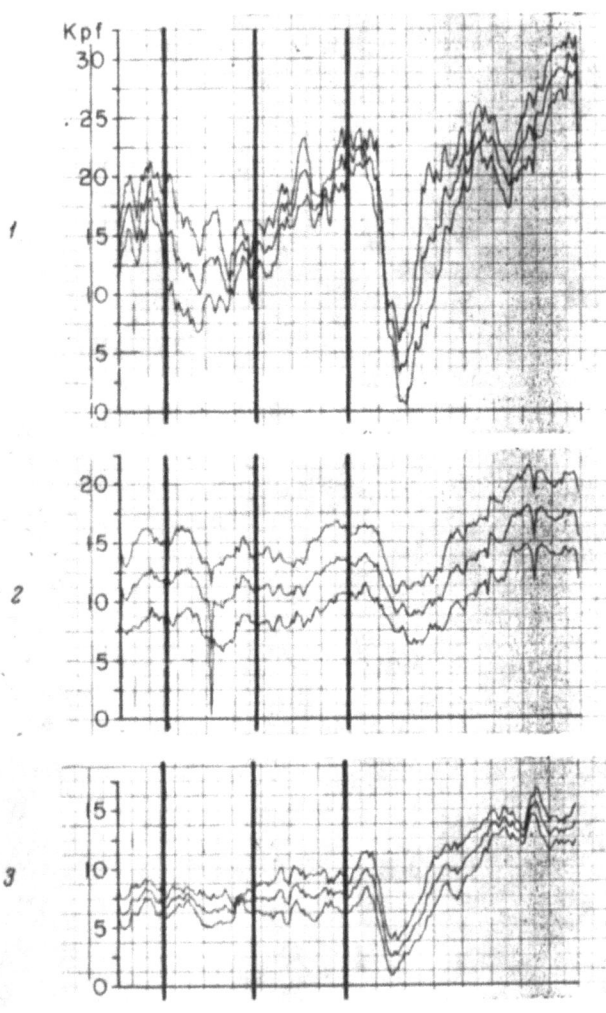

Fig. 5A

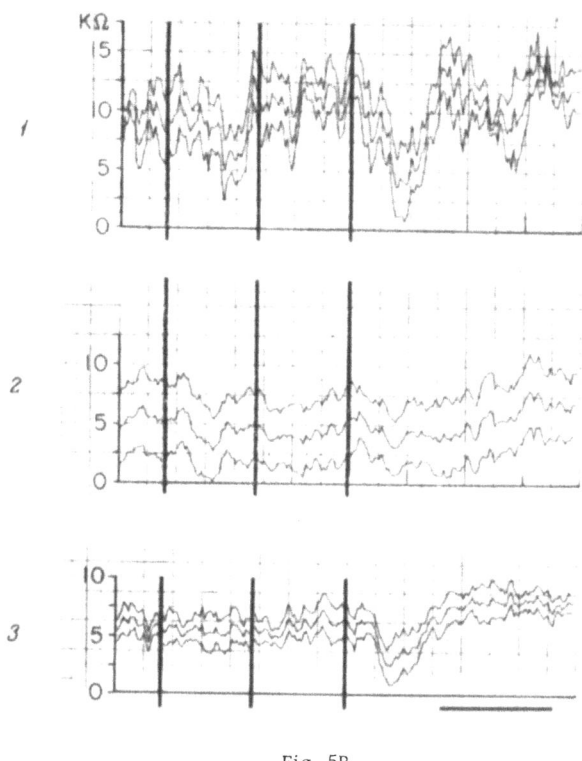

Fig. 5B

Fig. 5.   Calculations of means and variability in hippocampal
impedance over 5-day period at various levels of training. In
each graph, the middle trace indicates the mean, with upper and
lower traces showing one standard deviation from the mean. Cal-
ibrations indicate 50 pF, with mean baseline at 11.1 kpF throughout the training maneuvers; and 100 Ω, against a mean baseline
of 16.0 kΩ for the same period. Variability was low at 100%
performance (1), increased substantially immediately after cue
reversal (2), but decreased again after retraining (3). (From
Adey et al., 1966.)

visual discrimination, whereas immediately preceding alerting
and orienting responses were not accompanied by comparable impedance changes. Similar measurements in the rostral midbrain
reticular formation showed small responses during orientation and
discrimination, and less constantly during alerting responses. The
amygdala exhibited consistent responses only in the alerting epoch.
The magnitude of the responses in hippocampus and midbrain increased with the level of behavioral performance.

Unilateral visual cortical resection, leading to retrograde
loss of about 80% of lateral geniculate neurons, was followed by
perturbations in geniculate impedance baseline from 10 to 30 days
postoperatively. Subsequently, responses to a cyclohexamine drug
were reduced in the degenerated nucleus to about 20% of those in
the intact nucleus. These findings support the view that removal
of the majority of neurons from a cerebral nucleus greatly diminishes impedance responses to pharmacological manipulations, in
agreement with previous findings of diminished responsiveness to
alcohol in the cat, and decreased effects of hypocapnea in sclerotic
human hippocampal tissue (Porter et al., 1964).

Thus, the differential location of the impedance responses in
the course of a repertoire of alerting and learned responses, their
dependence on levels of learning, and their decrease in variability
at high performance levels all support an origin in the intrinsic
characteristics of cerebral tissue. They do not arise in simple
relationship to alterations in blood pressure, cerebral blood flow,
or brain temperature (Adey et al., 1965; Porter et al., 1964). The
relative conductivity of neurons, neuroglial cells, and intercellular
fluid has been reviewed elsewhere (Adey et al., 1966).

The extracellular space offers a high conductance pathway,
estimated to be at least one thousand times higher than through
neuronal membranes. Neuroglial conductance has been variously
estimated at values similar to extracelluar fluid (Hild and Tasaki,
1962), to substantially higher levels approaching neuronal membrane resistance (Kuffler and Potter, 1964). If the extracellular
space is the site of these impedance changes, a substrate must be
sought in a temporary and presumably reversible movement of
ions into it, a movement presumably initiated in neuronal elements, but capable of modulation by neuroglial cells, or of influencing neuroglia. Movements of ions in the extracellular space

will be substantially modified by the presence therein of macromolecules exhibiting ion-binding and fixed charge characteristics (Katchalsky, 1964; Weiss, 1964), so that it is not possible to model ionic behavior in the perineuronal environment from consideration of a mere aqueous solution permitting unimpeded flow. Identification of macromolecular material in electron micrographs of kidney and intestinal tissue led to a search for comparable intercellular substances in central nervous tissue (Pease, 1966). In sections of material prepared in ethylene and propylene glycol, and in the absence of calcium salts, Pease has disclosed substantial amounts of material in intercellular clefts in cerebral cortex not revealed in classical electron micrographs prepared from fixed material, and in staining strongly with phosphotungstic acid.

Such macromolecular polycarboxylic acids are susceptible to precipitation by alkaline earth cations. Remarkable shrinkage in volume is induced by exchanging part of the sodium counterions at the membrane surface for calcium ions. Neurons in tissue culture and dissected fragments of neuronal membrane both exhibit electrokinetic effects in the presence of a focal emf, in ways indicative of fixed negative membrane charges and apparently related to adherent macromolecular layers (Elul, 1965a, 1965b). Our current studies indicate that as little as 40 mequiv of intraventricular calcium solution changed impedance in structures adjoining the ventricle by as much as 25% for several hours. Bennett (1963) has termed these macromolecular envelopes "glycoalyces," and has suggested for them a direct role in determining differential entry of sodium and potassium to positions close to the plasma membrane.

Although the impedance responses seen here may well have origins in an intercellular fluid containing a matrix of macromolecules, assignment of such a role exclusively to an interface between neuronal elements and a series of glycoalyces may substantially oversimplify interchanges to which the neuroglial compartment may also contribute (Nicholson, 1966), either by direct changes in neuroglial membrane resistance, or by modulation of conductance in intercellular fluid (Nicholls and Kuffler, 1964). Location of enzymes by cytochemical techniques at neuronal–neuroglial and glial–glial interfaces (Barrnett, 1963) indicates that, if significant metabolic transactions occur in this way, they do so across a space in which the macromolecules may play a role, not only in modulation of ionic flow, but also in the coding of chemical signals that would underlie long-lasting changes in chemical structure of neuroglial cells interfacing with neuronal elements.

ASPECTS OF A MODEL OF CEREBRAL ORGANIZATION IN LEARNING

This study has emphasized the structural characteristics of cortical tissue that might underlie mechanisms of information storage: the palisade of cells in close proximity to one another; and the dendritic tree of one cell substantially overlapping that of one or more adjacent neurons. Concomitantly, wave-like electrical activity characterizes this tissue, and has its origins in an intracellular wave process many millivolts in amplitude, and apparently arising in the dendritic tree as an electrotonic phenomenon. This wave process has not been detected in neurons in other parts of the central nervous system, such as the spinal cord. Moreover, a propagated spike is not a regular concomitant of the depolarizing phase of the intracellular waves, even where this exceeds the threshold level for firing in some cases. The intracellular wave and initiation of a propagated impulse thus appear to involve processes that may occur in parallel in individual neurons, and may bear nonlinear interrelations to one another. As discussed above, computer analysis of the gross wave activity has established patterns relating to a wide variety of behavioral performances, and at the level of the individual neuron, the separate generators have been shown to meet the requirements of the statistical theorem of central limits, behaving essentially independently, and with gross EEG arising in the normal distribution ensuing from combination of many such independent or nonlinearly related neuronal generators. Both rate and regularity of these electrotonic wave processes may be modulated by changing impedance in perineuronal elements (Adey et al., 1963).

Our computed analyses of these EEG wave processes have indicated the possibility of a stochastic mode of operation in the sensitivity of cortical neurons to recurrent similar, but not necessarily identical, patterns of waves (Adey and Walter, 1963). This hypothesis of a probabilistic mode of operation would infer that the excitability of the cortical neuron would depend on the relation of the spatio-temporal pattern of wave phenomena at the cell surface

to an "optimal" pattern of waves for which its firing threshold
would be lowest. This "optimal" pattern of waves would be determined by the previous experience of the cell, with the wave phenomena intimately concerned in the physicochemical changes associated with the deposition of the memory trace. This proposed
model of the cerebral system would therefore exhibit nonlinear and
stochastic characteristics as local phenomena within a particular
cortical domain. Such schemes are far removed from simple con-
nectivity concepts of facilitated connections as the basis of learned
behavior.

Finally, we have come to recognize the critical difference
between the sensing of physiological processes that relate to transmission and transaction of information, as opposed to its storage.
Future developments in such areas as impedance measuring techniques may enhance knowledge of those delicate processes underlying storage, and establish the basis of structural and functional
organization which endow brain tissue with a uniqueness not found
in "nonlearning" neural systems of admittedly great complexity
(such as the spinal cord).

SUMMARY

The structural uniqueness of cerebral grey matter has been
reviewed, with discussion of a model having neuronal, neuroglial,
and extracellular compartments, and with neurons in cortical tissue
arranged with overlapping dendritic palisades. The intraneuronal
genesis of wave phenomena is discussed as the basis for the EEG
recorded grossly. The nonlinear interrelations between depolarizing phases of the intracellular waves and generation of propagated
spikes are evaluated. Studies of unit and EEG activity in conditional
processes in animals and man are described. Impedance measurements in cerebral tissue in alerting, orienting, and discriminative
responses are evaluated in terms of modified conductance in neuroglia and in an intercellular fluid containing a matrix of mucopolysaccharides and mucoproteins. The role of divalent cations,
such as calcium, in controlling conductivity in such macromolecules is considered. A model of cerebral organization in learning
is described, in which wave processes, organized on a stochastic
basis, might underlie both initial deposition and subsequent recall
of stored information.

ACKNOWLEDGEMENTS

It is a privilege to acknowledge the stimulus provided for many of the recent experiments performed in our laboratory by discussions in the Neurosciences Research Program of the Massachusetts Institute of Technology, and, in particular, by its director, Professor F. O. Schmitt.

Mechanisms of Control of Pyramidal Cell Activity*

P. Andersen and T. Lømo

Laboratory of Neurophysiology
Institute of Anatomy
University of Oslo
Oslo, Norway

Whatever the means employed to control a neuronal system, the ultimate effect must be measured by the number and pattern of discharges of the final neuron in the chain.

Largely through the work of Eccles (1957, 1963), our knowledge on the mode of activation of nerve cells is based upon a comprehensive study of the motoneuron. The basis of output control is a synaptic potential, called an excitatory postsynaptic potential (EPSP), created by the afferent volley to the motoneuron. The depolariztion caused by this EPSP will, when it reaches a certain threshold, initiate a spike in the initial segment of the axon (IS spike), which again will trigger off a larger spike from the soma dendritic membrane (SD spike). Thus, the initial part of the axon has come to be regarded as the trigger point of the motoneuronal membrane. However, there is increasing evidence that many synaptic influences may affect more remote parts of the neuronal membrane. Thus, Granit et al. (1963, 1965a, 1965b) have shown that adequate stimulation in the form of muscle stretching will lead to an alteration in the output of the neuron in spite of a small or undetectable change in the soma resting potential. Hence, this synaptic influence cannot be effected through the direct action of the synaptic current on the membrane of the initial segment. It is therefore possible that the model behavior of dorsal root af-

---

* Pages 5-13 in the Russian edition.

ferents upon motoneurons (Eccles, 1957) is only one of several possible patterns of excitatory neuronal control.

Turning to other types of nerve cells, there is even more evidence pointing to alternative or additive methods of control of their discharge patterns. In particular, the pyramidal cell, with its multitude of synaptic inputs from various sources ending upon very different portions of the cell membrane, represents an example of a cell which is structurally different from the motoneuron, and possibly functionally different as well. From the studies of Lux and Klee (1962) and Nacimiento et al. (1964), it is clear that the pyramidal cells of the motor cortex in cats can be activated by a thalamo-cortical volley in a manner very similar to that operating in the motoneuron. Thus, relatively large EPSPs can be recorded by an intrasomatic electrode. The EPSP brings the membrane potential to the critical level for firing and initiates this discharge in the cell. It is very likely that the initiation of discharge of this cell under these conditions takes place in the initial segment, since upon repetitive stimulation, an IS, SD spike could be seen, similar to the condition in motoneurons. However, when the nonspecific areas of the thalamus were stimulated, the ensuing EPSPs had longer latency and duration and a longer rise time. Furthermore, the EPSPs produced by nonspecific thalamic stimulation were less affected by intrasomatic injection of current than were the specific thalamo-cortical EPSPs. These data suggest that the nonspecific thalamo-cortical EPSPs were generated in a part of the membrane that is relatively remote from the soma. Yet the depolarization seemed to discharge the neuron through depolarization of the IS segment. The picture emerging from these studies shows that an afferent system can directly excite cortical pyramidal cells, much like a reflex operation in the spinal cord, with the motoneuron as effector.

On the other hand, the situation with another set of cortical pyramidal cells, the hippocampal pyramids, is different from the one outlined above. In the hippocampus, a single afferent volley does not excite the cells through direct excitation of the soma membrane by an EPSP. In order to excite the hippocampal pyramidal cells, the afferent bombardment must fulfill certain criteria, the description of which is the main purpose of the present article. The study of the mode of activation of hippocampal pyramidal cells and a comparison of this activation with the thalamic activation of sensori-motor pyramidal cells have led to a division of cortical synaptic activity into two classes, namely detonator and integrator synaptic activity.

Detonator Synapses

Detonator synapses can be defined as those synapses that can excite the cell through direct synaptic depolarization of the initial segment. Systems using this type of synapse require only moderate convergence to the target cell in order to reach the critical depolarization level. No potentiation of the synaptic activity is necessary in order to reach this level. Examples of such detonator synaptic activity are the dorsal root afferents to the motoneuron, and the very efficient excitatory synapses in the somesthetic projection system. These are synapses located in the dorsal column nuclei and the ventro-basal complex of the thalamus, together with the thalamo-cortical synapses on the pyramidal cells of the post-cruciate cortex. All these synapses of the somesthetic

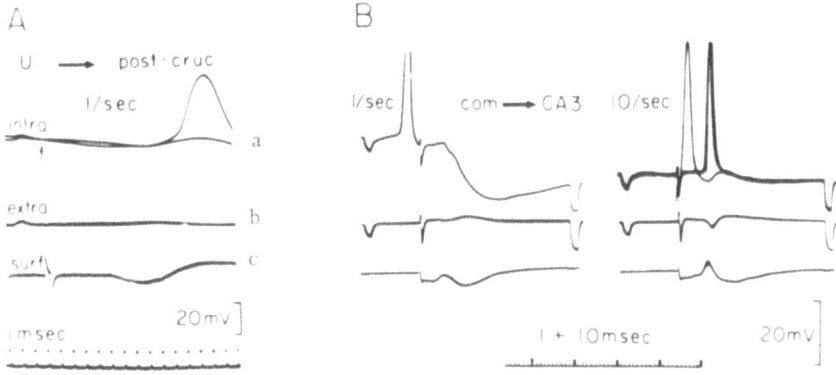

Fig. 1. Detonator and integrator synapses. A) Responses obtained intracellularly (a) and extracellularly (b) from a postcruciate pyramidal cell in response to single shock stimulation of the ulnar nerve. Intracellular response shows a subthreshold EPSP superimposed upon a large EPSP which gives rise to a full action potential. (c) Surface record. B) Responses of a CA3 hippocampal pyramidal cell in response to 1/sec stimulation(left) and 10/sec stimulation (right). Legend as for A. A 10/sec stimulation replaces the large hyperpolarizing potential at 1/sec with an antidromic or synaptically driven spike, the latter taking off without any obvious depolarizing prepotential.

projection path are characterized by large and unitary EPSPs, indicating that they are produced by a relatively small number of
axons only (Fig. 1A). Since the EPSPs produced by detonator synapses are relatively large and have a short rise time, it is possible that their location, electrically speaking, is not far from the
soma.

Such detonator synaptic activity gives the system a high degree of reliability, or a high safety factor for propagation along
the neuronal chain. Although possessing a high degree of reliability, the system has few possibilities for adjusting the activity of its
individual components, and will not be suitable for a process requiring integrative action, such as in higher mental processes.
It is, therefore, not unlikely that such synaptic activity is used in
cortical pathways subserving reflexes.

A neuronal path which has the possibility of integrative activity must have synapses where the synaptic transfer has a low or
moderate degree of transmissibility. In other words, the initial
failure factor must be reasonably high. If these conditions are not
fulfilled, any plastic changes of neuronal activity must depend upon
inhibition only. This concept can probably not be defended in view
of recent findings in the somatic projection system.

Integrator Synapses
_____

By contrast with the first class of synapses, the integrator
synapses stand out as a special type well suited for the integrative
function of the pyramidal cells. This type of synapse is exemplified by the dendritically located synapses on the hippocampal pyramidal cells (Fig. 1B). A single afferent volley very rarely depolarizes the soma at all. At any rate, any depolarization of the
soma is too small to reach the critical depolarization threshold
for initiation of the spike discharge. In order to discharge the
neuron, these synapses need a relatively large degree of convergence, that is, many volleys of the afferent system have to impinge
fairly simultaneously upon the same target neuron. Furthermore,
even if a large degree of convergence is granted, the probability of
discharge still remains low unless the activity of the excitatory
synapses involved is potentiated. As seen in Fig. 1B, neither a single volley nor a synaptic activity potentiated by an increase of the
rate of stimulation is capable of producing soma depolarization.
The spike which is triggered by the afferent volley must be initia-

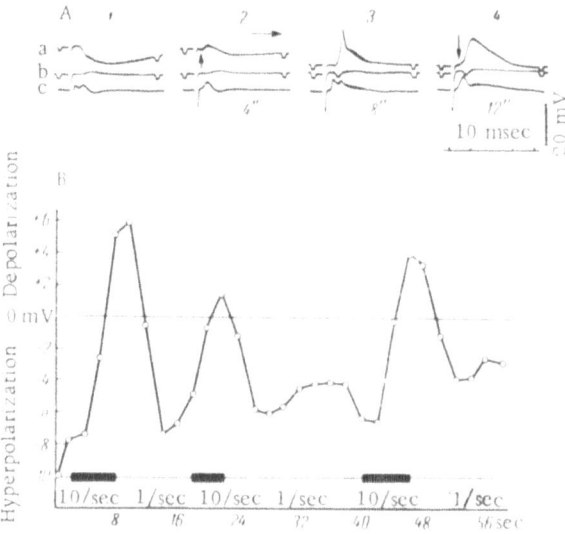

Fig. 2. Frequency potentiation of hippocampal synapses.
A) Intracellular (a), extracellular (b), and surface (c)
records in response to 1/sec and 10/sec stimulation of
the commissural pathway to CA1. The records in response
to 10/sec stimulation show progressive growth of a depolarizing potential, giving rise to discharge of the cell.
B) Size of synaptic potentials in response to stimulation
of the commissural pathway to a CA3 pyramidal cell. A
10/sec stimulation converts the ordinary hyperpolarizing
response to a depolarizing one.

ted by some other mechanism than direct depolarization of the initial segment. This integrator type of synaptic activity is dependent
upon terminal structures situated far from the soma, at least electrically speaking. As shown by Andersen et al. (1966), these excitatory hippocampal synapses are located exclusively on thin dendrites, most often on the small dendritic spinae which encrust dendrites with a diameter of about 1μ only. The dendritic location of
excitatory synapses seems to be an exclusive one, since very few,
if any, degenerating synapses can be seen on the soma after section
of known afferent excitatory pathways to the hippocampal formation
(Foss and Blackstad, 1966).

Single shock activation of integrator synapses rarely discharges the hippocampal pyramidal cells. In a study of the mech-

anism by which such activation can take place, Andersen and Lømo
(1966) concluded that a large increase in excitatory synaptic efficiency was produced by an increase in the rate of stimulation. As
can be seen from Fig. 2, an increase in the stimulation rate from
1 to 10/sec gave a slowly developing depolarization of the neuron.
This depolarization was due to formation of a large and long-lasting EPSP. Since we know that these EPSPs are generated by synapses located on the dendrites, the electrotonic attenuation along
the dendritic tree means that the locally created EPSPs are even
larger than that appearing in this figure. Much of the increase in
the EPSP is due to monosynaptically activated synapses. The most
likely explanation of the effect is that the later shocks in a train of
stimuli liberate more transmitter than the first shocks of the same
train. The effect can hardly be due to hyperpolarization of the terminals, since an extracellular negative dc shift at the level of the
excited synapses was observed under the same conditions. Furthermore, the results could be explained by hyperpolarization of
the cell membrane. However, the opposite was recorded and experiments with artificial hyperpolarization of pyramidal cell membranes (Nacimiento et al., 1964) have shown that it is very hard, if
not impossible, to increase the size of remote EPSP by hyperpolarization, just as for dorsal root motoneuron EPSPs (Eccles, 1957).
Thus, the most likely explanation of the potentiation effect is an increased liberation of transmitter due to the augmented rate of afferent impulses. We have called this process frequency potentiation.

Similar processes may occur in many neuronal systems,
notably in the autonomic nervous system. It has also been shown to
occur when pyramidal tract impulses impinge upon cervical motoneurons (Landgren et al., 1962) and by the action thalamo-cortical
volleys on sensori-motor pyramidal cells (Creutzfeldt, personal
communication). However, in these other locations, the degree of
potentiation is of the order of 2-10. In the hippocampal pyramids,
the factor by which the EPSP amplitude is potentiated is up to 50-
200. This formidable increase of the postsynaptic responses due to
an increased rate of stimulation must be of great importance for the
synaptic control of the cell discharge.

Dendritic Conduction

Since in the hippocampal formation excitatory synapses are

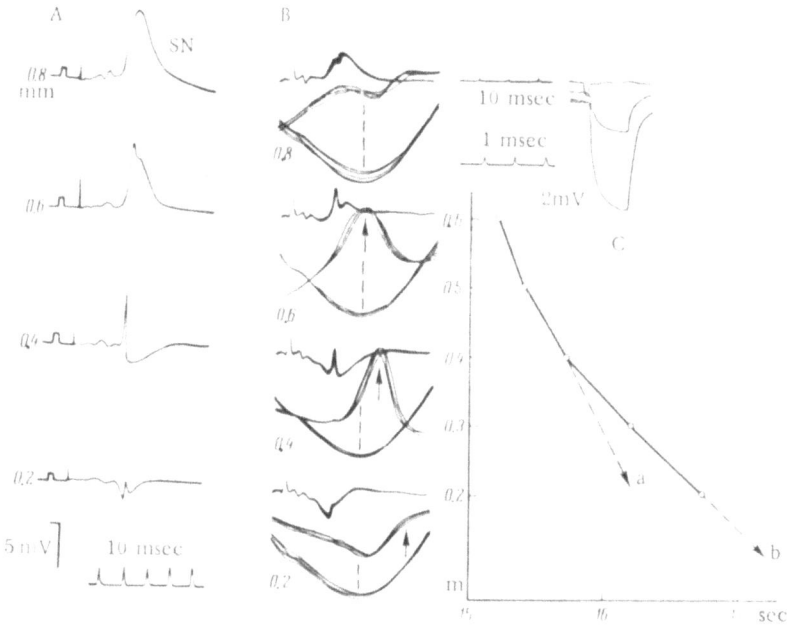

Fig. 3. Dendritic conduction. A) Extracellular responses obtained at the indicated
depths below the alveus of CA1 in response to entorhinal stimulation. The SN wave
(Schaffer negativity) is due to activation of excitatory synapses situated about 400 μ
from the soma. B) Records similar to those shown in A. In each assemblage of three
traces, the two lower show the thickened part of the upper trace displayed with an expanded time base. The arrow points to the peak latency of the spike. The stippled
line indicates the peak positivity of the surface spike which is the reference point.
The peak of the population spike occurs later with more superficial locations. This
latency increase is plotted to the right in the figure (C), giving an average conduction velocity of 0.4 m/sec along the apical dendritic shaft and 0.2 m/sec along the thinner
basal dendrites.

located exclusively on the dendrites, it is not surprising that
selective activation of one afferent path, like the Schaffer collaterals, produces a localized extracellular negative wave associated
with the synaptic activity (Fig. 3A). This negative extracellular
wave is no doubt due to the inward current of the activated synapses
on the dendrites at this depth. A spike can be seen, which is associated with the deep negative wave. In recordings from different
levels, the spike amplitude differed in its distribution from the am-

plitude of the SN (Schaffer negativity) wave. At the level of the
Schaffer collateral synapses, the spike is recorded as a purely positive wave, indicating a blockade of the spike invasion of the synaptic area. With increasing distances from the synaptic territory in
question, the extracellular population spike showed increasing amplitude and latency (Fig. 3B). This can only be explained by the
spike being generated at or very close to the activated synapses
and being conducted away from this territory toward the soma with
a conduction velocity in the neighborhood of 0.4 m/sec. Since the
spike is generated peripherally, the growth of the population spike
cannot be attributed to an increase in the number of activated neurons. The alternative explanation of a reduced inward sodium current due to an increased area is not likely either, since the relative
increase of dendritic membrane area decreases towards the soma.
This is due to the progressively increasing branching of the dendritic tree at more peripheral locations. Thus, the most likely explanation is that the individual spikes constituting the basis for the extracellular population spike are reduced close to the site of generation, due to the conductance increase of excitatory synapses acting
as a shunt across the membrane current. With more somapetal locations, the shunting current will be less and less important, resulting in a full spike amplitude.

Conducting and Synaptic Dendrites

On the basis of these results it is possible to define two
classes of dendrites: synaptic and conducting. Synaptic dendrites
are those dendrites upon which excitatory synapses end, causing an
inward current. Due to the great increase of transmembrane conductance, these dendritic areas act as a very efficient shunt for the
outward current that leaves the cell, making the conditions for initiation of a spike close to the synaptic territory not very favorable.
Neighboring areas of the dendrite will be in a more advantageous
situation, having a higher transmembrane resistance. Although attenuated by the cable properties of the dendritic tree, the synaptic
current will create here a sufficiently great voltage drop across the
dendritic membrane to initiate a spike through an increase of the
sodium permeability.

Summarizing, the conducting dendrites are depolarized by the
outward current produced by the more distant synaptic activity, and

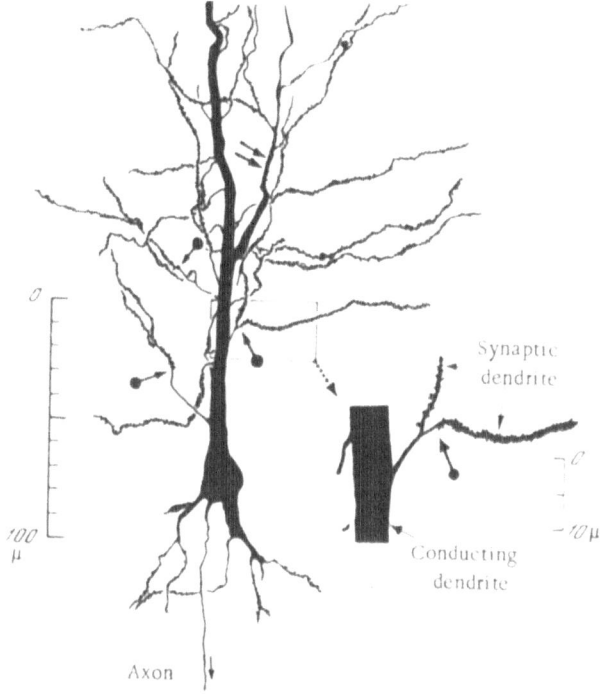

Fig. 4. Morphological basis of conducting and synaptic dendrites. Draw-ing of a Golgi-impregnated CA3 pyramidal cell from a rabbit. The thick apical dendrite and its main branches are smooth and show no spines. The secondary dendritic branches have a large number of spines, but none on the initial 10−30 μ. The transition from smooth to spine-encrusted secondary dendrites is indicated by an arrow connected with a filled circle. The double arrow indicates a transition zone in which a smooth dendrite starts to carry spines. The inset (lower right) shows the portion outlined in the rectangle with higher magnification, indi-cating the very large density of spines on a secondary dendrite.

serve as structures along which the spike may be propagated to-wards the soma.

The physiological classification has received morphological support. In impregnation pictures of hippocampal pyramidal cells, the diameter of different parts also allows division of dendrites into two types (Fig. 4). The first type is the thick, smooth portion of the dendrites, consisting of the apical dendritic trunk and its thicker

branches, and the thicker portions of the basal dendrites. The diameter of this portion of the cell ranges from $8\,\mu$ to $2.5\,\mu$. A dis-
tinctly thinner type of dendrite, never amounting to more than $1\,\mu$
in diameter, emerges from the thicker branches. The first $10\,\mu$
of these thinner branches is smooth, but from this point on, toward
their ends, they are heavily covered with small dendritic spines
morphologically resembling short match heads. The stems of the dendritic spines are very thin, measuring about $0.08\,\mu$ in diameter
and from 0.3 to $1.0\,\mu$ in length (Blackstad and Westrum, 1963).
The end of the spine consists of a small enlargement about $0.4\,\mu$
in diameter. The majority of excitatory synapses make contact
with these spines. Thus, synaptic activity obviously will be effected
through spines on the thin dendrites $1\,\mu$ in diameter.

POSSIBILITIES FOR CONTROL OF PYRAMIDAL CELLS THROUGH ACTIVATION OF EXCITATORY SYNAPSES

Choice of Line

The first means by which the discharge of a pyramidal cell
can be controlled is the choice of afferent line. In particular, are
detonator or integrator synapses being used? Using detonator
synapses, the depolarization will be reliable and effective. This
type of coupling is useful when reflex operations are demanded.
However, if integration and evaluation of different afferent information are going to take place, or if probability measurement of incoming information is to be secured, integrator synapses are the line of
choice. Through this system, it is possible to exert differential control
on the cell discharge according to the type of afferent information.

After the choice of the integrator synaptic line has been made,
the impulse traffic will be determined largely by the afferent volley
frequency. With very low afferent volley frequency, the individual
EPSPs will be of low amplitude and constitute not much more than
the random noise of the system. However, with an increase in the
rate of stimulation, the afferent system employed will increase in
importance in relation to other afferent lines and, eventually, it
will lead to discharge of the cell. The physiological basis for this

increase in importance of the afferent volley used is to be found in the potentiation of EPSPs produced by this particular pathway.

Provided the frequency potentiation of an integrator synaptic population has succeeded in creating a dendritic spike, the spike will propagate toward the soma. However, in order to determine the cell output, the dendritic spike must invade the soma and axon. At each branching point along the dendritic tree, the area to be invaded increases. Therefore, each of these points represents a possible point of blockade. Consequently, the safety factor for propagation along the dendrites will be very much lower than that for axons, and many of the initiated dendritic spikes will be blocked. If the propagation is assisted by additive depolarization exerted by synaptic territories lying along the path of dendritic propagation, the probability of dendritic spike invasion of the soma will increase. Therefore, in addition to the factors mentioned above, convergence of different synaptic lines will be of great importance in the determination of cell discharges.

Recurrent Inhibitory Control of
Pyramidal Cell Discharges

In the regulation of pyramidal cell discharges, purely excitatory control is not sufficient. In addition, the cell discharges must evade the very powerful inhibition that can be recorded from most pyramidal cells (Phillips, 1959; Andersen et al., 1964; Stefanis and Jasper, 1964). The inhibition can be seen in three forms: recurrent inhibition, forward inhibition, and competitive conductance shunt.

As initially reported by Spencer and Kandel, the very large IPSPs produced in the hippocampal cortex are due to a recurrent inhibitory pathway. Later the inhibitory interneurons and their synapses were identified as basket cells and their synapses upon the hippocampal pyramidal cell soma (Andersen et al., 1964). This recurrent inhibition is very long-lasting and powerful, completely swamping all pyramidal cell activity for as long as 200–300 msec after the initial discharge. However, a competitive state can be achieved in which the large inhibition is overwhelmed by the powerful depolarization exerted by the frequency-potentiated synapses. The recurrent inhibition is extensive, being able to spread for

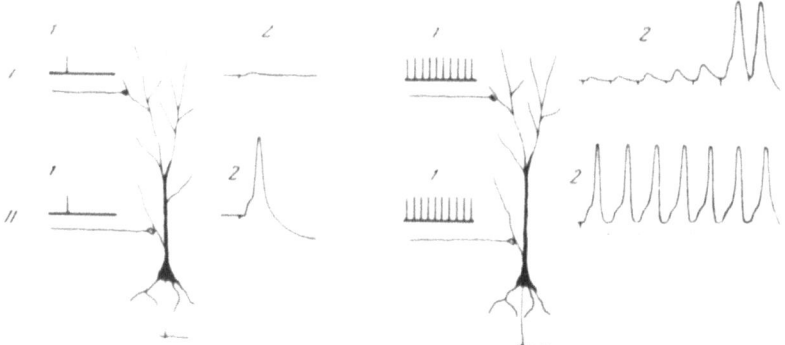

Fig. 5. Diagram indicating the choice of line of afferent input to a pyramidal cell.
Two types of line, the integrator (I) or detonator (II) lines can be used. Further explanation in text.

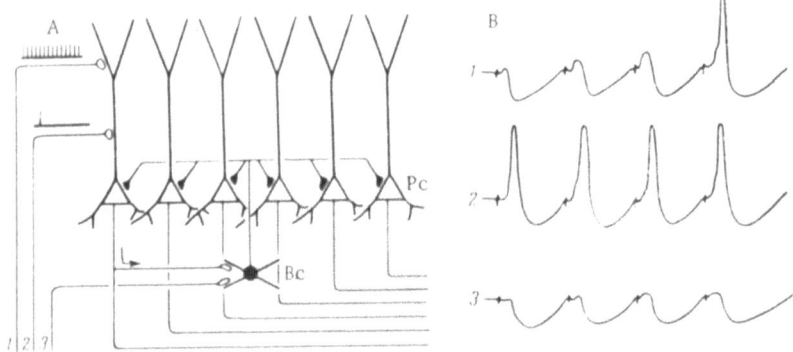

Fig. 6. Diagram showing inhibitory control of pyramidal cell discharge. A) The afferent volley can take three lines: 1, 2, or 3. The integrator line (1), on repetitive
stimulation, produces a progressively increasing EPSP which eventually may lead to
discharge of the cell. The detonator line (2) discharges the cell efficiently, even with
a single volley. Following discharge of the cell by line 1 or 2, a recurrent inhibitory
pathway will be operated through activation of basket cells (Bc) through the action of
pyramidal cell axon collaterals. An afferent volley may also take line 3, by which
the basket cells are excited directly, and inhibition exerted on many cells without
previous discharge of some pyramidal cells (Pc). The latter is an example of forward
inhibition. B) The responses to a train of afferent volleys in the three lines shown in A.

many millimeters in the cortex following excitation of a small
group of neurons. A single basket cell is estimated to connect
with 200-500 pyramidal cells. The situation will result in a fringe
of inhibited cells around a small nucleus of excited neurons (Fig.
5). Indeed, by direct stimulation of a single cell through an impal-
ing micro-electrode, it has been shown that the cell creates distinct, but slight, hyperpolarization of itself (unpublished observations). Thus, recurrent inhibition is distributed not only to the
neighboring cells, but also to the cell of origin. It should be
stressed, however, that the self-recurrent inhibition is much
smaller than that distributed to other neurons.

Forward Inhibition

A striking feature of the hippocampus is the great preponderance of inhibitory actions recorded in spite of a very small
stimulus strength. In fact, it is possible to record large and long-
lasting IPSPs in spite of any detectable field potential indicative of
an initial discharge of hippocampal neurons. Therefore, it is highly possible that large EPSPs can also be initiated, without a previous discharge of hippocampal neurons, by excitation of basket
cells by the afferent volley. This would be comparable to the situation in many other systems of the nervous system, notably the
cerebellar cortex (Eccles et al., 1966). The possibility of forward inhibition provides another means by which afferent control can be exerted on the discharge by hippocampal pyramids, possibly increasing
the "contrast" between intensively and less intensively excited cells.

Competitive Excitatory Shunts

The activation of a localized synaptic territory produces a
considerable conductance increase, particularly in the state of
frequency potentiation. Consequently, the efficiency with which a
neighboring synaptic territory can initiate a discharge will be reduced by this conductance increase. In this situation, the law of
algebraic summation of EPSPs produced by two afferent routes
does not hold true. In fact, the very intense synaptic currents produced by the frequency potentiated synapses may cut off the cell
discharge that they themselves initiated.

CONCLUSION

In conclusion, the physiological mechanisms available for control of pyramidal cell activity are many. The main point is whether detonator or integrator synapses shall be used. When integrator synapses are employed, the rate of afferent volley bombardment is of great importance, as well as the convergence pattern in that given situation. Further control can be exerted by restricted use of recurrent and forward inhibition. Although very powerful, these inhibitory actions can be counteracted by the frequency potentiation of excitatory synapses, leading to an increase in synaptic efficiency in cortical systems. It is likely that such plasticity will be found in integrator synaptic activity, whereas the detonator variety is not likely to show great pliability.

ACKNOWLEDGMENT

Supported by US Public Health Service Research Grant NB 04764, from the National Institute for Neurological Diseases and Blindness, and a grant from the Eli Lilly Research Foundation, which are gratefully acknowledged.

Correlation between Cortical Potentials in Patients with Obsessive Neuroses[*]

A. S. Aslanov

Institute of Higher Nervous Activity and Neurophysiology
Academy of Sciences of the USSR
Moscow, USSR

The application of electronic computer techniques to the analysis of electroencephaloscopic material has opened up extensive prospects for the study of human higher nervous activity under normal and clinical conditions. Research in this direction has shown that mathematical analysis of electroencephaloscopic data can not only provide a fuller assessment of cortical function as a whole, but can also reveal the relationships between different parts of the cortex.

In this paper I consider the possibilities afforded by determination of the correlations between biopotentials (Livanov, 1962b; Glivenko et al., 1962a, 1962b) for the study of changes in cortical functions in patients with obsessive neuroses.

METHOD

Cortical electrical activity was studied with the electroencephaloscope in 50 leads. Electrodes were arranged in five rows on the surface of the patient's head (the first two rows in the prefrontal areas). The reference electrode was placed on the chin. From the electroencephaloscope screen the potentials were photographed on motion picture film. For quantitative analysis of the potentials their amplitude was measured (visually) from frame to frame. The paired-sample correlation of bioelectrical activity of all 50 recorded points of the cerebral cortex was investigated. The coefficient of correlation was determined by the number of unidirectional changes in oscillations of the potentials over a fixed

---

[*] Pages 14-20 in the Russian edition.

time interval (1.5 sec). The correlation coefficients (numbering 1225 for 50 leads) were calculated on the M-20 Universal electronic computer. Phase changes of electrical oscillations were not considered.

The investigations were carried out on six patients with obsessive neuroses ranging in duration from 18 months to 5 years. The clinical picture was dominated by the obsessive states, manifested against a background of increased irritability, diminished ability to work, and insomnia. They appeared most frequently as obsessive fears and thoughts. Each patient was investigated at least three times. From 3 to 5 segments of the recording were processed in the computer and subsequently analyzed. Segments obtained at different stages of the investigation were analyzed.

Before describing the results, the basic results of study of biopotential correlations in the healthy human cortex must be considered. The principal role in causing synchronized activity in the healthy human cortex in a resting state belongs to those parts of the cortex whose potentials exhibit synchronous waves during 45–75% of the analyzed time (low correlation) (Livanov et al., 1964; Gavrilova et al., 1964a). Those parts of the cortex whose electrical processes follow an identical course for more than 75% of the time of analysis (high correlations) are few in number in healthy persons in a resting state. Besides the fact that there are few such areas and that they can be recorded in widely different regions of the cortex, each of them correlates as a rule with one, two, or occasionally, more areas usually located close together. The number of high correlations in healthy human cortical background bioelectrical activity usually did not exceed 2% of the total number of possible combinations.

The distribution of high correlations between potentials in the healthy subject B.E. in a state of relative rest is shown in Fig.1, which is a spatial scheme of the human cerebral cortex. Clearly, the number of areas giving a high correlation coefficient is exceedingly small in B.E. in a resting state. Synchronized waves are found only at points located side by side or a short distance apart, each correlating with one or two others. However, as soon as subject B.E. is given a mental problem to solve, the pattern described changes sharply. Performance of mental work is accompanied by a clear increase in the number of areas giving a high biopotential correlation coefficient. Most areas now correlate not with one or

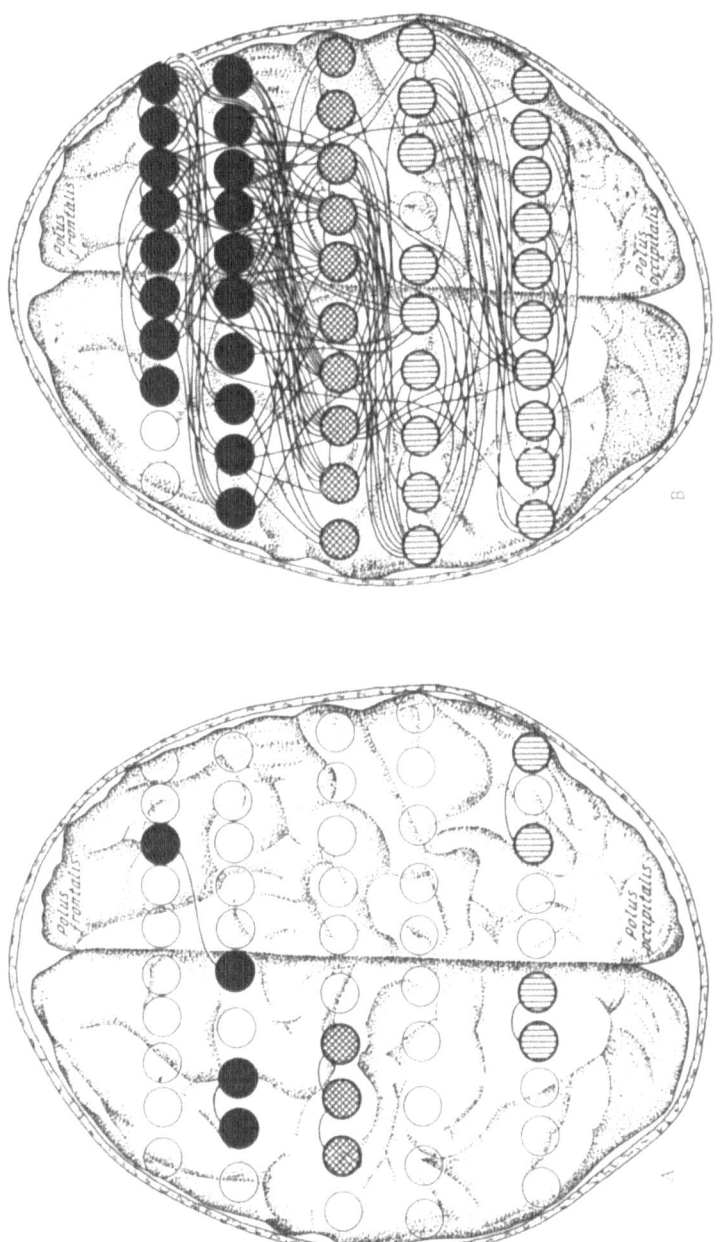

Fig. 1. Temporal correlation between potentials in different parts of the healthy human cerebral cortex. A) Resting state; B) during mental work. The circles denote locations of recording electrodes. Areas whose potentials give a high correlation coefficient are joined by lines. Circles in prefrontal area are shaded black; those in the motor area are cross-hatched, those in the posterior portions are shaded vertically.

two, but with many (10 or more) cortical areas. Compared with the resting state, the total number of high cortical correlations in B.E. increases several times during mental work. The number of correlations reaches a maximum in the prefrontal areas of the cortex (Fig. 1b).

Results similar to those described above were obtained with the other healthy subjects investigated. They showed that the degree of spatial synchronization of potentials depends on the state of cortical function. In healthy subjects in a state of relative rest, synchronization of cortical potentials is slight, whereas during mental work a sharp increase in processes of synchronization of electrical activity is observed.

For the results obtained on patients with obsessive neuroses, the following case is typical.

Patient Ch. V., a woman aged 37 years, was admitted to the psychiatric hospital complaining of an obsessive fear of death, a tendency to tire rapidly, irritability, and disturbed sleep.

Mental State. Consciousness was lucid, correctly oriented. Complaints of obsessive fear of death from "heart disease," which no physician has yet succeeded in finding. Fixation on occasional attacks of palpitations and dyspnea. She herself states that these symptoms arise as soon as she thinks about them. Later, fear of death from heart disease gave way to cancerophobia. Next these obsessive thoughts were relegated to the background and she began to be afraid of "losing her reason." The patient is critical and understands that there are no grounds for her fears, yet she cannot get rid of them. Diagnosis: obsessive neurosis.

The results obtained during examination of patient Ch. V., as in the preceding case, are inscribed on a diagram of the cortex (Fig. 2). This shows that almost two-thirds of all the recorded cortical areas give a high biopotential correlation coefficient. Each area correlates simultaneously with many others, their number exceeding 10 or more. Waves correlate not only from areas located side by side, but also from areas some (occasionally a considerable) distance apart. For example, the cortical area lying beneath the 10th electrode correlates with areas in both the ipsilateral and contralateral hemispheres. The number of high correlations in the case of Ch. V. is 212, more than 8% of the total number possible.

TABLE 1. Number of High Correlations in Cerebral Cortex of Healthy Subjects and Patients with Neurosis of Obsessive States

Healthy subjects	Mental work	Patients	Background
N. G.	233	M. L.	307
S. E.	238	Ch. V.	212
B. E.	229	E. G.	142
V. A.	209	A. Z.	118
A. S.	136	K. T.	62
A. M.	44	N. L.	54
Yu. Ch.	35	—	—

So far as their topographic distribution is concerned, analysis shows that most correlations are found in the prefrontal cortex. Synchronization processes are weaker in the posterior portions.

Comparison of the data given in Figs. 1 and 2 clearly shows two facts: first, spatial synchronization in patient Ch. V. in a resting state was much stronger than normal; second, it is very similar in degree to the picture of biopotential correlations discovered in the healthy human cortex during mental work. This is confirmed by an examination of Table 1.

From a comparison of the data in Table 1 it will be obvious that the number of high correlations in patients in a resting state is mainly on the same level as the number of high biopotential correlations in the healthy cortex during mental work. However, the spatial synchronization of the potentials showed individual differences in each of the healthy subjects investigated. In some, in Yu. Ch., for example, the number of high correlations was measured in tens, while in others, in S.E., for example, it was measured in hundreds. Similar individual variations in the degree of biopotential synchronization were also observed in the patients with obsessive states. It may be asked, if the difference in the degree of spatial synchronization of potentials in healthy subjects, as our investigations showed (Livanov et al., 1964), was due mainly to difficulty of the tasks presented to them, what is the explanation of this difference in the patients with neurosis? Help in answering this question was obtained from a comparison of the results ob-

tained with the clinical picture of the disease. Spatial synchronization of bioelectrical processes taking place in different parts of the
patients' cortex was found to be closely connected with their clinical state, changing along with a change in the course and severity
of the clinical manifestations of the disease. This is clearly demonstrated by the example of patient A.Z.

Patient A.Z., a woman aged 35 years, was admitted to the
psychiatric hospital complaining of an obsessive fear of death from
"paralysis of the heart," of being readily fatigued, and of insomnia.

Physical State. No evidence of any organic heart disease.

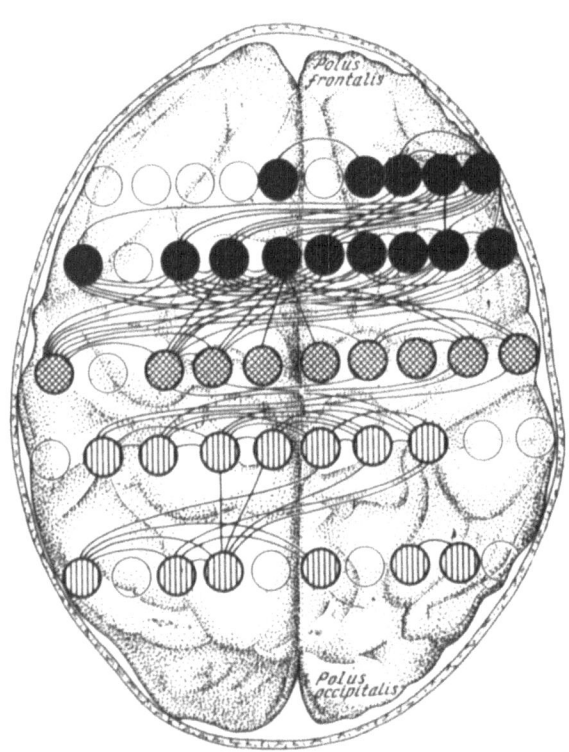

Fig. 2. High correlations between potentials in various
parts of the cerebral cortex of a patient with obsessive
neurosis at rest. Legend as in Fig. 1.

Mental State. Consciousness lucid, correctly oriented. Presents various complaints, mainly an obsessive fear of death from infarction. Constantly tormented by the thought that "her heart is in a bad way." Listens attentively to her heart beat. Counts her pulse all the time. If any sensations develop in the region of her heart, she develops an uncontrollable fear. Attempts to overcome her obsessive fears are of no avail. Diagnosis: obsessive neurosis.

During examination, patient A.Z. could be seen in two different states: one in which she was clinically relatively well, and another, of severe apprehension and alarm, when she could no longer control herself; at such times her pulse was rapid, dyspnea developed, and she perspired freely.

The trend of high cortical correlations in the patient A.Z. when she was clinically relatively well (a) and during an exacerbation of her obsessive phenomena (b) is shown in Fig. 3.

It will be seen from the results in Fig. 3 that when this patient was clinically relatively well, 36 of the 50 recorded cortical areas show synchronized activity. Each of these areas correlates with between 2 and 5 other areas situated side by side or a short distance apart. The number of high correlations counted in the cortex of patient A.Z. was 118, or 4.5% of the total number possible. The distribution of correlations over the cortex was fairly regular, but rather more were found in the anterior portions of the cortex. The pattern of spatial synchronization of the potentials was different when the patient's obsessive fears were aggravated (Fig. 3b). There was a definite increase in the number of areas with high correlation between their potentials. Now more than 53 such areas were counted, 33 of them each correlating with 6 or more other areas. Areas located at the opposite poles of the cortex now began to correlate. The number of high correlations was three times greater than their number at rest and amounted to more than 13% of the total number of possible correlations. The highest "density" of high correlations was observed in the frontal cortex. Here more than half the total number of high correlations were found. The number of high correlations also rose considerably in the motor area.

These investigations revealed the following pattern of corre-

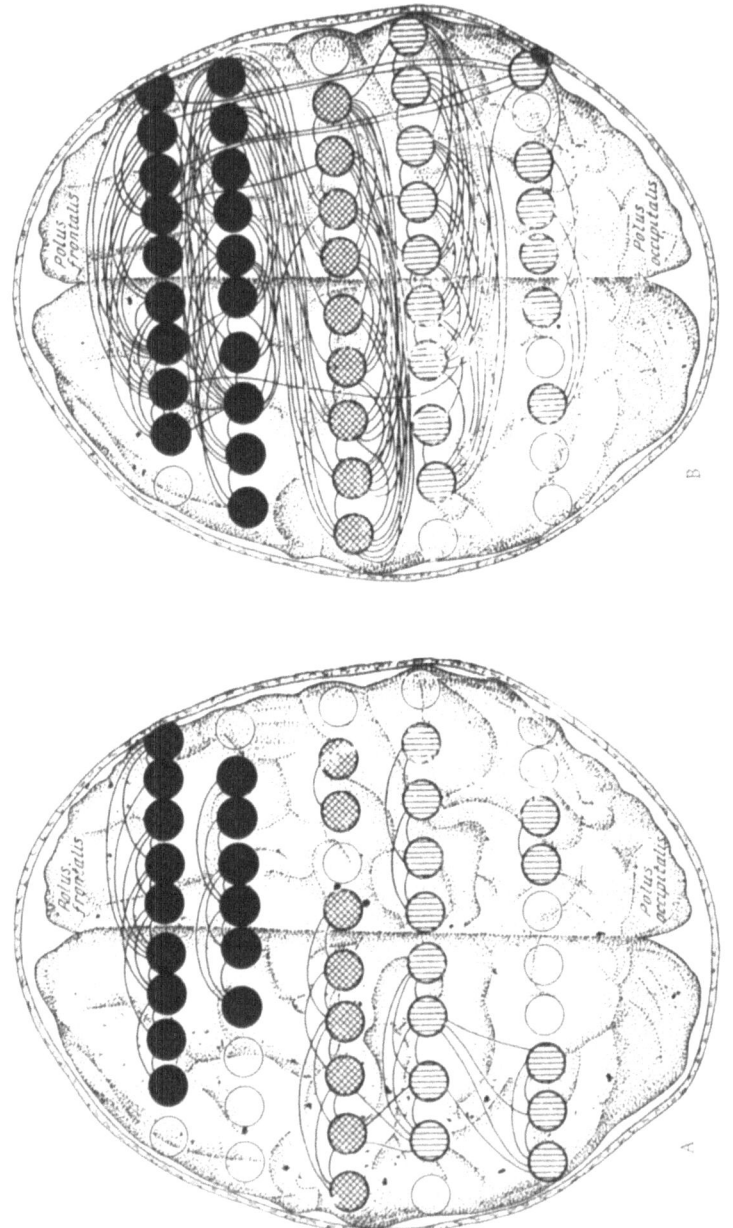

Fig. 3. High correlations between potentials of various cortical regions of a patient with obsessive neurosis. A) When clinically relatively well; B) during sharp exacerbation of obsessive phenomena. Remainder of legend as in Fig. 1.

lation between bioelectrical processes taking place in different parts of the cortex in patients with obsessive neurosis.

In contrast to the healthy cortex, in the cortex of the patient at rest synchronization is more marked. The level of synchronization of potentials in the patient fluctuates constantly, sometimes rising and sometimes falling, although in the latter case it usually remains above the level observed in healthy subjects in a resting state. Such fluctuations are closely connected with the patients' clinical condition. The increase in the number of high correlations observed in this group took place in all parts of the cortex, but mainly in the frontal regions. This distribution of correlations was very similar to that observed in the healthy subjects during performance of intellectual operations, and it may be evidenced that these correlations are connected with complex associative processes taking place in the cerebral cortex. From this standpoint the investigations of N. A. Gavrilova (1965) on patients with the paranoid form of schizophrenia are interesting. She found that in this disease the frontal cortex, compared with all the other regions, has the smallest number of high correlations. In patients with obsessive neurosis, as shown here, as a rule no such distribution of high correlations is found.

Changes in Electrical Phenomena in the Human Subcortex in Relation to Operative Memory[*]

N. P. Bekhtereva, V. M. Smirnov,
and A. I. Trokhachev

Institute of Experimental Medicine
Academy of Medical Sciences of the USSR
Leningrad, USSR

What is the role of the subcortex in the mechanisms of mental activity and, in particular, in the mechanisms of operative memory? One way in which this problem can be investigated is by a rational combination of psychological and electrophysiological methods, using mathematical analysis (Walter and van Leeuwen, 1961; Bekhtereva et al., 1965).

When electrical stimuli are applied to deep brain structures through implanted electrodes (stimulation, polarization, and electrolysis) for diagnostic and therapeutic purposes, observation must be kept not only on neurological and autonomic effects, but also on psychological manifestations.

To assess the state of the higher mental functions we have used a special test, a test of the range of operative memory, derived from the old method of Jacobs. The patient is instructed to repeat a random series of number immediately after the physician or to reproduce it after a pause of 10-15 sec, at the examiner's request (the delayed response). The test is applied several times under normal conditions (the control experiment) and during stimulation of deep brain structures — the main experiment (current 0.1-0.5 mA, pulse duration 1 msec, frequency 1, 4, 10, 25, and 50/sec, for 1 sec). In the second test the patient had to add (or subtract) a certain number to (or from) a given number and to say the result

---

[*] Pages 31-40 in the Russian edition.

aloud only if the sum (or difference) is even or (depending on the instruction) odd (this test was proposed by V. M. Smirnov).

When conducting tests of short-term (operative) memory, various aspects of the electrical activity of the deep brain structures were studied: the electrosubcorticogram (ESCG), changes in the level of the steady potential, and cell activity.

Analysis of the ESCG included preliminary manual and subsequent computer analysis of the data. The distinctive characteristics of each ESCG (seven characteristics in all) were given code numbers. The program for the electronic computer (Moiseeva and Orlov, 1965) was devised so that the characteristics of the ESCG correlated with any variables.

Another method of ESCG analysis was based on asymmetry of the durations of the ascending and descending phases of wavelike fluctuations of the ESCG (Genkin, 1962), or on delta t of a parameter equal to $(g - d)/(g + d)$, where $t = 0.5 - 2.0$ sec, g is the duration of the ascending phase, d the duration of the descending phase, and $g + d$ the whole period of the wave.

Changes in the level of the steady potential were investigated during presentation of memory tests without additional stimuli and in conjunction with trigger stimulation by flashes, the ESCG to the trigger attachment being recorded from the same structure as that from which the steady potential was recorded.

Cell activity was recorded: a) in the form of a "natural" motion picture, b) through the filter-trigger device on an ink-writing electroencephalograph, and c) by special counters, also through a filter-trigger device. The cell activity was recorded by means of the same gold electrodes (diameter 100 μ) as those through which electrical diagnostic and therapeutic stimuli were applied and the ESCG recorded. Cell activity was recorded in patients before electrical treatment and after stimulation, polarization, and electrolysis, during a waking state at rest, and during various function tests, including tests of operative memory.

As a rule in the ESCG in patients with hyperkinesia no profound pathological changes were found. During electrical stimulation moderate, general changes in activity were detected (when a current of minimal strength was used), whereas performance of tests of operative memory were often accompanied by more marked and widespread changes in the ESCG.

Changes in biolelectrical activity during these psychological tests, when applied alone or in conjunction with electrical stimulation, were observed in various deep brain structures. An increase, or conversely, a decrease in intensity of various types of electrical activity, depending to a large extent on the original background, could be observed. Under these conditions the only "stable" feature was the actual presence of convincing changes in potentials in the deep brain structures (Table 1).

In individual cases the character of the bioelectrical effect depended on the character of the test (correct or incorrect answer, etc. — Table 2), as well as on the original background and method of presentation. Differences of this type were found in a very demonstrative form in patient M. (parkinsonism) when the electrical response was investigated during the second psychological test.

In patient N., with unilateral epileptiform hyperkinesia of myoclonic type, with major and minor epileptic fits, the background ESCG was dominated by high-voltage slow activity with single or multiple high-voltage pointed waves. Stable pointed waves and slow activity were recorded from the underlying white matter of the motor area in the region of the epileptogenic focus. In some cases neither visual assessment of the ESCG nor analysis by the "specialist + computer" method could yield convincing enough results for distinguishing between changes in electrical activity associated with electrical stimulation and changes accompanying electrical stimulation of the deep brain structures in conjunction with psychological tests. All that could be noted was that inhibition of pathological activity was more or less probable depending on the conditions of investigation. Generalized inhibition of pathological activity developed, for instance, during stimulation in the hippocampal region in about half of all the cases. When electrical stimulation was combined with the psychological test, this phenomenon was observed much more frequently — in about 9 out of 10 cases.

Analysis of the ratio between the durations of the ascending and descending phases of the waves (by Genkin's method) showed that in a waking, resting state these ratios between activity of the hippocampus and the cortical epileptogenic zone varied sometimes

TABLE 1. Changes in Bioelectrical Activity after Isolated Electrical Stimulation of White Matter and after Electrical Stimulation Combined with Psychological Tests (Data Analyzed by "Specialist + Computer" Method)

Conditions of stimulation	Region from which potentials recorded													
	Ponds		Cerebral peduncles			Red Nucleus	Cerebral peduncles–globus pallidus		Globus pallidus		Putamen		Ventro-lateral nucleus of thalamus	Cortex (posterior part of middle frontal gyrus)
	15/I	17/I	2/I	15/I	17/I	2/I	15/I	17/I	15/I	17/I	2/I	15/I		
Isolated electrical stimulation	θ — 18	Burst of slow waves–18	β — 20	0	0	0	0	0	0	0	β — 18, θ — 25, Δ — 20	θ — 15	0	α — 25, θ — 25, Δ — 25; θ — 23, Δ — 15, α — 33
Stimulation combined with psychological test	Burst of slow waves–35, Pointed waves–25; Δ — 55; θ — 18, α — 18	Burst of slow waves–70, Pointed waves–15; θ — 30, Δ — 42	Burst of slow waves–50; β — 50	Burst of slow waves–67, Pointed waves–33; β — 18, Δ — 50; θ — 18, α — 18	Burst of slow waves–57, Pointed waves–15; θ — 15, Δ — 30	θ — 31, Δ — 31	Burst of slow waves–67, Pointed waves–50; θ — 18, Δ — 50; α — 18	Burst of slow waves–33, Pointed waves–18; θ — 18, Δ — 18; α — 18	Burst of slow waves–40, Pointed waves–40; θ — 20, Δ — 60; α — 20	Burst of slow waves–65; α — 15, β — 15, θ — 50, Δ — 50	Δ — 25	Burst of slow waves–25; θ — 23, Δ — 50; Pointed waves–25, α — 75	0	Burst of slow waves–70; α — 30, θ — 30, Δ — 30; θ — 30, α — 15

↑ Increase in amplitude

↓ Decrease in amplitude

α, β, Δ, θ alpha-, beta-, delta-, theta-like activity

Numbers denote percentages of presentations of tests and stimuli evoking these changes.

TABLE 2. Changes in Electrical Activity in Subcortical Structures Depending on Psychological Test and its Performance (Analyzed by the "Specialist + Computer" Method)

Answer		Region from which potentials recorded							
		Substantia nigra—subthalamic nucleus	Amygdala—putamen	Lateral-ventral nucleus of thalamus	Putamen	Internal capsule	Hippo-campus (uncus)	Hippo-campus—internal capsule	Medial part of uncus—amygdala
Correct	Spoken aloud	25 40 / 50	75 / 75 55 45	20	55 / 50	76 / 57 30	25	40 15 / 65	57 / 57 50 57
	Not spoken in accordance with instructions	40 57 80	40 65 65	45 25 57	40 50 45	57 25 65 50	15 57 70	45 45	50 57 65 57
	Incorrect	50 / 50	25 50	25			25		50 50

← Increase in activity
→ Decrease in activity
Numbers denote percentages of presentation of tests evoking these changes, different figures represent different subjects.

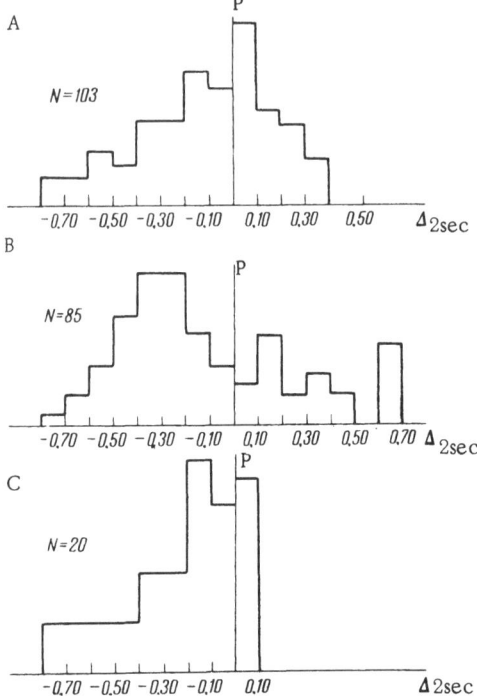

Fig. 1. Distribution of asymmetries of duration
of ascending and descending phases of hippocampal ESCG waves in 2-sec intervals at rest (A),
during electrical stimulation of the hippocampus
(B), and during a test of operative memory (C).
Abscissa: asymmetries of duration of ascending
and descending phases. Negative values correspond to long duration of the descending phase,
positive to longer duration of the ascending phase.
Ordinate: relative frequency (P) of occurrence of
definite values of asymmetry between durations
of ascending and descending phases of waves.
Number in left upper part of figure denotes number of 2-sec cuts of ESCG analyzed.

identically, sometimes independently. During electrical stimulation of the hippocampus the character of these relationships varied
and sometimes they were opposite in direction. Repeated stimulation of the hippocampus could give a different concrete type of relationship, always different from the background pattern.

By this method of analysis it was possible to detect additional features of the changes in activity evoked by electrical stimulation of the hippocampus and by stimulation combined with psychological tests (Fig. 1).

Presentation of a psychological test was accompanied by changes not only in the ESCG, but also in the level of the steady potential (LSP). Changes in the LSP under these conditions were obvious in most cases even by simple examination of the electrograms.

Superposition of the data showed that in certain structures, under ordinary conditions of recording and without the use of a strong, additional stimulus (notably trigger stimulation), the deviations in LSP with repeated presentations of the short-term memory test followed a stable pattern. This stable pattern of LSP was observed in the region of the ventro-lateral nucleus of the thalamus, the globus pallidus, the hippocampus, structures of the cerebral peduncles, and elsewhere. In many cases disturbance of the pattern arose if the test was performed against a background of trigger stimulation: The dynamics of LSP during the psychological test differed from that of the background LSP, losing the characteristic pattern reproducible in repeated tests.

Besides this type of dynamics, in five of 27 cases a clear and reproducible pattern of change in LSP developed against the background of trigger photic stimulation. This phenomenon was seen most clearly in steady potentials recorded from the cerebral peduncles and the amygdaloid nucleus. Only in one case was the repeatability of the pattern convincing when accompanying correct replies and very severely disturbed during complete refusal to reproduce the sequence of numbers.

A characteristic change of the LSP during presentation of the psychological test under normal conditions and against a background of trigger stimulation was not, however, always observed. In some cases the initial background of the LSP was stable and this made analysis of the changes in LSP in connection with the psychological test difficult. In some cases they likewise could not be detected, even when the initial background was sufficiently variable, and in certain brain structures, on the other hand, they were found under normal conditions and also against a background of additional trigger stimulation (especially in the region of the ventro-lateral nucleus of the thalamus).

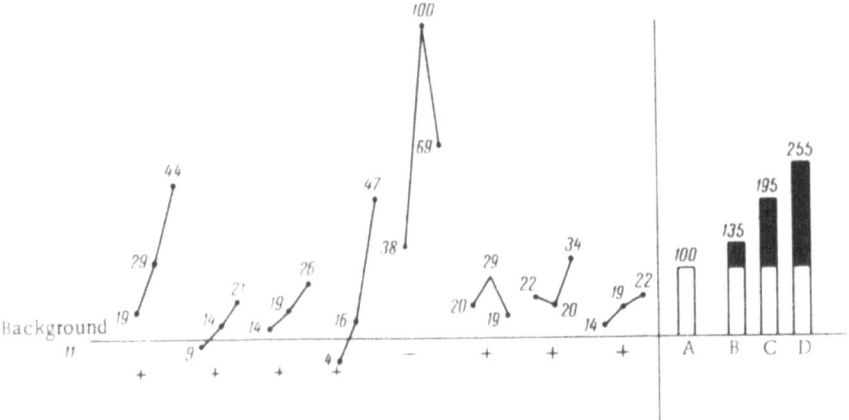

Fig. 2. Dynamics of cell activity, stage by stage, in ventral posterolateral nucleus of thalamus during tests of operative memory: +, correct answer; -, incorrect answer. Numbers denote number of cell discharges in 30 sec. On the right, composite diagram of dynamics of cell discharges in the same structure at various stages of the tests: A) background number of cell discharges; B, C, and D, at stages I, II, and III of the tests, respectively. Numbers denote percentages.

Cell activity was recorded in the ventrolateral, postero-lateral, ventral posterolateral, and ventral posteromedial nuclei of the thalamus, in the globus pallidus, the cerebral peduncles, and the caudate nucleus. The dynamics of cell activity in certain of these structures showed characteristic features during perfor-mance of the operative memory test: either a progressive in-crease in frequency of the discharges, or an increase followed by a decrease in frequency, and so on. Cell activity during presenta-tion of the operative memory test was recorded in three stages (each for 30 sec); at the end of the first stage the examiner pro-nounced a series of 6 or more numbers, during the next stage the patient kept the series of numbers in his memory, and in the last 30 sec cell activity was recorded while the patient recited the se-ries of numbers.

The stage-by-stage dynamics of cell activity was particu-larly characteristic and regular in the ventral posterolateral tha-lamic nucleus, the globus pallidus, and the caudate nucleus. In these structures the dynamics of cell activity was definite, re-producible, and identical not only during repeated investigations of the same patient, but also during investigation of the same

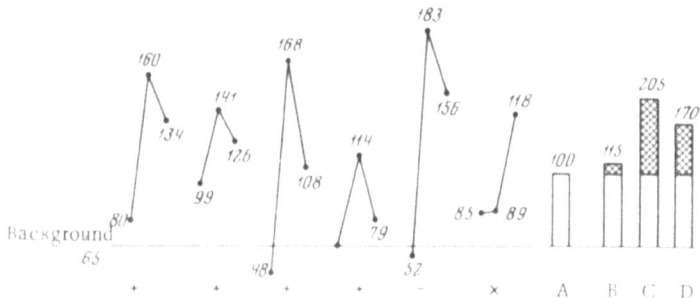

Fig. 3. Dynamics of cell activity, stage by stage, in ventro-lateral nucleus of thalamus during tests of operative memory on patient K. X: a disturbance arose during performance of the test. Remainder of legend as in Fig. 2.

structure in another patient. In the ventral posterolateral thalamic nucleus, possessing the most distinct lateralization in relation to movement (increase in frequency of cell activity during flexion of contralateral limbs only), the number of discharges from the cells increased regularly during successive stages of the test of range of operative memory (Fig. 2). In the globus pallidus and caudate nucleus, and also in the ventrolateral thalamic nucleus (possessing less marked lateralization in relation to movement) the number of discharges from the cells changed in the form of a "hump," with maximal increase of frequency during the second stage, when the subject was retaining the series of numbers in his memory (Fig. 3).

The number of cell discharges in the three stages of the test of range of operative memory changed in the opposite direction in the posterolateral thalamic nucleus: during retention of the series of numbers in the memory (second stage of the test), slowing of the discharges in this structure was observed.

Cell activity during this test in other structures was irregular and indefinite.

In only one of the structures investigated (cerebral peduncles) were the dynamics of the cell discharges completely absent: in this structure their background number remained unchanged at all stages of the test, affording some evidence of the uninvolvement of this structure, at least in this particular type of human mental activity.

Nearly all the subcortical structures investigated were characterized by a sharp increase in the number of cell discharges in stages II and III of the test if the answer was incorrect (mistakes in reproduction of the series of numbers).

The need to apply repeated anodal polarizations for diagnostic purposes and for therapeutic electrolytic destruction presented a favorable opportunity for investigating operative memory actually during these electrical procedures, immediately after their end, and at various times thereafter. The range of operative memory was found not to change as a result of temporary exclusion and destruction of small areas of brain tissue within the limits of the following subcortical structures in one hemisphere: medial part of the globus pallidus, ventrolateral and posteroventral thalamic nuclei, head of the caudate nucleus, subthalamic region, and hippocampus. Pinpoint foci of destruction within the limits of the ventrolateral thalamic nuclei or globus pallidus of both cerebral hemispheres likewise did not change the range of operative memory.

On the assumption that the integrity of operative memory is an important prerequisite of certain long-term memory processes, for example of recollection, special investigations were undertaken. Long-term memory (recollection and retention) were studied by S. Ya. Rubinshtein's method, widely used in clinical practice. This method can be used to investigate individual differences in learning 10 words (or other objects) by heart and retaining the learned material for a certain period of time (1 hr or longer). Material for memorizing is selected afresh for each experiment in accordance with special rules. The first series of investigations was carried out before the operation, when the patient showed no deviation from the accepted normal. Similar results were obtained also after the operation. Investigations of long-term memory in the period of diagnostic and therapeutic electrical procedures showed that after these procedures directed toward various subcortical structures (globus pallidus, certain thalamic nuclei, caudate nucleus, hippocampus, etc.) the curve of memorizing was essentially unchanged from the control: memorizing and retention of the material in the memory were not affected. Obviously temporary exclusion or destruction of brain tissue in these subcortical

structures in one hemisphere, or of the thalamus and globus pallidus in both hemispheres, had no significant effect on the mechanisms of long-term memory.

Investigations of patients in the late follow-up period $(1-2$
years after the end of treatment) revealed no deterioration of operative and long-term memory. In fact, results were obtained suggesting some improvement of long-term memory, possibly because
of a general tendency toward normalization of brain activity as a
result of eradication of pathological manifestations by the treatment given.

The practical significance of the results described above for
the subsequent fate of the method of implanted electrodes is perfectly obvious, because had the result been opposite, it would have
been a contraindication to this method of treatment. On the theoretical plane, the results must be discussed from the standpoint of
the extremely wide, universal importance of operative memory for
many psychological processes and for mental activity as a whole.
The evidence of these results is in support of a general cerebral
rather than a narrowly localized structural and functional organization of the mechanisms of operative and long-term memory.

The appearance of changes in the ESCG during presentation
of psychological tests under our conditions depended primarily on
the underlying pathological process. The tests applied had the significance of a function test. On the other hand, the consistency with
which changes in electrical activity were found in many brain structures during these psychological tests may be ascribed to mechanisms responsible for the ability, inborn in the great majority of
people, to perform not more than one type of complex mental activity at the same time. Naturally, in time the process may, and
does, develop into something more complex. We stress this aspect
because the distinction between cortical regions mainly related to
a given activity (to exaggerate the importance of the cortex) could
apparently serve as the basis for the simultaneous occurence of
several activities. A closely similar view has been expressed by
the mathematician A. K. Kolmogorov (1963), who claims that the
process of thinking consists of a series of elementary intellectual
acts following each other in rapid succession. More or less similar opinions on this question are held by G. I. Polyakov (1964) and
others.

Is the participation of all (or many) of the deep structures necessary for performance of a relatively simple test? The results of temporary exclusion (polarization) of various structures (hippocampus, etc.), during which this type of mental activity is undisturbed, and also the absence or the vagueness of the dynamics of cell activity in various structures during this test enable us to answer this question in the negative. Even if investigations show that the exclusion of any part of the brain does not lead to disappearance of a particular phenomenon, this is still not evidence against its participation in these processes under normal conditions.

The results of the investigations described above show that deep portions of the brain evidently play a more important role in the mechanism of mental phenomena than has hitherto been stressed. Their participation creates a background representing the resultant of complex interactions of ascending and descending influences, whereas the role of the cerebral cortex is predominant in determining the specific character of the activity, and also, evidently, in the organization of this background. The wide variety of changes in the ESCG of different deep brain structures during the performance of psychological tests and their dependence on the initial background emphasize the reason for failure of attempts to find correlations between the character of the ESCG and the type of mental activity.

The discovery of a regular pattern of changes in LSP in the course of this psychological test and of a regular dynamics of cell activity indicates once again the involvement of many deep brain structures in this mental activity. Disturbance of the characteristic pattern of LSP, in several of these structures during application of additional trigger stimulation evidently demonstrates that, if not these structures themselves, then at least a rigid characteristic of this parameter in these structures, is not essential for performance of mental activity of this type. The appearance of a regular, reproducible pattern of LSP during trigger stimulation, on the other hand, perhaps indicates that against a certain functional background, additional involvement of at least some deep structures in this type of mental activity must take place.

The stability of the changes in steady potentials which is found in a number of subcortical structures during presentation of a psychological test under different conditions (including against

the background of trigger stimulation) has now received unique confirmation in the results of investigation of cell activity in these
structures. A severe disturbance of reproduction of a numerical
series could be combined with considerable changes in the LSP pattern. The change in pattern of the dynamics of cell discharges under these conditions was much more regular and obvious.

By using the criterion of cell activity, and by using it in particular in man, information has been obtained confirming the structural and functional characteristics of the phenomena investigated,
and the dynamics of changes in the character of cell activity of
certain deep brain structures under these conditions has been studied. Further accumulation of facts in this direction must define
more precisely the participation of the various deep brain structures in mental activity and assess the degree to which this participation is necessary in each individual case.

Ephaptic Transmission of Excitation as a Factor in the Synchronization of Neuronal Activity[*]

N. Yu. Belenkov

Department of Physiology
Kirov Medical Institute
Gor'kii, USSR

Synchronized rhythms recorded in large or small areas of the central nervous system occupy an important place in electroencephalographic patterns of brain activity. Many of the more important physiological functions are connected with their appearance. This accounts for attempts to discover the origin of these rhythms and the mechanism of their formation from separate elements of cell activity.

During his many years of study of the encephalographic and encephaloscopic manifestations of conditioned-reflex activity, M. N. Livanov has paid much attention to the investigation of synchronized electrical oscillations. In one paper (1962) he describes an observation made by his collaborator L. I. Kaburneeva. She found that if microelectrodes are gradually pushed deeper into the cortex, cells firing at the same rhythm can be detected. Assuming that one type of synchronization is dependent on the ability of large numbers of neurons located close together and possessing identical ability to begin to act in unison, Livanov writes: "It seems to me that besides synaptic influences, influences exerted by closely situated neurons on each other through their electric fields may also be responsible for the appearance of spontaneous synchronized activity." Believing, like Livanov, that the extrasynaptic factor in excitatory transmission plays an important role in brain activity and,

---
[*] Pages 21-30 in the Russian edition.

in particular, in the synchronization of neuronal activity, I am devoting this paper to a discussion of this problem.

Electroencephalographic waves recorded from the brain by macroelectrodes are the resultant of addition of currents from many neuronal elements. Consequently, the amplitude and duration of the waves composing the encephalogram must depend on the more rapidly flowing currents arising in brain cells, the frequency of their appearance, and their coincidence in time. With an increase in the number of units firing simultaneously, the potential in the region of the recording electrodes will naturally increase. From this is derived the concept of the phenomena of "synchronization" and "desynchronization" of electrical activity reflecting two extreme aspects of the combined activity of neurons.

When a nerve cell is excited, as well as the rapid and high potential generated by its soma and axon, and discharged in accordance with the "all or nothing" principle, a slowly increasing postsynaptic current is generated in the neuron before this spike potential. Many investigators consider that the encephalographic waves are the resultant of summation of the postsynaptic potentials of the neurons, although this idea still remains nothing more than a hypothesis. The original view put forward by Adrian and Matthews (1934) on the importance of summation of spike potentials in formation of the surface waves of the brain cannot be fully rejected. Papers are continually being published in which the close connection between these waves and the manifestation of cell spike activity is described (Baumgarten and Schaefer, 1957; Jung, 1953, and others).

In support of Adrian's view, Li (1959) cites observations which he made during simultaneous recording of the activity of cortical cells not more than 1 mm apart. Their discharges were extremely similar and consisted of bursts of 5 - 10 spikes. These spikes from a particular pair of neurons were recorded almost simultaneously, with a discrepancy of 2 - 40 msec. In the state of "arousal," evoked by stimulation of an afferent nerve, the synchronized character of the discharges became less marked, but even under these conditions there was a tendency for both cells to fire after equal intervals. The most marked synchronization was observed after application of strychnine to an area of the brain, when the discrepancy between spikes was 1 - 8 msec.

It must not be forgotten, however, that spikes are formed when the postsynaptic potential reaches a critical level. Accordingly, although these potentials may constitute the basis for formation of surface slow waves, there are evidently not sufficient grounds for a unitary interpretation of their origin. The evidence, in fact, suggests rather that they reflect cell currents of different genesis and localization. If, of course, the postsynaptic potentials do not reach threshold intensity, the encephalogram will be formed by these currents alone.

Cell potentials generated in an electrically conducting medium are shunted and cannot therefore spread widely. At the same time, these potentials are not short-circuited, and they evidently influence processes taking place not only in the cell itself, but also in other nearby elements. If cells are closely packed, their extrasynaptic interaction will be very significant both in the cortex and in other structures.

Interactions of this type probably occur in pyramidal neurons which lie parallel to one another (Fessard, 1961).

Excitation of living structures by the direct action of biological currents on them has been called ephaptic, from the Greek word "ephapto" meaning to set on fire. The existence of this mechanism in the central nervous system is largely based on indirect evidence, although some direct evidence has now been described. Some of this direct evidence consists of facts obtained in experiments on nerve cells of invertebrates.

In lower animals electrical synapses are found in which the transmission of excitation takes place by an electrotonic mechanism (for example, the synapses of the giant cells of the crayfish). Facts are accumulating which show that the passage of excitation from one neuron to another may take place by direct action of the electric field on the cell membrane. These conclusions have been made principally on the basis of absence of presynaptic potentials during excitation of neurons. Ephaptic transmission has been found during investigation of the cells of the cardiac ganglion of the lobster (Hagiwara et al., 1959), the ganglia of Aplysia (Arvanitaki, 1942; Tauc, 1959), the supramedullary cells of the crayfish (Furshpan and Potter, 1959), and in certain other cases.

Hagiwara and Morita (1962) investigated the relationships between two "colossal" cells in the segmental ganglion of *Hirudo medicinalis*. In the absence of synaptic connections between these cells, these workers found definite electrotonic connections. By introducing a microelectrode into each cell and an additional electrode into one of them to create artificial depolarization, they observed that a spike potential developed in that cell when the depolarization reached a certain value. At the same time, the second cell also became depolarized and discharged with a spike. When the electrical stimulus acted on the root of the ganglion, an antidromic spike was recorded in one cell, and before its arrival, slow depolarization was observed. Similar phenomena took place simultaneously in the other cell. Conduction was in two directions. When the spontaneous activity of these "colossal" cells was recorded simultaneously, their excitation was seen to be synchronized. The existence of purely electrotonic connections was thus demonstrated between the investigated nerve cells. One year later similar experiments with identical results were published by Eckert (1963).

In experiments on the cardiac ganglion cells of the lobster, Watanabe and Bullock (1960) showed changes in the activity of one neuron when a subthreshold slow potential acted on another. This effect took place electrotonically. Fast potentials of the second cell began to arrive synchronously with spikes of the first. They conclude from their observations that subthreshold, electrotonically spreading currents may bring about widely synchronized bursts of activity of a whole group of neurons. Bennett et al. (1963) investigated electrotonic connections between spinal electromotor neurons of electric fishes. When the fibers of one cell were stimulated, a slow depolarizing potential and spike were generated simultaneously in the neighboring cell. Artificial depolarization of one cell caused depolarization of the other. Transmission was electrotonic. The existence not only of chemical synaptic transmission, but also of electrical transmission between the investigated cells was confirmed by electron microscopy. Areas were found in which the membrane of one cell was in close contact with that of the other, and vesicular elements and mitochondria were absent. In other regions a synaptic space was present, together with typical synaptic plaques and vesicles. Bennett and co-workers consider that their findings are of particular interest because the cells they studied are related to the vertebrate central nervous system.

Investigation of the ephaptic factor in a system of neurons possessing ordinary synaptic connections is made difficult by the fact that the experimenter has as yet no reliable method of isolating the cells from synaptic excitation. However, on the assumption that nicotine completely blocks transsynaptic transmission, Libet and Gerard (1938) used it for this purpose. Having applied nicotine to the olfactory bulb of the frog and observed synchronized electrical activity in it, they concluded that an ephaptic factor of cell excitation is present.

Hence, the results described above show that, at least in lower animals, cells may become involved in synchronized activity not only through synaptic connections, but also through purely electrical interaction. The phenomena examined must be regarded as a possible mechanism of internal synchronization (Fessard, 1961), by which is meant the generation of a focus of isorhythmically functioning neurons.

The other variant of synchronization, which must now be examined further, may be called external, because in this case the isorhythmia of the cells is achieved by the influence of a preformed focus of rhythmic activity lying outside that region, and thus acting as pacemaker or trigger, on them. External synchronization in a clearly defined form is seen, for example, during the spread of epileptiform discharges in the brain, discharges whose origin is attributed to hypersynchronization of cell activity. The trigger role in this case is played by the epileptogenic focus of activity. This classification, of course, is to a certain extent artificial.

By injecting strychnine intravenously, Bremer (1941) produced high-amplitude rhythmic discharges in cats in all parts of the spinal cord. After transection, intrinsic rhythms were recorded in the two separate parts, but when the edges were placed in contact, a common rhythm was restored as the result of restoration of physical conduction of the currents. Jasper (1958) considers that even during normal brain activity interaction takes place between adjacent groups of neurons through their electric field. In the brain of the epileptic, when the discharges in the focus of maximal activity are 10-20 times greater than those in

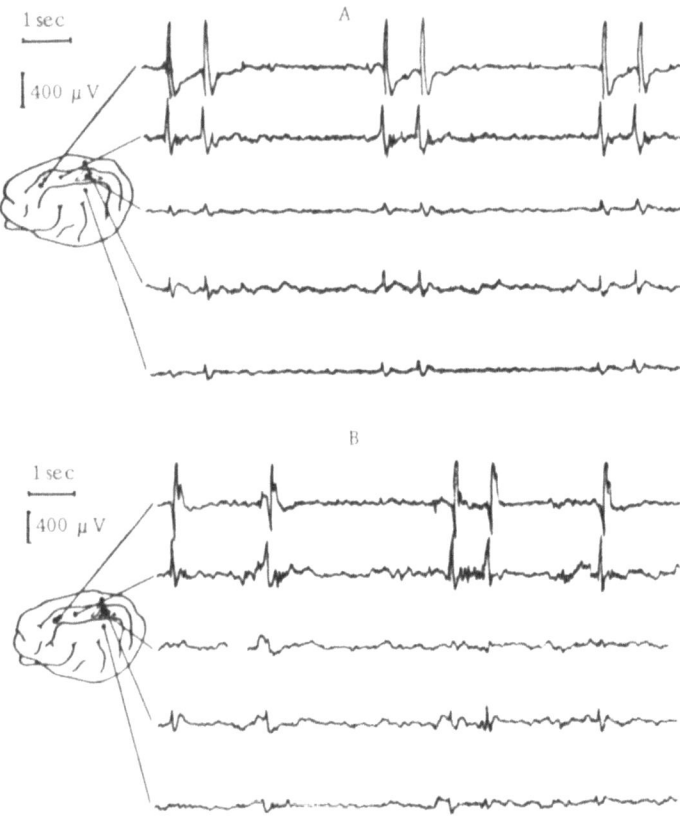

Fig. 1. Effect of epileptiform discharges evoked in the cerebral cortex
on electrical activity of functioning and nonfunctioning cortical islands.
A) Paroxysmal discharges recorded from a functioning cortical island,
smaller in amplitude than discharges of adjacent areas (cortical island
shown by a broken line); B) paroxysmal discharges recorded from nonfunctioning tissue of the island, reduced in amplitude to an even greater
degree (this area is shaded).

normal activity, in his opinion this factor acquires an important
role in the synchronization of cell function. In investigations of
the spread of excitation induced in the cortex of cats by local application of strychnine (Belenkov and Chirkov, 1961), it was found
that the generalized epileptiform discharges thus arising were recorded in parts of the cortex separated from the rest of the brain
by horizontal or vertical incisions. Later experiments (Belenkov
and Chirkov, 1964) showed that epileptiform discharges evoked by

strychnine in the region of the suprasylvian gyrus spread not only across the incisions described above, but also to an isolated island of cortex. This island, 0.5 cm^2 in area, was formed at a distance of 0.5 - 1 cm from the point of strychnine application by making a subpial incision in the brain with a special thin curved scalpel. Neuronal isolation of the island was verified in macroscopic and microscopic brain preparations. When paroxysmal discharges of the order of 600 μV or more appeared in the region of strychninization, analogous discharges, although considerably smaller in amplitude, began to appear synchronously in the island (Fig. 1A). It was thus concluded that high-voltage discharges, recordable on the surface, may develop in brain tissue on account of its physical conduction. This was confirmed also by the fact that after extirpation of the island followed by its replacement at its original position, discharges continued to be recorded from it (Fig. 1B).

Soon after, N. Yu. Belenkov and T. E. Kalinina (1965) showed that not only epileptiform discharges, but other synchronized potentials also, such as high-amplitude primary responses, can spread physically through brain tissue. They found that in the presence of an active focus evoked by clicks in the auditory cortex, an identical primary response was recorded not only in the neighboring intact area of the cortex, but also in a neuronally isolated island formed at that place. It was also recorded from a wad of cotton wool, soaked in physiological saline, and placed in the position of the extirpated island (Fig. 2). Hence, since the electric fields created by the primary responses spread for some distance over the cortex as along a conductor, it may be postulated that the true projection areas of the analyzers in fact occupy rather smaller areas than those determined by monopolar or bipolar recording of primary potentials. Perl and Casby (1954), who forestalled the passive spread of current over the cortex by the use of a Laplacian recording method, in fact obtained a more restricted localization of primary responses in the temporal lobe to acoustic stimulation.

Our observations are in agreement with the findings of Green and Shimamoto (1953), who also showed that currents can pass across incisions in the brain. In this way they explained the spread of discharges evoked in the hippocampus to the temporal region after division of all pathways between these regions. By the use of

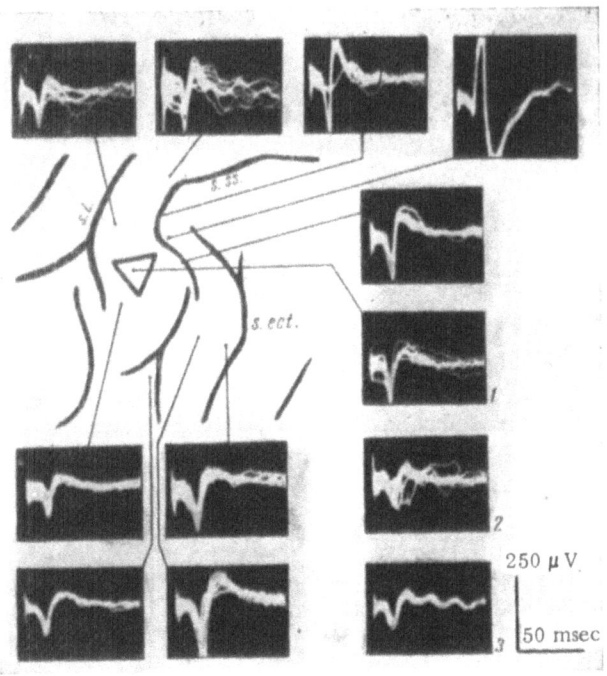

Fig. 2. Recording of evoked responses to clicks. 1) Evoked response of intact cortex; 2) from isolated island formed at that
place; 3) after replacement of island tissue with moist cotton
wool. In all cases a potential of the same type is recorded.
s. l., sulcus lateralis; s. ss. suprasylvian sulcus; s. ect., ectosylvian sulcus.

vector-electrography, Euler et al. (1958) showed that after-dis-
charges pass from the hippocampus to other structures as a result of electrotonic spread along pathways not corresponding to
their normal conduction.

Hence, irradiation of epileptiform discharges and also of other high-amplitude potentials may take place by physical spread as
well as by the well-known and intensively studied neuronal connections. The question thus arises: does a biological current spread
purely passively over the brain tissue without affecting the course
of physiological processes in it, or is it capable of acting on cells
and perhaps of exciting them into synchronized activity?

A preliminary answer to this question was obtained in experiments with the isolated island of brain tissue, which showed that after death of its cells, the paroxysmal discharges recorded from it became appreciably smaller (Fig. 1). However, more conclusive evidence was obtained in experiments in which the neuronal activity of this island was recorded. The cells of the island were found to increase their activity as epileptiform discharges evoked in the intact cortex spread into it (Belenkov and Chirkov, 1964).

We next (Belenkov and Chirkov, 1965) studied in detail cell activity in a focus of epileptiform discharges (i.e., in a strychninized area), at a distance of 0.5 - 1 cm from it, and in a neuronally isolated island formed at the same place. Observations were made on more than 2000 neurons in 50 experiments on cats immobilized with muscle relaxants. Simultaneously with extracellular recording of their spike activity, surface paroxysmal waves were recorded from the same region by another electrode. It was found that the cells behaved differently relative to the discharge recorded on the surface, consisting usually of three waves.

In the strychninized area during one of the waves of the paroxysmal discharge, 49.1% of all recorded cells were in a state of excitation. Of this number, 23% fired during the first positive short wave, 60% during the longer and higher negative wave, and 17% during the third positive wave. Meanwhile, 20% of all cells investigated which were silent before and during the epileptiform discharge began to show activity at its end, and another 20%, previously active, ceased their activity during the paroxysmal wave. Only 10.9% of neurons functioned independently of the waves recorded on the surface (Fig. 3A).

At a distance of 0.5 - 1 cm from the focus of activity, the amplitude of the slow discharges was slightly lower. The cell activity here also was depressed. However, the general ratio between the number of cells manifesting different activities was the same as in the epileptogenic focus itself. Only the number of excited cells was slightly reduced and the number of independently functioning cells increased (Fig. 3B).

The distribution of the numbers of cells with different types of activity in the isolated area of cortex, where the surface discharge had the lowest amplitude, differed from that in the intact

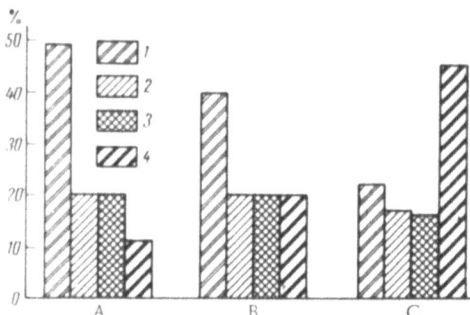

Fig. 3. Distribution of number of cells with activity correlated with surface epileptiform waves,
in % of total number of cells investigated. A) In
area of strychninization; B) 0.5 - 1.0 cm from
point of application of strychnine; C) in isolated
cortical island. 1) Cells working synchronously
with waves of epileptiform discharge; 2) cells
firing after epileptiform wave; 3) cells ceasing
activity during one wave of epileptiform discharge;
4) cells discharging independently of epileptiform
waves.

cortex, yet nevertheless 22% of the investigated cells showed increased activity during this discharge. The number of cells whose
bursts arose after the epileptiform wave and which were inhibited
during the discharge showed no appreciable change (17 and 16%
respectively). The number of cells functioning independently increased to 45% (Fig. 3C). The sharp decrease in the number of
cells excited simultaneously with the epileptiform discharge in the
island was evidently responsible for the increase in asynchronously functioning neurons. Some manifestations of the activity of neurons in an isolated cortical island are illustrated in Fig. 4.

Thus the main conclusion that can be drawn from experiments with an isolated cortical island is that excitation of the cells
in it takes place under the influence of physically spreading potentials arising outside its borders. This suggests that in the intact
cortex, an epileptiform discharge spreading extrasynaptically over
the cortex may initiate synchronized activity in cells in different
areas. As the distance from the focus of maximal activity increases, the amplitude of the paroxysmal discharges decreases

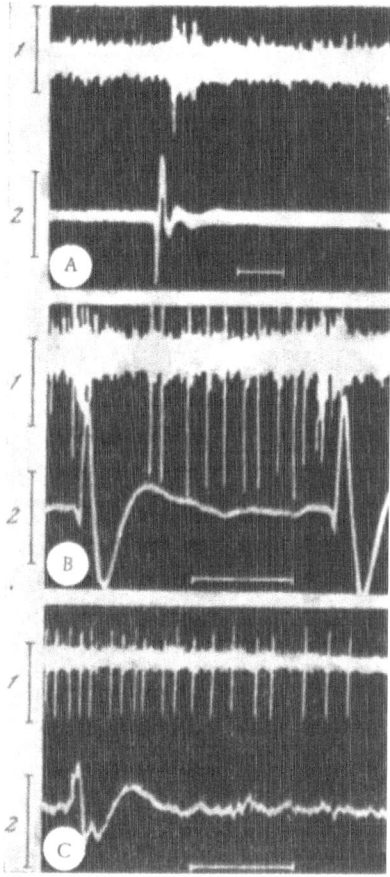

Fig. 4. Certain manifestations of activity of neurons
in a cortical island some distance from a focus of
epileptiform activity. A) During 3rd wave of paroxysmal discharge; B) cessation of spike activity of cell
during surface epileptiform wave; C) activity of cell
independent of this wave. 1) Cell activity in cortical
island; 2) surface epileptiform waves recorded from
strychninized focus. Calibration: 1) 250 μV; 2) 1 mV.

(evidently on account of the ohmic resistance of the tissue), and
this is accompanied by a decrease in the number of cells involved
in this synchronized activity. Ephaptic excitation does not, of
course, rule out the possibility of the synaptic conduction of im-

pulses. So far as other cerebral manifestations are concerned, it may be postulated on the basis of the experimentally based conclusion regarding epileptiform discharges that during their formation the activity of cell populations is influenced by the electric field. These influences, of course, are more probable when regions with a high-voltage pacemaker appear in the brain, or in association with evoked responses, slow "sleep" waves, and other electrical waves with considerable amplitude. Whereas extrasynaptically spreading currents are of too low an intensity to evoke spike discharges in the cells, they probably have some influence on their excitability, facilitating the rise of postsynaptic potentials up to the threshold level.

Tasaki et al (1954) studied currents in individual cells of the visual cortex and lateral geniculate body in cats and concluded that the extracellular potentials generated by the neurons do not exceed 3 mV. However, subsequent experiments showed that extracellular action potentials may be much greater, and according to Martin and Branch (1958), these potentials generated by pyramidal neurons may reach 17 mV. Very probably this amplitude of current is sufficient to excite cells in that particular electric field.

By stimulating the cat cortex with single pulses, Chirkov (1965) determined the magnitude of potentials spreading physically into a neuronally isolated island and causing excitation of the cells. Potentials were recorded extracellularly. Preliminary results showed that this potential under these conditions amounted to 8 - 16 mV. Consequently, the ephaptically acting factor is commensurate with the threshold values at which excitation of the neuron takes place. This potential may in fact be smaller, especially for cells of the intact cortex. After application of strychnine to the island the magnitude of a physically spreading potential causing the generation of cell spikes fell to 2 - 8 mV, and this must naturally be connected with an increase in excitability of the cells (Fig. 5).

It follows from what has been said that electrical waves recordable on the surface, even epileptiform, are much smaller in magnitude than those exciting individual neurons. It is evident, however, that if the intensity of encephalographic gradual slow waves passing across a tissue incision is high enough, they will

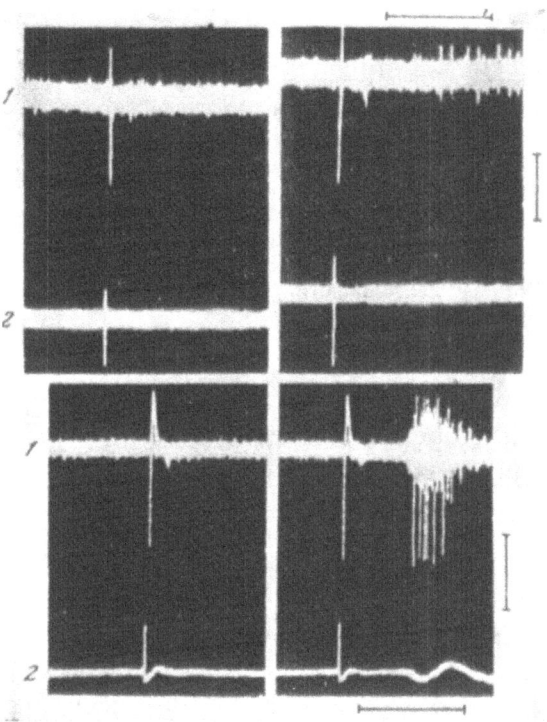

Fig. 5. Determination of excitation threshold of cells
in an isolated island of the cortex during single electrical stimulation of a cortical point outside this island.
1) Extracellular recording of cell discharges; 2) recording of magnitude of stimulating potential spreading into
island. Moment of application of discharge can also be
seen in 1 (artifact). Calibration 10 mV applies only to
curve 2. Frames on left represent subthreshold stimulation, on right, threshold. Lower frames after application
of strychnine (0.1%) to cortical island. Time marker
200 msec.

excite ephaptically the cells of an isolated cortical island. It may
be postulated that cells in the island located next to the incision,
with highest excitability on account of injury, will be excited electrotonically before the others. Further, synchronization of discharges of the remaining cells in the island is dependent on the
activity of these cells. From this point of view, previously ex-

pressed by Granit et al. (1944), an artificial synapse is formed at the site of the incision, possessing increased excitability because of the partial depolarization taking place in its neighborhood.

Finally, if the gradual, slow brain potentials are inadequate to excite the cells of either the intact cortex or the isolated island, very probably they increase their excitability, thus facilitating the conversion of their synaptic potentials into spikes. The solution of this problem, of course, requires special investigation including the recording of intracellular potentials.

CONCLUSION

Generation of electricity by nerve cells and their special sensitivity to this type of energy suggests the possibility that biopotentials may have a direct influence on neuronal function. The special significance of the ephaptic factor is seen in the activity of the nervous system of lower animals, in which synaptic connections have not attained a high level of development and sometimes are completely absent. However, there are definite grounds for considering that the action of an electric field on cells may also take place in the brain of higher animals. The evolution of special synaptic mediator mechanisms in the central nervous system is aimed at improving the coordination and integration of its activity. Meanwhile, the generalized and general tonic regulation of brain activity, with no specific features of any kind, may possibly be maintained in larger or smaller regions of the brain by the ephaptic factor. If the role of this factor is revealed by the generation and spread of epileptiform discharges, reflecting the result of cell hypersynchronization, the problem of the significance of the extrasynaptic influences of biopotentials under normal conditions becomes less hypothetical. Investigation of the mechanisms of synchronization of cortical unit activity is one of the fundamental problems in modern neurophysiology. The study of the ephaptic factor must broaden concepts of these elementary, yet essentially significant, electrical manifestations of brain activity.

The Biologist and the Mathematician: A Necessary Symbiosis*

M. A. B. Brazier †

Brain Research Institute
University of California, Los Angeles

To gain knowledge about the internal organization of living systems is the goal of every research biologist and is especially tantalizing to those who work in the neurophysiology of the brain. It is only comparatively recently that the approach through mathematical models and artificial simulation has been added to the classical scientific methods of analysis and synthesis, of stimulus and response (with the brain in between as a black box).

Much is owed to the influence of figures such as Norbert Wiener, who insist on similarities between living and nonliving systems in terms of information transmission and control, and it is not mere chance that the evolution of this type of biological research should develop in the age of computers. It is easy to forget that only 15 years ago it was rare for a biologist to have access to a computer, let alone to have thought out his problems in the form of questions that could be put to a computer.

Successful modeling requires a close symbiosis of mathematician and biologist, each with his special knowledge and with a common language that allows them to understand each other exactly. For the biologist the practical test of a model is matching it to some aspect of physiological reality, yet often the material the mathematician is given to work with is not physiological reality at all, but the hypothetical concept of the biologist, a hypothesis that may be erroneous.

---

* Pages 41-46 in the Russian edition.

† Work supported by #5-K6-NB-18,608 and Grant #NB 04773 from the USPHS.

There have been and still continue to be, many intriguing mathematical models proposed for various facets of neuronal mechanisms. Possibly the most widely used model in neurophysiology is the one for the conduction of the nerve impulse, which we owe to Hodgkin and Huxley. Their mathematical model was originally published in 1952 (Hodgkin and Huxley, 1952) and more explicitly stated in 1958 (Hodgkin, 1958).

This model describes the movements of potassium ions out of and sodium ions into the nerve, when impulses pass along the fiber in the form of electrical action potentials. These ionic fluxes result from a sequence of changes in permeability of the membrane of the nerve to potassium and sodium ions (Hodgkin, 1964). The mathematical treatment developed by these workers is based on the probabilities that these charged particles are concentrated at the relevant place on the membrane at the critical time. The equations developed from this model for expression of the membrane current density and the velocity of the propagated action potential along a nerve fiber will be found in their publications. Essentially, the equations consist of a partial differential equation describing the current distribution along the nerve fiber, and some additional differential equations describing the current-voltage-time characteristics of the nerve membrane. Hodgkin and Huxley have published an example comparing a curve calculated numerically with one recorded experimentally from a biological preparation, in this case the nerve axon of a squid, an animal with an unusually large nerve fiber that has proved a delight to the neurophysiologist.

The Hodgkin-Huxley equations for propagation of nerve signals along the axon, although they have met their challengers from time to time, have been used very extensively by others in the field, and have in some cases been treated to further refinements, bringing them closer to biological reality [some quite recent models bringing in the spatial element are, for example, those of Dodge and Cooley (1965), and of Harris and Weaver (1965)].

Now the model just discussed is only for the initiation of the nerve impulse in the shaft of the neuron. It is not intended to apply to the cell-body or its dendrites, and is not intended to encompass the organization of any complex of neurons or nerve net. It represents one important link in the propagation of activity but does not describe, and is not intended to describe, how impulses are transmitted across junctions, i.e., across synapses, between different neurons.

There are many types of these synapses within a nerve net, that is, within a biological nerve net, and the universally descriptive mathematical model has proved elusive to its searchers.

When the first model for a nerve net that made any great impact was proposed, i.e., the model of McCulloch and Pitts (1943), physiologists had not yet realized that all parts of the neuron did not share the same properties of conduction and transmission as those that had been found for the long fiber.

Twenty-two years ago, when that model was first proposed, physiologists really knew very little about the behavior of neurons in the brain, and were basing their concepts on the hypothesis that the brain shared the same electrical properties as the long axons of the peripheral nerves in the limbs. About the latter a great deal of information had been gathering for over a century — in fact, ever since Du Bois-Reymond finally provided incontrovertible evidence that the nerve impulse was accompanied by an electrical signal, the action potential. In his famous book on animal electricity (Du Bois-Reymond, 1848) he said:

> "If I do not greatly deceive myself, I have succeeded in realizing in full actuality (albeit under a slightly different aspect) the hundred years' dream of physicists and physiologists, to wit, the identification of the nervous principle with electricity."

The outstanding electrical property of the long nerve shaft of a peripheral neuron is that it either fires or it doesn't — a beautiful digital signal and one which lends itself to the logical notation: "if and only if." This property, christened by Adrian (1914) the "all-or-none" principle, states that provided an impulse is strong enough to be propagated at all, the amplitude of its action potential and the velocity of its conduction will be independent of the strength of the stimulus. The occurrence of an all-or-none action potential in the conductile part of the neuron carries the message that an adequate stimulus has been received, but is not in itself proportional in size to the stimulus strength.

At the time when the McCulloch—Pitts model was first formulated, the hypothesis generally held was that this was how neurons in the brain worked, and it was to match logic based on this hypothesis that their model was originally designed. In other words, it was an expression in logical notation, not of factual events but of a conceptual model. Unfortunately, to this day, many biomathematicians, not in close symbiosis with neurophysiologists, pursue this same concept in designing models for nerve nets, though the originators are among those who have recognized that a two-state switch cannot mimic a neuron (McCulloch, 1959). The scope and limitations of neural network models based on all-or-none behavior and stereotyped refractory periods have been well summarized by Minsky (1964).

In 1943, when the model was first proposed, neurophysiologists were only just beginning to get their electrodes in contact with the nerve cells of the brain and only later were they successful in penetrating the cell bodies with ultrafine electrodes through which they could record the potential changes associated with activity actually within the cell itself.

Hypotheses of transmission between neurons in the brain then began to give way to direct observation, and one of the more important revelations for the model-maker was the fact that the digital nature of impulse transmission is a characteristic only of the long shaft of the neuron and that at both ends, at the input and at the output, the electrical events are graded, i.e., analog in form, and presumably should be represented in mathematics by continuous functions. The direct electrophysiological evidence for this came from many laboratories, including that of Bishop, who summarized it well in his review (Bishop, 1956). That the nervous system was a "mixed" system with both analog and digital characteristics was recognized by Von Neumann (1958) in his Silliman lectures on "The Computer and the Brain." One should realize that computer development owes much to Von Neumann's interest in the field of brain science, a fact which we neurophysiologists are proud to recognize.

The demonstration that ultimate activity, i.e., output, was influenced and modulated by graded changes in levels of polarization in multiple neuronal units (that did not necessarily lead to a discharge in these units themselves), has led several workers to an exploration of probabilistic models — ones in which the concept that the probability of units firing, as a result of these multiple influences, is the computation one should seek. In fact, the evidence that the change in state (not necessarily firing) may influence the activity of neighboring neurons, imposes on us the necessity for recognizing not only the probability of a neuron firing but also the probability that a subthreshold change in its state may influence the ensemble.

One must realize, therefore, that the ingenious model of McCulloch and Pitts in its original form was based on so extreme a simplification of neuronal interaction as to be misleading to both physiologist and mathematician. It had, however, the valuable effect of stimulating the interest of mathematicians in the operations of the central nervous system.

The challenge that remains and which succeeding generations of biomathematicians have begun to tackle is for models of information flow in the brain that are not restricted to interactions based on a stereotyped concept of synaptic transmission rigidly determined by the refractory period of nerve. Among the other influences at work are the time-varying nature of the threshold of discharge for a neuron, the lack of uniformity between the individual neurons composing a net, and the possible field effect of activity in neuronal aggregates in the neighborhood, not necessarily concerned with the particular information-flow under study. And of course there are the many environmental and extra-neuronal influences brought to bear by the bloodstream, the hormones, the transmitter substances, and the glial cells.

It would be absurd to demand from our mathematical colleagues a model which encompasses all these properties in one package, and this is not the purpose of a model but it is a reminder that the frequently used term "a model of the brain" is a misnomer.

In any case, models, whether conceptual, methematical, or three-dimensional, should be thought of as one in a series of steps toward more accurate and appropriate models, each of which may be an approximation to some facet of nervous activity and each of which may suggest a new hypothesis or a new experiment. The McCulloch—Pitts model for a nervous net, essentially a finite automaton expressed in logical notation, has now been rejected by neu-

rophysiologists, though it has led to the development of other con-
structs, each attempting to correct for one or more of its failings.

We will give only three examples from many that might be
cited. There is Beurle's (1956) model, in which the connections
between neurons are considered statistically instead of determin-
istically and in which the threshold for discharge of a neuron is
lowered sequentially each time it fires. A second example is im-
plied by Chapman's (1959) mathematical model of a self-organizing
classification system in which the degree of connection between
units is a function of the past use of the net. Conditional probability as part of a model for a net has been considered by many people
and notably by Uttley (1955), who built a hardware version capable
of counting, storing, and acting on the various sequences of connections that occurred, thus calculating the probability of activity in
any one or more parts of the net. The importance of this model
lies in the introduction of conditional probability. The application
of probabilistic theory to neural nets has received extensive consideration from Rapoport in a series of papers published in the
Bulletin of Mathematical Biophysics (Rapoport, 1950). Many other
examples could be cited were there a necessity to give an all-in-
clusive list, which is not the intention here.

A whole other field of model-making that has invaded neurophysiology is represented by the approach through the study of information processing. One great advantage for the novice entering
this interface between neurophysiology and mathematics is that the
language is couched in terms generally more familiar to the biologist than is the logical notation used by many for models of neural
nets.

This approach has led directly to a wealth of investigation of
the coding systems used by the sensory systems of the brain in receiving information and rather less, as yet, of investigation of the
codes by which it sends out its signals to the peripheral effector
mechanisms. The coded message in the input as represented by
activity in neurons undergoes many transformations before it results in output — transformations which, in neurophysiological
terms, are imposed by the properties of synapses and by convergence and divergence of the multiple neuronal pathways within the
brain, as well as by the fact that the message does not enter a passive network but one which is continually active and unpredictably

variable. Information is being processed, not only sequentially, but in parallel circuits that may have different properties. It is presumably through these parallel circuits that associations and ultimately, one supposes, memory traces are mediated.

As neurophysiologists came more and more to realize that the one-to-one kind of connection of a purely deterministic sort is largely foreign to the nervous system, the multiplicity of chain re- actions set up by a stimulus began to receive attention. Some of these chains may well contain weak links, components rendered unreliable by some transient states of the organism, such as wan- ing of alertness or loss of attention, which shut off some of the physiologically demonstrated parallel paths for input to the brain.

This need for the nervous system to be able to assess the reliability of the input channels is one of the factors that have led to concepts of the neurons reacting on a basis of probability, for failure of a neuron in a deterministic model would be disastrous. For example, it has been suggested that in a multiple-routed input, when the ratio of messages received to those lost exceeds a criti- cal value, the probability of the stimulus having occurred is recog- nized by the nervous system and processed as an input on the basis of a statistical decision.

Models that recognize the probability of failure of individual elements in the net and the need for error correction have been de- signed by several workers, including the well-known one of Von Neumann (1956), which essentially made use of redundancy.

The brain has to deal not only with the assessment of arrival or non-arrival of a signal, but with the arrival of noise simultane- ously with the signal and the continual presence of background acti- vity. Again there would appear to be a probabilistic assessment of significance based on stored association.

It is now well known that the neuronal mechanisms for selec- tion of messages at the level of input at the sense receptor, for ex- ample, the eye or the ear, exist in anatomical form as feedback routes from the brain (Brazier, 1956). The initiation in the brain of controlling impulses flowing out to the receptors implies a matching, on a probabilistic basis, of the significance of the signal at the expense of the noise.

Now let us consider briefly the problem of coding in the ner-

vous system. Mention has been made of transformations in the
message reaching the brain caused by convergence and divergence
at synapses, i.e., at the junction points between neurons in the pathway. This is only one of the complications that defeats any attempt
at a simple wiring diagram.

A few biological facts may be quoted as examples. In the
visual system, for example, excitation in the mammalian retina
has to pass through at least two synapses before it reaches the
cell body of the neuron whose fiber reaches into the brain. It is
only at this last stage that the message, originating as a graded
potential, is passed on in a digital, all-or-none code. This coded
message reaches another relay station in the thalamus where it
has to cross yet another synapse to continue on to the cortex.

The point to be emphasized is that neurophysiologists who
have explored the response of single units have found individual
neurons that respond in the same way to wavelength changes as
they do to changes in position of a stimulus in the receptive field.
Thus the coding in single units, as MacNichol expresses it (Mac-
Nichol, 1965; Wolbarsht et al., 1961), is ambiguous, although the
total message is obviously not.

Findings such as this have led neurophysiologists to conceptual models in which the code is envisaged as being carried, not by
one-to-one connections, but by patterns of excitation formed by the
combined responses of several, or even very many units. The
challenge to the model-makers is that this patterning is both temporally and spatially distributed (Bullock, 1961).

Baffled as one may be by this evidence for ambiguity of message coded in single neuronal units when derived from the same
receptor organ, an even greater challenge is presented by the evidence for multisensory convergence onto cortical neurons.

It has been known for some time that several sensory modalities (visual, auditory, tactile, etc.) converge onto individual neurons in the nonspecific systems of the depths of the brain [for example in the reticular formation (Scheibel et al., 1955), in the nucleus center median of the thalamus (Albe-Fessard and Gillett,
1961) and in the hypothalamus (Cross and Green, 1959), as well as
in the hippocampus (Green and Machne, 1955)] but more complicating still, in terms of coding, is the more recent evidence in the

primary receiving areas of the cortex for convergence from different sense modalities onto the same neurons (Jung et al., 1963).

It is not so surprising to the electrophysiologist to find long-latency responses from a second modality evoked in a neuron in an area of the cortex primarily specific to a different modality, for this may have travelled by a long polysynaptic route, though even this must complicate the decoding process. More puzzling are the clearly bisensory neurons which give similar short-latency immediate responses to two different modalities, for example visual and acoustic, or visual and vestibular, or vestibular and acoustic. All of these bisensory neurons have been found by Jung and his colleagues (1963) in Freiburg, who have even found (but more rarely) trisensory neurons responding with short and constant latencies to visual, acoustic, and vestibular stimuli.

Possibly the existence of neurons in primary cortex that are multisensory in their responses may be one of the causes underlying what is usually called "spontaneous" activity by the experimenter who is controlling only one modality of sensory input. Spontaneity has always seemed rather a mystical term for a scientist to use.

In presenting a biologist's view of some of the problems encountered when attempting mathematical descriptions of biological systems, it is not the intention to disparage these; in fact the exact opposite is intended. What is needed is a closer symbiosis of the biologist with the mathematician.

It will be noticed that allusion has been made to only a few of the many constructive models that the mathematicians have contributed to neurophysiology. This is only because there are so many. Anyone who is familiar with this field must realize that even to give an adequate list of the references is an impossibility, for this is currently an area of great activity. Indeed, the mathematician is probably going to find the neurophysiologist turning to him more and more often, especially as he faces up more courageously than he has in the past to the nonlinearity of the nervous system. A step in this direction is consideration of quasilinear models, possibly along the line of those suggested by Licklider (1960) for manual tracking by the human operator. This is not to say that no linear relationships have been found in the biological

preparations. Within a considerable range of normal physiological
activity, linear functions are found, but there is little doubt that
the physiologist will present the mathematician with more nonlinear than linear examples for which he desires a solution. Biological systems do not give identical responses to identical stimuli.
They are not time-invariant. And study of the nervous system
makes it clear that it is not a network of linear components.

What we ultimately need for the mathematics of the central
nervous system is a nonlinear statistical theory. Perception, to
use the psychologists' word, is the result of the decoding process.
Clearly the earlier concepts of point-to-point connections from
receptor to cortex were oversimplifications and fell far short of
explaining "perception." Data are now available from deeply implanted electrodes in the brain of man, an experimental subject
who can report the result of his decoding of the message. From
such observations it becomes clear that many regions of the brain
at considerable distances from the primary receiving cortex are
involved in perception of the stimulus.

To extract significant information from the multitude of signals entering the brain through noisy channels there would appear
to be the need for some probabilistic selection process based on
comparisons made with previously encountered messages and
their associations. The amount of storage equipment required for
such a system would be immense, but not beyond the capacity of
the human brain.

Inhibition of the "Escape Reaction" in Rabbit: An Analysis of the Mechanism of Hypnotic Akinesia[*]

P. Buser, G. Viala, L. Chertok, and G. Fontaine

Laboratory of Comparative Neurophysiology
Faculty of Sciences, Paris, France

The mechanism of hypnotic "akinesia" or "catalepsy" (or tonic immobility) still remains a matter of discussion. Before the use of electrophysiological explorations, various authors suggested interesting, although sometimes complex and controversial, hypotheses on both the origin and the significance of this phenomenon. In these various explanations of hypnosis, a confusion sometimes exists between the general biological meaning of the phenomenon and its physiological mechanism per se. To consider only the latter aspect, it has been suggested that hypnosis is a "generalized inhibition" (Czermak, 1873), a paralysis caused by fear (Preyer, 1878), an inhibition of defense reactions caused by fear (Danilevski, 1890), a modification of postural reflexes (Verworn, 1897), a tonic inhibition of movements creating a state "close to sleep" (Mangold, 1914), etc. Pavlov's ideas on the problem are of considerable interest. According to him, hypnotic catalepsy represents an intermediate state between wakefulness and sleep, in which activity is partially "inhibited," while some aspects of wakefulness persist. This concept was mainly if not entirely focused on events at the cortical level. Some more recent studies have localized the origin of "inhibition" at subcortical levels — this in accordance with the growing interest in the functional significance of brain stem reticular structures (see Svorad, 1957).[†]

---

[*] Pages 47-53 in the Russian edition.

[†] For literature on the subject, see Chertok (1963).

Modern electrophysiological studies on mechanisms of animal hypnosis have been mainly based upon analysis of electrical activity of the cerebral cortex or rhinencephalic structures like hippocampus (Ectors, 1936; Gerebtzoff, 1941; Schwarz and Bickford, 1956; Svorad, 1957; Silva et al., 1959; Lievens, 1960; Lieberson, 1961; van Reeth, 1963; Ruckebush, 1964 etc.). According to the majority of these studies, a hypnotic state is not characterized by any specific electrocortical activity, nor does it necessarily coincide with sleep-like activity, the latter usually being characterized by slow waves or spindles. There is however, a general consensus that hypnotic immobilization may favor the development of sleep. Finally, it seems that EEG analysis, in animals as well as in man,[*] has provided little basis for the understanding of the real mechanisms of hypnosis.

The experiments which we are presenting were not, at first, concerned with mechanisms of hypnotic immobilization, but with the so-called escape reaction in rabbits. Results have appeared to us, however, as being of some interest for a better understanding of induced akinesia. Our observations were performed under different experimental conditions, either on preparations immobilized with curare (flaxedil) and only submitted to local anesthesia, on decerebrate preparations with surgery performed under ether, or on freely moving subjects, with implanted electrodes. Depending on particular cases and conditions, electrical activity was recorded from various points on the neocortex, in bulbar or pontine reticular nuclei, and from the central stump of cut branches of the sciatic nerve. In some cases, nerve discharges as well as fast reticular discharges (recorded with semimicroelectrodes) were fed into a frequency-to-amplitude converting device.

RESULTS

Somatic Escape Reactions in Acute
Preparation — Their Inhibition by
Cutaneous Stimulation

In rabbits, the familiar escape movements to prehension, which consist of strong simultaneous flexions and extensions of

---

[*] Human hypnosis will not be considered here.

both posterior legs, may be considered a somatic component of the defense reaction. Defense reactions also include visceral components, such as vascular or cutaneogalvanic, as shown by the work of Sokolov (see Sokolov, 1963). Only this somatic defensive reaction will be considered below.

Motor Escape Movements in Decerebrate Animals. In decerebrate unanesthetized preparations (by intercollicular transection), the escape reaction (ER) may appear "spontaneously" (i.e., in absence of any identified stimulus); it consists in rhythmic flexions, released from the hyperextended rigid state. More often, it is necessary, in order to elicit it, to apply one or another cutaneous stimulus, like touching the foot pad, pinching, tapping, subcutaneous repetitive electrical shock on one leg, etc. (Fig. 1).

On the other hand, ERs are completely abolished by small amounts of various anesthetics (chloralose, barbiturates, or ether) or by transection below the medulla at C_1 level. These preliminary observations indicated that ERs can only be obtained in the absence of anesthesia and are likely to be organized in the brain stem.

Finally, another important observation is that ERs are suppressed by a slight hand pressure on the dorsolumbar skin of the animal (Fig. 1). As long as pressure is exerted, there is no ER; this can be seen when the excitatory stimulus outlasts the suppressive one.

"Escape Reactions" in Paralyzed Preparations: Rhythmic Efferent Discharges. In preparations paralyzed with curare (with only local analgesia) efferent rhythmic discharges can be recorded from the central end of cut branches of one sciatic nerve. As will be seen, these discharges behave very similarly, to the ERs (Fig. 2). They will be considered, from now on, as the electrophysiological equivalent of such rhythmic escape movements (escape discharges, EDs).

EDs may be elicited by stimuli belonging to all modalities, although somatic stimulation is the most effective (tapping one leg, or the nose; repetitive subcutaneous electric stimulation of one leg, etc.). Auditory and visual stimuli have less effect. "Habituation" of these EDs to a constant stimulus sets in very quickly with nonsomatic stimuli; this phenomenon will not be considered here.

Fig. 1. Decerebrate rabbit. Mechanographic recording of movements of the right (upper trace) and left (lower trace) hind limb. Upward deflection, extension; downward deflection, flexion. Inhibition of "spontaneous" escape movements by slight dorsolumbar pressure on the skin (horizontal line).

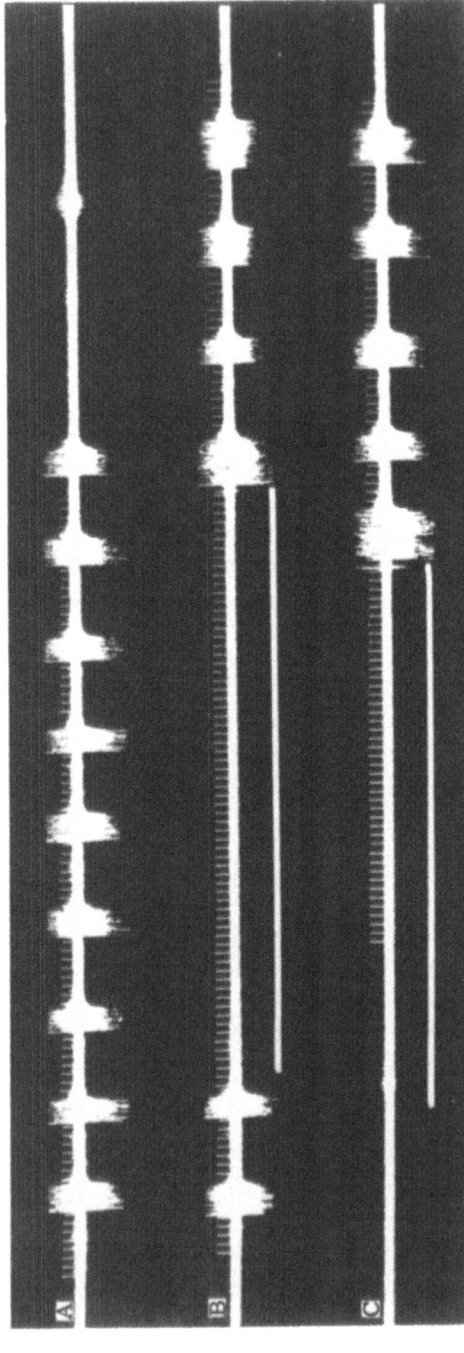

Fig. 2. Rhythmic "escape" discharges (EDs) recorded from the right sciatic nerve, elicited by repetitive subcutaneous stimulation of the right forelimb (stimulus artifacts are visible).
A: uninterrupted EDs to forelimb stimulation. B: same stimulation; suppression of EDs during dorsal pressure (horizontal line): C: dorsal pressure first; excitatory stimuli start second; EDs only appearing when pressure ends.

Fig. 3. Suppression of EDs during reversible pharmacological transection of the spinal cord at C_1 (through injection of small amount of xylocaine). A: before transection; B: after transection; C: after recovery. Single frames on the left: the flexion reflex to ipsilateral stimulation of the leg is maintained (while late components of supraspinal origin, which were not considered in this paper, are suppressed). Time: 20 msec.

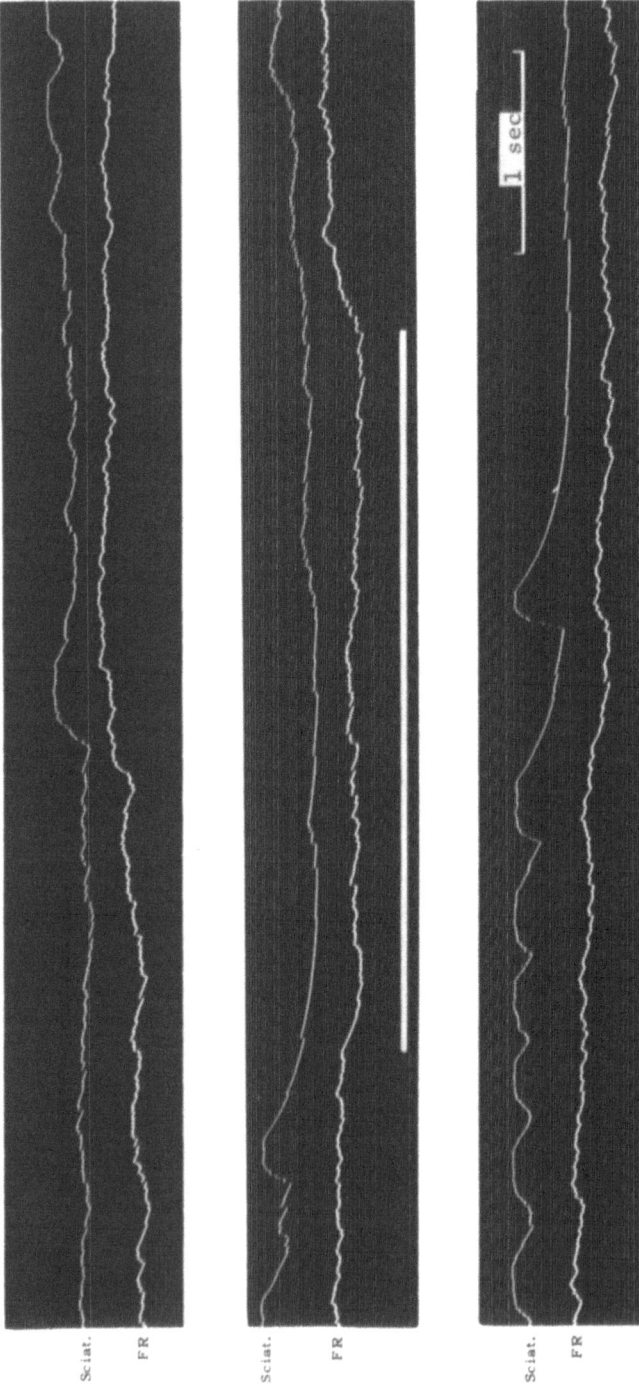

Fig. 4. Comparison of "integrated" neuronal activities recorded from the bulbar reticular formation (FR) with a semimicro-electrode (50 μ) and from sciatic nerve (Sciat.). Notice that increase in FR precedes onset of discharge in sciatic. Both activities are depressed during dorsal pressure (horizontal line).

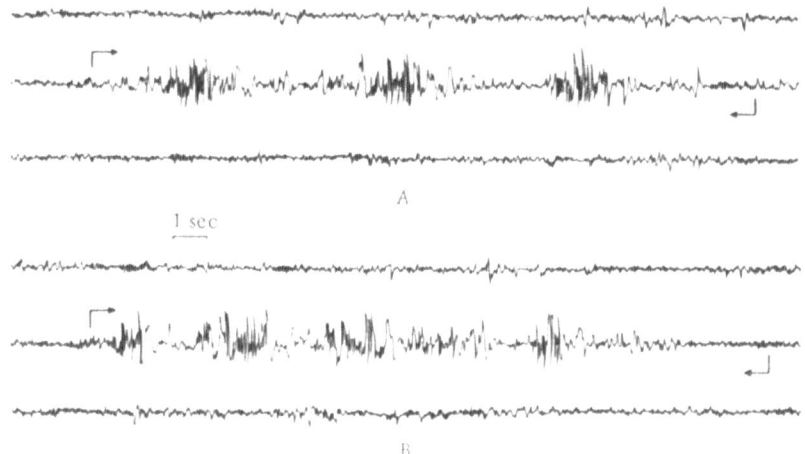

Fig. 5. Electrocortical recordings from sensorimotor cortex of rabbit (single channel). A: acute curarized preparation. Between arrows, dorsal skin pressure, produces slow cortical patterns. B: same experiment, performed on chronic animal with implanted electrodes.

EDs are elaborated at the bulbo-pontine level and seem independent of upper levels including the cortex. They are not modified during a cortical spreading depression, or even during ablation of the cortex. They reappear after inter-collicular decerebration. On the other hand, a transection between medulla and spinal cord is followed by their complete and irreversible disappearance (Fig. 3).

EDs can also be suppressed by pressure on the dorso-lumbar skin (when applied during an excitatory stimulus). Removing skin pressure causes a rebound, which strongly suggests that the suppression is actually due to a neuronal inhibition.

Together with recording of peripheral discharges, an evaluation was made of the "tonic" sustained neuronal activity in various points of the bulbar and pontine reticular formation (n. gigantocellularis, n. pontis caudalis, and n. pontis oralis). Points were thus found showing significant variations during ED sequences. In such points, the brain stem activity usually precedes the onset of the ED (Fig. 4). Depending upon the recording point, bulbar activity either oscillates, like the successive bursts of discharges, or

appears at the onset of the sequence and then dies progressively
away. Whether these different patterns of reticular responses are
topographically distinct or not, has not been investigated yet. At
any rate, lumbar skin pressure, which inhibits peripheral discharges, also blocks the bulbar increase (Fig. 4). These last observations suggest — although they do not demonstrate the fact —
that not only the elaboration of EDs but also their inhibition is likely to take place at the supraspinal–bulbar or pontine level.

A further observation which deserves mention concerns electrocortical activity. In paralyzed preparations, this activity recorded from the neocortex oscillates between a drowsy state (spindles and/or slow waves) and an "aroused" state (fast activity). It
has been regularly observed that in such preparations, a lumbar
skin pressure applied when the preparation is aroused induces a
drowsy pattern within 3 to 4 sec. (Fig. 5). Similar observations
have been made by Takagi et al. (1959) and Kumazawa (1963). In
free animals with implanted electrodes, a dorso–lumbar pressure
can induce similar slow patterns (Fig. 5).

DISCUSSION

A Possible Mechanism of "Hypnotic"
Akinesia in the Rabbit

As we mentioned above, the observations just reported are
not directly related to "hypnosis." We consider, however, that
they may contribute to the understanding of hypnotic akinesia in
this species.

It is well known that at least two methods can be used to induce hypnotic catalepsia in rodents: by submitting the animal to a
compulsory, monotonous, and possibly strong visual stimulation
(visual immobilization or fixation) or by forcing the animal into an
abnormal position (turning it onto its back).

Our data evidently concern the second procedure for immobilization.

Assuming that one necessary condition for immobilization of
the animal is inhibition of its escape reactions (somatic component
of the defensive reflex) and that such an inhibition may be produced

by pressure on the lumbar skin, then we may understand that, when the animal is lying with its back on the table, the resulting dorsal pressure plays a major role in suppressing movements. In other words, a subcortical (but supraspinal) neuronal inhibition of the escape reaction (and possibly also of other movements) seems to be the prime mover for induction of hypnotic catalepsia in rabbits. It is likely, moreover, that once inhibition has started, immobilization is maintained through a positive feedback. That is to say, skin pressure determines immobilization, which in turn favors maintenance of pressure.

Our study also included electrocortical recording during hypnosis in free implanted animals. These experiments, however, do not deserve further mention, as they could confirm previous observations. Namely, it has appeared to us that hypnotic catalepsy induced by abnormal body position is not necessarily accompanied by sleep. As already mentioned by other workers, phases of cortical slow activity may well develop during such hypnotic immobilization but intercurrent stimuli can arouse the animal without producing any change in the cataleptic attitude. It seems clear that the state of tonic motor inhibition provides favorable conditions for the development of sleep, although the latter is not a necessary condition nor a necessary consequence of immobilization.

On the other hand, from our observations under curare paralysis, it appears that lumbar skin pressure is always accompanied by sleep patterns. The dissociation which exists between descending and ascending effects in the normal animal thus seems to be lacking in paralyzed preparations. This lack of dissociation may be due either to paralysis, or to the pharmacological action of curare itself.

Pavlov's suggestion that, in hypnotic catalepsy, the "motor analyzer" is inhibited while other areas (of the cortex) may continue to function normally, deserves some final comments.

Our observations that a motor inhibition is the essential first step for tonic immobilization, and the well-accepted assumption that generalized drowsiness or sleep is but a secondary consequence of immobilization, is in good accordance with the general idea that a dissociation takes place during animal hypnotic states. On the other hand, the data just presented tend to considerably minimize the importance of the cortical level in the basic mechanism of immobilization. Of course, our observations do not necessarily apply to other species, nor can they be extended to other means of immobilization, such as "visual fixation;" in other words, they should not be extrapolated to cases which may indeed involve higher levels of control.

SUMMARY

In unanesthetized paralyzed rabbits (with only local anesthesia) rhythmic bursts of efferent discharges can be recorded from the central cut end of branches of the sciatic nerve. These discharges appear either spontaneously or following cutaneous stimulation of given parts of the body (legs or face); they mimic the "defensive escape" reaction which exists in the normal animal and may also appear in decerebrate preparations (simultaneous rhythmic flexion and extension of both hind limbs). These discharges are completely suppressed by any of the anesthetics tested. They also disappear after bulbo-spinal transection, a fact suggesting their supraspinal origin. Finally, they can be completely inhibited by a slight pressure on the dorso-lumbar skin and muscle area.

Latter observations allow some hypotheses regarding the mechanism of tonic immobilization (so-called "hypnotic catalepsia" or "akinesia") in the rabbit. It is assumed that this inhibition of the escape reaction, originating from some body areas, and likely to take place in the brain stem, is the basic mechanism for akinesia produced by placing the animal in an abnormal position (placing it on its back). It is further considered, in accordance with conclusions reached by previous workers, that drowsiness or sleep, as evaluated from inspection of the EEG, although it is favored by immobilization, is only a secondary and unnecessary consequence of motor inhibition.

On Butterflies in the Brain[*]

R. W. Doty

Center for Brain Research
University of Rochester
Rochester, New York

The external world is revealed only by processes that occur in molecules constituting the brain. Thus in a very real sense the brain creates the world, and an abnormal brain, or one which is manipulated, can create objects which exist nowhere but in the brain itself. A dramatic example of such creation was reported by Foerster (1936). Electrical stimulation of what was believed to be area 18 in an unanesthetized patient created such a strong hallucination of a flitting butterfly that the patient reached out quickly from the operating table to catch it. This butterfly obviously was nothing more than a pattern of neural activity.

To understand how neural activity can be a butterfly is a problem that challenges philosopher and physiologist alike. Just 100 years ago Simonoff (1866) first utilized a procedure which will have increasing usefulness in analyzing the mechanism of such creation. Simonoff apparently originated the technique of applying localized electrical stimulation to the brain of unanesthetized animals through chronically implanted electrodes. Using Simonoff's technique it has now been possible to create "butterflies" in the brain of a macaque and determine that this hallucination can be transferred from one cerebral hemisphere to the other in the absence of the corpus callosum, psalterium and, probably, the anterior commissure. These observations arose fortuitously from a series of experiments in which electrical stimulation is being utilized to investigate mnemonic and perceptual processes in the brains of various animals (Doty et al., 1956; Doty and Rutledge, 1959; Doty and Giurgea, 1961; Rutledge and Doty, 1962; Doty, 1965a).

---

[*] Pages 96-103 in the Russian edition.

METHOD

Seventeen macaques have been studied so far using procedures detailed elsewhere (Doty, 1965a). They are first trained to
obtain a banana-flavored food pellet by pressing a lever with their
noses by turning left when a green light or clicks at 3/sec are presented; and to avoid a shock or, in some cases obtain food, by -
pressing a lever with their right hands upon presentation of a red
light or clicks at 30/sec. Intertrial lever-pressing is punished.
Seventeen pairs of platinum-iridium electrodes with tip separations of 1-2 mm are then implanted in various loci. In animals
in which the corpus callosum has been cut a postoperative recovery period of six months has been allowed for the skull flap to knit
prior to implantation of electrodes. Electrical stimulation of the
brain (0.5-msec rectangular pulses, 50/sec, 0.5 mA for 4 sec) is
then used as a conditional stimulus (CS) at different loci to establish lever-pressing shock-avoidance (SA) or food-reward (FR)
conditioned responses (CRs). By pressing one or the other lever
the monkey can communicate which brain stimulus is similar to or
different from another, and thus allows himself to be questioned
concerning the perceptual qualities of the effects evoked by the
stimulation.

RESULTS

Figure 1 shows the location of 65 of the more than 100 points
in the cerebral cortex where electrical excitation has been effective as a CS. No unresponsive points or areas have been found,
either cortically or subcortically. With electrodes in white matter
or placed transcortically, macaques are able to detect a single
0.5-msec pulse of 0.02-0.4 mA (usually about 0.2 mA) applied apparently any place in the brain. At some loci the threshold is much
lower for tetanic than for single stimuli whereas at other points
such differences are minimal (Doty, 1965a).

The macaques are able to detect the location of stimuli at
points <1-3 mm apart, i.e., within wide limits of random variation
of intensity and frequency in stimulation randomly presented to one
or another point, >90% of the time they make the CR appropriate
to the locus of the CS. This has been true in area striata (Fig. 1,

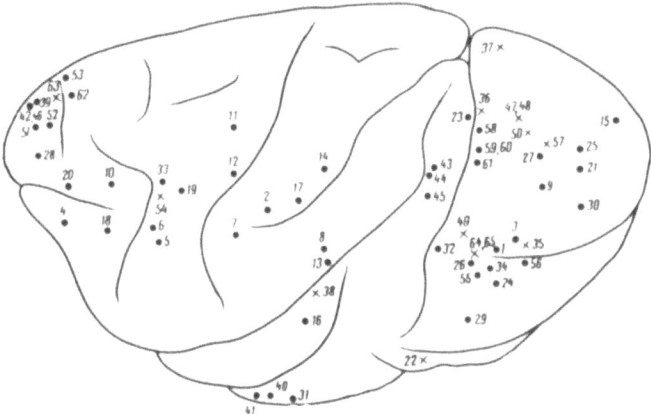

Fig. 1. Location of first 65 cortical points from which CRs were
elicited by electrical stimulation. Summary from 10 macaques,
data from right hemisphere here represented as though it were
also from left. "X" indicates penetration of electrode >3 mm
in from surface.

points 24, 26, 29), area 18 just posterior to the lunate sulcus
(points 58 vs 59, 60 vs 61), area 19 in the prelunate gyrus (points
43, 44 vs 45), rostral area 9 medial to sulcus principalis (points
39 vs 42, 51 vs 52, 62 vs 63) and area 20 in the inferior temporal
gyrus (points 40 vs 41). In one animal tested with electrodes in
area striata about 10 mm apart a single stimulus pulse, again within a wide range of intensities, to either electrode pair, was sufficient to elicit the appropriate CR.

Following the establishment of CRs to stimulation at two
usually widely separated loci (e.g. stimulation of area striata as
FRCS, frontal cortex as SACS) stimuli were applied to other loci
for the first time in the training situation to test for stimulus generalization. Such tests with new stimuli were given aperiodically
during sessions in which the monkey continued making CRs normally to the established conditional stimuli. FRCRs to test stimuli were never rewarded, and the animal was prepared for such
omissions by witholding reward for up to 20% of the FRCRs to the
usual FRCS. Failure to make a CR upon presentation of a new
stimulus can, of course, be considered significant only if it is subsequently shown that the stimulus can be effective.

Nine intact macaques have been utilized so far in these exploratory studies on stimulus generalization. When the animal is trained to shock-avoidance for stimulation at several points (monkey 4, Doty, 1965a) or for widespread stimulation as of the internal capsule, or when only shock-avoidance is used (Doty and Rutledge, 1959), SACRs occasionally are produced by stimulation at several cortical loci having no clear relation to the originally stimulated point. Despite this tendency, however, and especially if differential training is used, the most common finding is that the animal makes no CRs to stimulation in areas unrelated to that of the original CS. Equivalence of points on the surface of the occipital operculum (area striata) has been found in six of seven animals tested, and it is bilateral. In other words, if excitation of area striata serves as FRCS and of area 9 as SACS, stimulation of area striata in either hemisphere elicits an unhesitant FRCR. Stimulation of area 18 in such a situation sometimes elicits occasional FRCRs, but just as often does not. In the face of this definitive generalization within area striata, the two cases of clear-cut failure to obtain any CRs for stimulation of new loci in area 9 in the above situation may be significant.

One macaque, taught to discriminate between two loci about 10 mm apart in area striata, consistently made appropriate CRs without reinforcement or further training when the corresponding points in the contralateral hemisphere were stimulated. Data consistent with this possibility of interhemispheric generalization of a cortical spatial discrimination have been obtained in four other animals, although because of technical or procedural flaws they are not as definitive. A frequency discrimination, e.g., 4 versus 10/sec, is readily generalized between points in area striata and, on one animal, from optic tract to lateral geniculate nucleus. With stimulation in some other cortical loci in posterior parietal and superior temporal areas there is not even any transfer of training in such frequency discriminations. One animal was trained with 4/sec as FRCS in the left mid-striate area and 10/sec as SACS in right lateral striate area. Tested with both 4 and 10/sec at the left lateral striate area and right medial striate area, only FRCRs were made to mid-striate stimulation and SACRs for lateral striate stimulation. In other words, location was prepotent over frequency.

In an effort to trace the paths by which such interhemispheric

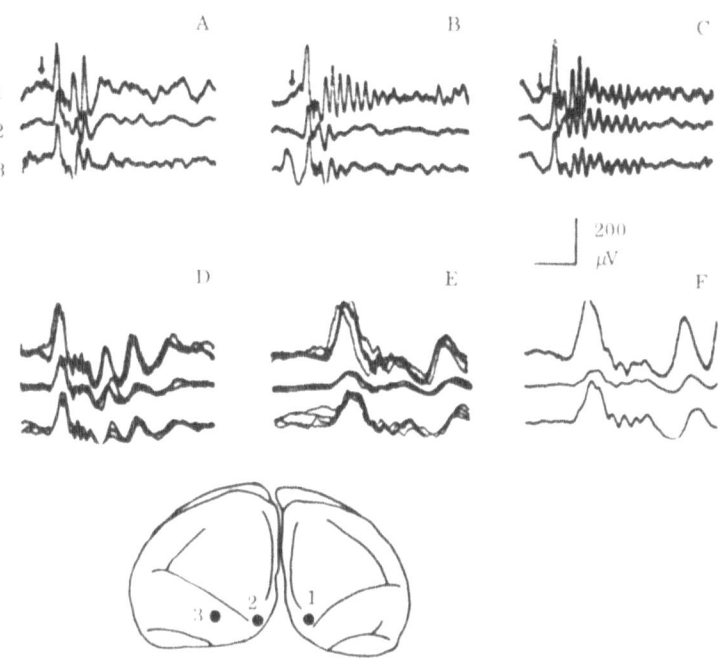

Fig. 2. Responses elicited from indicated points on occipital pole of M-18 by intense stroboscopic flashes, 0.3/sec. Monkey was alert and able to move head or eyes away from flash. Corpus callosum and psalterium had been severed nine months before these recordings were made with chronically implanted, bipolar electrodes located approximately transcortically. Note that highly complex waveforms may or may not appear symmetrically in the two hemispheres. Flash at start of trace in D, E, F arrow indicates flash in A, B, C. Time: A, B, C — 100 msec; D — 40 msec; E, F — 20 msec.

generalization may be effected, the corpus callosum and hippocampal commissure were severed in one macaque (M-18, histologically verified) and these structures plus the anterior commissure in two animals still living (M-24 and M-26).* After establishing FRCRs to stimulation of point 2 (Fig. 2) in M-18, stimulation at point 1, 3, or a point in the right hemisphere homotopic to point 3 also consistently elicited FRCRs. The animal was then trained to stimulation at point 3 (Fig. 2) as SACS. When retested at point 1

*Subsequent histology has shown that the sections in M-24 and M-26 were as intended.

on the contralateral hemisphere, two FRCRs and a few movements toward the nose lever were observed; but aside from this SACRs were elicited from stimulation at all striate points in either hemisphere except for point 2 yielding FRCRs. Differentiation was then established at point 2 so that 2/sec produced FRCRs; 10/sec, SACRs. Testing with 2/sec at point 1 contralaterally produced 15 FRCRs in 25 trials whereas 10/sec gave two SACRs and two FRCRs in 25 trials. The fact that 10/sec produced so few FRCRs is highly significant, and the paucity of SACRs is readily understood from the fact that they were made with the right hand which the animal does not readily control with its right brain (although it can learn to do so). Stimulation at point 3 gave five FRCRs in ten trials at 2/sec and two SACRs in ten trials for 10/sec. Except for nine SACRs to 10/sec for 15 trials at a deep striate locus, stimulation at the other striate loci now gave no response. The paucity of errors is remarkable.

It seems clear that interhemispheric generalization occurred to a considerable degree despite absence of the corpus callosum. Such generalization, however, in the two animals with the anterior commissure also sectioned, is so far completely absent. M-24 was trained with stimulation of the right arcuate area as SACS and right mid-striate as FRCS. An occasional SACR was made to stimulation of the left striate area and even these "fear" responses disappeared after the right arcuate excitation was also made a FRCS. The same paradigm was used with M-26 except that each point was a FRCS from the beginning and no pretraining had been given. No CRs have been elicited in 100 stimulations at eight loci in the left, "untrained" hemisphere.

In the animals with the anterior commissure sectioned there is no particular symmetry in the EEG of homotopic points on the two hemispheres. Unfortunately, the photically evoked potentials are poor. In M-18, on the other hand, where photically evoked potentials were excellent (Fig. 2), a high degree of symmetry in the responses, and occasionally in the background activity, could often be observed.

The failure of interhemispheric generalization in M-24 is particularly interesting because this monkey had previously demonstrated an ability to catch hallucinated "butterflies" with either hand to stimulation in either hemisphere! Under visual control

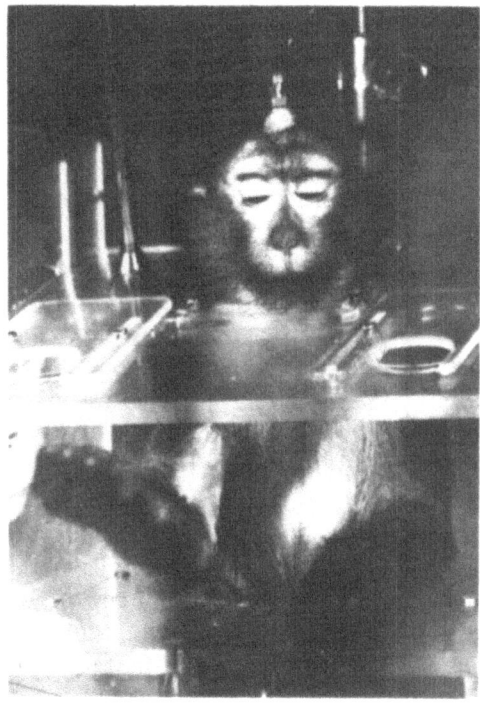

Fig. 3. M-24 looking intently at his fist and carefully opening it just after having "caught a butterfly" hallucinated upon stimulation of left area 18
just posterior to the lunate sulcus.

three pairs of electrodes had been implanted in a line in each occipital lobe 18 mm from the midline. One pair was in the prelunate
gyrus (probably area 19), another in area 18 just posterior to the
lunate sulcus, and a third in area 17 about 5 mm posterior to the
second. The first time the left area 18 was stimulated, at 0.2 mA,
50/sec, there was a fixation of the gaze. The eyes moved steadily
down. Then there was a sudden catching movement with the hand
and, staring intently as if to see what it had captured, the monkey
carefully opened its fist (Fig. 3). The movements imitate exactly
those made by macaques in catching flies.

I was so startled by the results of this first stimulation that
I was subsequently unable to be certain which hand had been used

in this first catching movement. On the second stimulation, however, a perfectly skillful, explosively sudden capturing movement
was made with the hand and arm ipsilateral to the stimulation of
area 18! The response to stimulation about four minutes later
was made with the contralateral hand, as were all subsequent ones
to stimulation here on that day. Only a slight eye movement was
made to the fourth stimulation so the current was increased to
0.4 mA. This produced two vigorous catching movements and
careful hand openings during the 4-sec stimulation. The threshold
continued to rise upon repetition of the stimulation every few minutes. Stimulation of the corresponding point in the left hemisphere
elicited the same response, and again on one occasion the ipsilateral hand was used. Stimulation at the point in left area striata
could also elicit this response but not as reliably or as vigorously
as from either area 18. Cinematographic analysis shows that the
response is not stereotyped, but rather has a wide variation in
starting and terminal postures and in the latency of its components. Frequent repetition has apparently caused a permanent
disappearance of the response. The monkey's eyes still occasionally fixate and move down, but perhaps his lack of success in
catching that elusive "butterfly" has palled his appetite for the
game.

DISCUSSION

The pathway for interhemispheric generalization will obviously depend upon many factors in the experimental situation. On
simplistic anatomical and physiological grounds, however, it might
even be expected that it would not occur via the corpus callosum
when stimuli are applied to area striata in mid-operculum. Myers
(1962, 1965) has shown on macaques that there are no transcallosal connections from one area striata to the other, and that it is
only that portion of area striata representing the vertical meridian
which has access to the transcallosal pathways in area 18. Whitteridge (1965) offers physiological evidence in support of this
idea, that it is only the cortex related to the vertical meridian
which participates in interhemispheric relations. A priori this
seems unlikely, but the ability of M-18 to effect interhemispheric
transfer of a frequency and to some extent a spatial discrimination
also suggest that some rather complex operations of area striata
can be transferred by noncallosal paths.

Whether this path involves the anterior commissure (M-24 and M-26) is uncertain, but the dramatic ability of M-24 to pursue with either hand objects hallucinated through stimulation of either brain indicates some extremely complex information can pass by lower subcortical paths. This could also be perceived in the general behavior of these animals that were able to jump several meters with unerring accuracy, and by the lack of disturbance of eye-hand coordination in "unlearned" pursuit movements by split-brain monkeys (Myers, 1965). While many sites may serve for this remarkable integration, and for the transfer of brain-created objects (!), the mesencephalic reticular formation seems a likely candidate on two grounds. First, it has been shown to play a prominent role in elaborating the interhemispheric delayed response (Rutledge and Kennedy, 1961) and, even more important, this pathway undergoes a specific increase in its excitability as a consequence of the establishment of CRs to direct cortical stimulation (Rutledge, 1965). Second, in so far as the visual system is concerned, the mesencephalic reticular formation plays an extremely important role in primates. Electrical stimulation in this region of the brain stem and/or stimuli which alert a monkey can almost turn synaptic transmission on or off in the lateral geniculate nucleus (Doty et al., 1964; Wilson et al., 1965; Pecci–Saavedra et al., 1966). In this last sense the mesencephalic reticular formation, or perhaps the entire centrencephalic system (Doty, 1965b), has extraordinary relevance to the problem of how the world of mind is created from the world of matter.

SUMMARY

1. From tests at 100 cortical points in 17 animals it can be concluded that macaques can learn to respond to electrical stimulation applied to any neocortical location.

2. They are able to discriminate between stimuli applied to points 1-3 mm apart in frontal, temporal, or occipital cortex.

3. Following training in which electrical excitation of area striata served as a conditional stimulus, excitation at other striate points in either hemisphere was immediately effective in eliciting the same conditioned reflex. Such equivalence has not been observed for excitation in the frontal lobe.

4. A spatial discrimination established for stimulation of two points in area striata about 10 mm apart is transferred without training to corresponding points in the other hemisphere. Elements of such interhemispheric generalization, together with interhemispheric generalization of discrimination between cortical stimuli at 4/sec vs 10/sec still occurred following histologically confirmed section of the corpus callosum and psalterium. Adding to this the section of the anterior commissure may forestall such interhemispheric transfer.

5. Stimulation of area 18 on either side just posterior to the lunate sulcus in one macaque with corpus callosum, psalterium, and anterior commissure severed elicited a downward movement of the eyes, and a sudden catching movement with either hand, and then intent observation of the carefully and slowly opened fist. This behavior and movements imitated exactly that of macaques catching flies, and would also seem to be identical with that made by Foerster's patient who attempted to catch an hallucinated butterfly upon stimulation of the occipital lobe. The locally elicited hallucination in the macaque must be accessible to either hemisphere via deep subcortical pathways.

Electroencephalographic Investigation of Cortical Relationships in Dogs During Formation of a Conditioned Reflex Stereotype*

V. N. Dumenko

Bogomolets' Institute of Physiology
Academy of Sciences of the Ukrainian SSR
Kiev, USSR

It was found in Livanov's laboratory in 1952-53 that synchronism of the course of electrical potentials in the cortical representations of combined stimuli developed in rabbits during formation of temporary connections (Livanov, 1952; Dumenko, 1953b, 1955). Later investigations in the same laboratory (Knipst et al., 1959; Glivenko et al., 1962a; Gavrilova et al., 1964a; Livanov et al., 1964, and others), using an encephaloscopic technique, confirmed and considerably expanded the previous findings and showed that synchronism between the elctrical activity of different cortical regions appeared only at certain stages of conditioned reflex formation in rabbits and also in a period of intensive mental work in man. In this way, for the first time, an attempt was made to compare physiological processes such as the formation of temporary connections with the character of the relationships between electrical activity of different cortical regions. In the West, work in this direction has been carried out by Adey (1961a).

Before the investigations mentioned above, the phenomenon of synchronism (especially bilateral synchronism in homologous regions or synchronism of potentials in closely situated areas) had been observed by many authors (Adrian and Yamagiwa, 1935; Beritov, 1943; Garoutte and Aird, 1958; Barlow and Freeman, 1959; Tunturi, 1959; and others). However, in the reports of their inves-

---

* Pages 104-112 in the Russian edition.

tigation the problem of a possible physiological interpretation of this phenomenon was not mentioned.

By synchronism (in contrast to the widely used term "synchronization") is meant similarity of direction of changes in the potential of comparable areas in time. For several years I have investigated the relationships between electrical activity of different cortical regions in dogs during the formation of motor–defensive conditioned reflexes (Dumenko, 1960, 1961). I have shown that under these conditions distant synchronism becomes prolonged and stable.

METHOD

Experiments were performed on five dogs (No. 1, Zhuk; No. 2, Galka; No. 3, Zhuzhu; No. 4, Norka; No. 5, Med) in a shielded, conditioned reflex chamber. Eleven electrodes were implanted in each dog, into the auditory, cutaneous, and visual analyzers and into the cortical representations of the fore- and hind limbs (two electrodes into each of the areas mentioned above; one electrode, placed on the animal's nose, could be used as reference electrode).

A stereotype of motor–defensive conditioned reflexes of the hind limb was formed to the following stimuli: bell (+), tone of 600 cps (+), tone of 200 cps (-), regular flashes (+). The unconditional stimulus was electrical stimulation of the skin of the leg by square pulses for 0.5 sec.

Quantitative assessments of synchronism were undertaken manually (Dumenko, 1965). The signs of the first derivative in time of each EEG curve were compared with all the rest after each 9 msec. The duration of the segment chosen for analysis was 1.5 sec. The background electrical activity in periods between stimulation was analyzed for all the dogs. For two dogs (Nos. 4 and 5) responses to conditional stimuli also were analyzed. The EEG curves obtained both with bipolar (dogs Nos. 1, 2, and 3) and with monopolar recording (dogs Nos. 4 and 5) were compared in pairs. In the first case there were ten possible pairs for comparison, in the second, 45. Since the values of synchronism were close for each pair in the intervals between stimuli, the mean level of synchronism was calculated for each pair throughout the experiment.

TABLE 1. Comparison of Synchronism of Two
Points within One Analyzer

Points compared	Degree of synchronism, %			
	Dog No. 4		Dog No. 5	
	$\bar{\chi}$	σ	$\bar{\chi}$	σ
Representation of hind limb	87	5	94	3.7
Visual analyzer	86	8	80	2.8
Cutaneous analyzer	76	3.7	78	5.8
Auditory analyzer	75	2.5	74	10.0

The criterion of synchronism (level of significance 0.027) was
constructed in such a way that values of synchronism lying outside the limits of $50 \pm 6\%$ indicated positive or negative correlations.

RESULTS AND DISCUSSION

Comparison of EEG curves from different cortical areas in
pairs in the intervals between stimuli before the animals developed a stereotype shows that very high synchronism is seen in
closely situated areas (for example, on comparing two points 4-5
mm apart in the same analyzer; Table 1).

Areas of the cortex lying in the motor analyzer (representations of the fore- and hind limbs) also were interconnected, although in this case the degrees of synchronism were less (60-70%)
than in the other analyzers.

Comparison of the electrical activity of the different analyzers before stereotype formation revealed either the complete
absence of synchronism in the periods between stimuli or a very
low level (between cutaneous and motor analyzers). After stereotype formation, on the other hand, very clear distant synchronism
was established in the periods between stimuli (Table 2).

In Table 2, two rows of numbers correspond to each pair of
regions compared. Since two electrodes were implanted into each
cortical region, by comparing the electrical activity of two analyzers four comparable pairs were obtained, and as a rule the levels of synchronism were similar in these pairs. Accordingly, the
results for only two pairs are given in this table, i.e., the results

TABLE 2. Changes in Synchronism after Stereotype Formation
in Dog. No. 4

Areas compared	Degree of synchronism, %				
	Before formation		After formation		Significance of difference
	$\overline{x}$	σ	$\overline{x}$	σ	
Representation of hind limb (two points) and auditory analyzer (one point)	55 51	4.5 4.2	67 67	5.5 4.7	$P \ll 0.001$
Representation of hind limb (two points) and visual analyzer (one point)	52 50	5 5.5	66 68	4.7 5.6	$P \ll 0.001$
Representation of hind limb (two points) and cutaneous analyzer (one point)	62 59	6 2.5	71 71	5.5 5.5	$P \ll 0.001$
Auditory (one point) and cutaneous analyzers (two points)	53 58	7.8 4.5	69 63.	5.9 6	$P \leqslant 0.001$ $P \leqslant 0.05$
Auditory (one point) and visual analyzers (two points)	54 55	6.4 5	68 70	5.6 5.5	$P \ll 0.001$
Cutaneous (one point) and visual analyzers (two points)	52 54	— —	70 69	4.7 5.1	$P \leqslant 0.001$

of comparison between two points in one analyzer with one of the
two points in the other. It is clear from this table that the levels
of synchronism in both rows, both before and after stereotype formation, are very similar. For brevity, the values of synchronism
between the cortical areas mentioned and the representation of the
forelimb are not given in Table 2. A statistically significant increase in synchronism likewise was observed in these pairs, but
it was much smaller, and when the levels of synchronism were
compared before and after stereotype formation, the level of significance of the differences was much lower (0.1; 0.05).

Hence, during formation of a stereotyped system of conditioned reflexes, distinct interaction appeared between the centers
of the conditional stimuli composing this stereotype and the center
of the unconditional stimulus.

Similar results were obtained with dog No. 5. It is interesting that a stereotype was formed initially in this animal to conditional acoustic stimuli only (without subsequent photic stimulus);

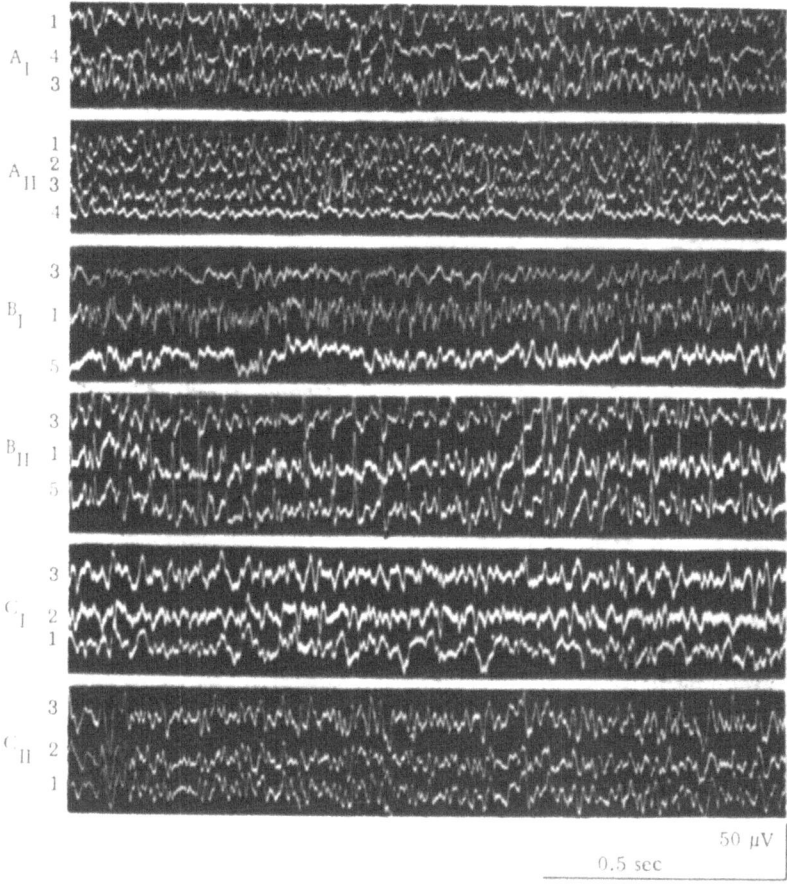

Fig. 1. Examples of synchronism between EEG curves of different dogs. A) Dog No. 1; B) dog No. 4; C) dog No. 5. I) Before stereotype formation; II) after stereotype formation. 1) Auditory analyzer; 2) cutaneous analyzer; 3) representation of hind limb; 4) representation of forelimb; 5) visual analyzer (in A the recording of the EEG is bipolar, in B and C monopolar).

the electrical activity of the visual analyzer under these circumstances was not involved in the synchronous activity. Only after the photic stimulus was included in the stereotype of stimuli did the oscillations of potential in the visual cortex become synchronous with the electrical activity of the remaining "involved" areas of the cortex.

The high degree of synchronism observed within the limits of one analyzer (Table 1) showed no statistically significant changes after stereotype formation.

Electroencephalographic examples of the synchronism described above are given in Fig. 1.

Similar changes in relationships between potentials of different cortical regions, indicating the appearance of distant synchronism, were observed in dogs Nos. 1 and 2 (during bipolar recording of the EEG). In dog No. 1 the stereotype was formed comparatively quickly (experiments 8-10). By this time a clear interaction appeared between the auditory and cutaneous analyzers and the representation of the hind limb (to stimulation of which the reflexes were formed), on the one hand, and the auditory and cutaneous analyzers on the other. At the same time, no such interaction was observed in dog No. 1 with the representation of the forelimb (Fig. 1A). With a change to nonstereotyped conditioned reflexes, a gradual decrease in distant synchronism took place and ultimately the interaction disappeared completely.

In dog No. 2 stereotype formation took much longer and synchronism appeared initially with the representation of both fore- and hindlimbs, and not until the period of stabilization of the stereotype (experiments 49-50) was the representation of the forelimb excluded (Dumenko, 1965). The transfer of unconditioned reinforcement from the hindlimb to the forelimb led to disturbance of synchronism with the representation of the hind limb, and synchronism was formed very gradually (parallel with the appearance of stereotyped conditioned reflexes of the forelimb) with the representation of the forelimb (Dumenko, 1965).

Consequently, distant synchronism appeared in those parts of the cortex which were connected functionally in the course of formation of the stereotyped system of conditioned reflexes.

Various disturbances of the stereotype (effect of irrelevant stimuli, changes in the frequency of the conditional stimuli, exclusion of unconditional reinforcement from the stereotype, reducing the intensity of illumination in the chamber, changing the duration of the periods between stimuli) led to weakening of synchronism or ultimately to its complete disappearance. However, the factors listed above were not consistent in their influence on different ani-

mals at different stages of stereotype formation and the degree of
weakening of synchronism which they produced varied from case
to case. In dog No. 1, for instance, exclusion of electrodermal reinforcement from the stereotype led to the disappearance of synchronism. In dogs Nos. 4 and 5, on the other hand, this procedure
caused no significant changes in synchronism; the decisive factor
for these animals was the constant duration of pauses between
stimuli. As Fig. 2 shows, a sharp weakening of distant synchronism took place when the intervals between stimuli were lengthened from 6 to 7-8 min (experiments 63, 67, 99). Meanwhile an
extraneous stimulus such as the presence of a strange person in
the chamber, evoked much less weakening of synchronism in this
animal (experiment 89). It should be emphasized that in all animals the factors causing weakening or disappearance of synchronism did not always cause disturbances of conditioned-reflex activity. For instance, the weakening of synchronism in experiments
63 and 67 (Fig. 2) was not accompanied by disturbance of conditioned reflexes, but in experiment 89 de-inhibition of differentiation took place. The synchronism in the course of electrical activity arising during stereotype formation proved to be more "frag-
ile" than the conditioned reflexes.

The distant synchronism remained at this level for many
months during work with the animals on presentation of the stimuli

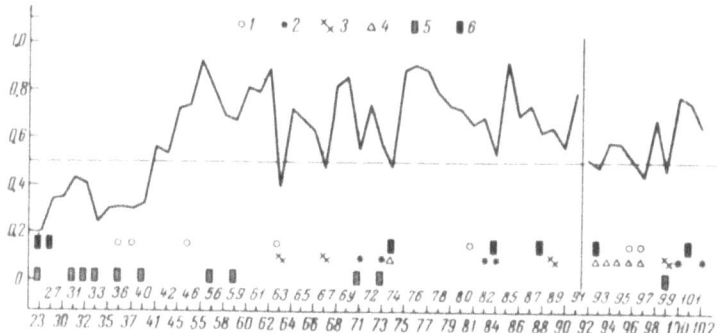

Fig. 2. Comparison of synchronism with conditioned-reflex activity and general condition of the animal (dog No. 4). Abscissa: No. of experiments; ordinate: coefficients of synchronism. Right vertical line denotes interval of two months in investigation. 1) Considerable dimming of illumination in chamber; 2) increased excitability
of animal; 3) disturbance of stereotype; 4) inhibited state of animal; 5) disinhibition of differentiation; 6) disturbance of positive conditioned reflexes (lengthening
of latent period or disappearance of conditioned responses).

of the stereotype. It appeared when the dog was placed in the chamber and all preparations for the experiment were complete. It is very interesting to note that the animal responded not only to the beginning of the experiment, but also to its end. This indicates that dogs are to some extent capable of measuring precisely the passage of time. If, after presentation of the last stimulus of the stereotype (flashes), the animal was kept in the chamber longer than the usual 5-6 min, the synchronism observed in the experiment either weakened or disappeared (Table 3).

In dog No. 3 the formation of a statistically significant distant synchronism was not observed. This was evidently because no stereotyped system of conditioned reflexes could be formed in this animal.

It may be concluded that the distant and long-lasting synchronism observed in the intervals between stimuli is the electrographic manifestation of the dynamic stereotype. In its physiological nature, this synchronism is evidently an expression of cortical tone and it ensures the general state of preparedness of the animal for responding with its various systems in a given situation. Investigations using the classical conditioned-reflex technique (Asratyan, 1941; Kupalov, 1933; etc.) showed that the whole experimental situation gives rise to a definite functional state in the cortex, with no external manifestation, giving it the proper "tuning" for forming conditioned-reflex connections.

In connection with the possible tonic nature of the distant synchronism I have described, it may naturally be asked how this synchronism changes under the influence of conditional stimuli. The electrographic responses of two dogs (Nos. 4 and 5) to conditional stimuli were analyzed. This showed that the synchronism between the potentials in two regions of the cortex arising during intervals between stimuli was considerably increased under the influence of positive and inhibitory conditioned stimuli. The mean results of 10 experiments with dog No. 4 are given in Fig. 3. The increase in synchronism during the time of action of the conditional stimuli can be seen both in the EEG pairs where it was well marked in the background (1/8, 3/8, and 7/8), and in those when it was weaker (7/9). Soon after discontinuing the conditional stimulus, the synchronism fell sharply. Consequently, each conditional stimulus included in the stereotype increased the distant synchronism still more at the moment of its action, pointing to an increase in

TABLE 3. Decrease in Synchronism after End of Experiment

Areas compared	Degree of synchronism, %			
	Experiment 95		Experiment 100	
	I	II	I	II
Representation of hind limb and auditory analyzer	68	60	65	54
Representation of hind limb and cutaneous analyzer	72	63	68	60
Representation of hind limb and visual analyzer	62	56	64	59

I) Mean values of synchronism for experiment; II) values of synchronism 7-10 min after presentation of last stimulus of stereotype.

interaction between the regions to which the conditioned and un-conditional stimuli are directed. These results show that the synchronism described above reflects not only the animal's general response to the experimental situation, anticipating the presentation of a fixed system of stimuli, but also its direct response to the stimuli.

It is of definite interest to compare changes in synchronism of the potentials with changes in conditioned-reflex activity throughout the period of the investigation. This relationship is shown graphically in Fig. 2. The coefficient of synchronism was calculated as the ratio between the number of positive correlations between pairs, indicating synchronism, and the total number of correlations between pairs in that particular experiment. Calculation of such a coefficient became necessary so that the general "gross" assessment of synchronism could be made over the whole period of the experiment. It is clear from Fig. 2 that before experiment 40 the coefficient of synchronism was low (0.2-0.4), i.e., synchronism was observed in approximately only one-third of the compared EEG pairs. This fraction of synchronism was provided by closely situated pairs within the nuclear zone of each analyzer (Table 1), and distant synchronism was absent. In this period the stereotype had not yet been formed, and correspondingly, disturbance of both positive and inhibitory conditioned reflexes (black and shaded columns) occurred frequently. Starting with experiments 41-42, the coefficient of synchronism increased and remained at a high level (0.7-0.9) until experiment 91, indicating the onset of distant synchronism. Exceptions were given by a few experiments in which sharp weakening of synchronism was observed (experi-

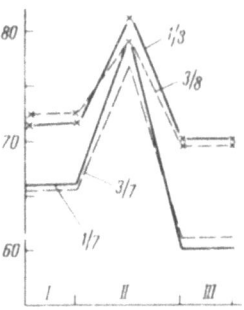

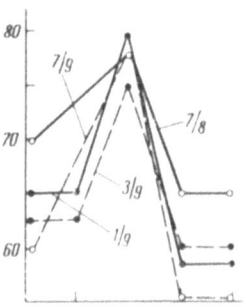

Fig. 3. Changes in degrees of synchronism of different EEG pairs under the influence of conditional stimuli. Ordinate: degrees of synchronism in percent; abscissa, time: I) 1.5-2 sec before action of conditional stimuli; II) period of action of conditional stimuli; III) immediately after their discontinuation. 1, 3) Representation of hind limb; 8, 9) cutaneous analyzer; 7) auditory analyzer (monopolar recording of EEG).

ments 63, 67, 74). In experiments 63 and 67 this weakening of synchronism was due to disturbance of the stereotype; in experiment 74 the animal (for certain unknown reasons) was in an inhibited state (elevation of the threshold of the unconditional response, drowsiness), and this evidently caused weakening of distant synchronism. This dog, it will be noted, when used in the experiments after an interval of two months (to the right of the vertical line), showed signs of a neurotic state, (elevation of the threshold of the unconditional response by 3-4 times, lengthening of the latent period of the positive conditioned reflexes or their disappearance, drowsiness in intervals between stimuli), and the level of synchronism was correspondingly lowered.

It is clear from the graph that in the period after formation of the stereotype (experiments 41-91), the disinhibition of differentiation taking place in a balanced animal (probably for accidental reasons) was not accompanied by weakening of synchronism (experiment 56, 59). If, however, besides disturbance of its conditioned-reflex activity, the animal showed a state of increased excitability (severe restlessness, yelping, frequent raising of the limbs in the intervals between stimuli), in these cases a distinct decrease in synchronism was observed (experiment 71, 73, 84).

A direct relationship thus exists between the normal, balanced state of the animal, correct stereotyped conditioned-reflex responses, and a high level of distant synchronism. Disturbance of the animal's balance (whether toward excitation or towards inhibition), like disturbance of the stereotype, led to a sharp weakening of synchronism.

A similar relationship was obtained with dog No. 5.

SUMMARY

1. During formation of a stereotype of motor-defensive conditioned reflexes in dogs, synchronism of the potentials in cortical regions corresponding to the stimuli combined in the stereotype appeared in intervals between stimuli. The synchronism appeared only in those areas which were connected functionally.

2. This distant synchronism persisted for several months during work with the animals (on presentation of the stimuli of the stereotype), and it appeared regularly at the beginning of each experiment, as soon as all preparations for it were complete.

3. During the action of the conditional stimuli included in the stereotype, the level of synchronism rose considerably.

4. The synchronism was greatly weakened if the stereotype was disturbed.

5. Distant synchronism was observed in those animals in which a stereotyped system of conditioned-reflex responses could be formed. If the animal's balance was disturbed (whether toward excitation or toward inhibition), the degree of synchronism fell clearly.

6. The appearance of lasting and selective synchronism in certain cortical areas is one of the electrographic manifestations of stereotype formation, indicating organization of a fixed system of connections in the cortex ensuring that the relevant systems take part in the animal's response.

Correlation Analysis of Spike Activity of Adjacent Cortical Neurons in Rabbits*

T. M. Efremova, M. N. Zhadin, L. I. Kaburneeva,
I. N. Kondrat'eva, A. M. Melekhova,
and G. I. Shul'gina

*Institute of Higher Nervous Activity and Neurophysiology
Academy of Sciences of the USSR
Moscow, USSR*

At the present time neurophysiologists are concentrating their efforts on finding a solution to the problem of the functional organization of the central nervous system. Very probably both the temporal and the spatial parameters of activity of individual neural units are used for coding information in the central nervous system. The numerous investigations demonstrating changes in discharge frequency in response to stimulation are evidence of the first. The role of spatial parameters is shown by investigations revealing integration of impulses of different modalities at one neuron. This was demonstrated, in particular, by the fact that the activity of a neuron during simultaneous action of stimuli of two modalities differed from the responses to each of them (Baumgartner, 1958; Lomo and Mollica, 1960; Skrebitskii and Voronin, 1965; and others). The second hypothesis is also supported by the existence of neurons possessing complex receptor fields (Hubel and Wiesel, 1962). According to these workers, the activity of such neurons is the result of their integration of activity of neurons with simple receptor fields. Purposive activity of the living organism is undoubtedly possible only with precise coordination of the activity of individual neurons, maintained by integration of the activity of the many neurons forming complex networks.

How are these functional interrelationships between individual neurons in the central nervous system established in the pro-

---

* Pages 113-118 in the Russian edition.

cess of activity? This question can be answered by analyzing the activity of several neurons recorded simultaneously. This can be done by recording with one ordinary microelectrode (as most of the investigations in this field have been performed); by recording with one specially designed electrode capable of recording activity from 3–5 neurons simultaneously (Gerstein and Clark, 1964), and also by recording with several microelectrodes (Verzeano and Calma, 1954; Verzeano, 1956, 1962; Verzeano and Negishi, 1960; Verzeano et al., 1965).

Research in this direction has been carried out on various systems: the faceted eye of Limulus (Hartline et al., 1956), the ganglion of the edible snail (Kogan and Chorayan, 1965), the tectum mesencephali of the frog (Kogan, 1965; Chorayan, 1963, 1965), the rabbit medulla (Keder–Stepanova et al., 1966), the visual and motor cortex, thalamus, and lateral geniculate body of the cat (Verzeano and Calma, 1954; Verzeano, 1956, 1962; Mountcastle, 1957; Amassian et al., 1959; Li, 1959; Verzeano and Negishi, 1960; Hubel and Wiesel, 1962; Creutzfeldt and Meisch, 1963; Gerstein and Clark, 1964; Griffith and Horn, 1963; Murata et al., 1966), and the cortex and thalamic nuclei of the monkey (Evarts, 1965; Verzeano et al., 1965). Most of these investigations were carried out without the use of special methods of analysis for detecting correlation between the activity of neurons, and the corresponding mathematical operations were used in only a few of them (Griffith and Horn, 1963; Keder–Stepanova et al., 1966; Kogan, 1965; Chorayan, 1965).

The overwhelming majority of investigations was performed on immobilized, anesthetized animals or on encéphale isolé preparations, and neurons of a given brain region were investigated in the same animal using stimuli of identical modality and strength.

The object of the present investigation was to study the simultaneously recorded activity of closely adjacent neurons in different parts of the cortex of the unanesthetized, nonimmobilized rabbit during the action of adequate and inadequate stimuli of different strengths and modalities.

To analyze neuronal activity the correlation method was used. The cross–correlation coefficient was calculated by the formula:

$$R = \frac{\sum\limits_{i=0}^{n} (a_i - \bar{a})(b_i - \bar{b})}{\sqrt{\sum\limits_{i=0}^{n} (a_i - \bar{a})^2 \sum\limits_{i=0}^{n} (b_i - \bar{b})^2}},$$

where n may be from 20 to 400. The cross-correlation coefficient
was determined for the number of spikes in every 100 msec.
The analysis epoch ranged from 2 to 40 sec. To assess the results obtained, mathematical statistical tables (Bol'shev and Smirnov, 1965) were used to determine whether the coefficient of
cross-correlation differed from zero with a probability of 0.99.

To judge the degree of synchronism it was necessary to
know the shape of the distribution curve of cross-correlation coefficients for each pair of neurons. If only single measurements
of the cross-correlation coefficients were available, it could only
be stated that the degree of synchronism differed from zero.

To record spike activity, glass microelectrodes filled with
2.5 M NaCl or KCl solution (diameter 1-5 μ, resistance 20 MΩ)
were inserted into the brain by means of a micromanipulator designed by A. M. Melekhova and V. L. D'yakonov, fixed to the skull
with phosphate cement. The indifferent electrode was placed on
the rabbit's ear. The animal was tied to a frame by its limbs.
Impulse activity of two or three neurons was recorded simultaneously by one microelectrode. The neurons were identified from
the shape and amplitude of the spikes. Potentials were fed into a
type UBP 1-01 amplifier and recorded from a cathode-ray or loop
oscilloscope on photographic film moving at a speed of 10 and 50
cm/sec. Spike activity of units in the visual, auditory, and sensorimotor cortex was investigated in the absence of stimulation and
also in response to a single flash and flashes with a frequency of
5/sec (in the visual and sensorimotor areas); during a continuous
tone of 500 cps (in the auditory, visual, and sensorimotor areas);
during electrical stimulation of the skin (in the sensorimotor area);
after subcutaneous injection of caffeine or morphine in doses of
1.5-2 and 3-4 mg/kg respectively, or administration of bromide
in a dose of 170 mg/kg by mouth (in the visual and sensorimotor
areas). Altogether 213 pairs of neurons were analyzed, 73 in the
visual, 15 in the auditory, and 125 in the sensorimotor cortex.

RESULTS

Investigation of synchronism in the activity of simultaneous-ly recorded units in the absence of special stimulation showed that of 62 pairs of neurons in the sensorimotor area four pairs (9.6%) displayed some degree of synchronism. In the visual cortex 6 of 40 recorded pairs (i.e., 15%) showed some degree of synchronism, while in the auditory cortex, only one of the 15 investigated pairs of neurons fired synchronously.

During a continuous tone of 500 cps no change took place in the relationships between the neurons in the investigated areas: of 8 pairs of asynchronous somatosensory neurons, the coefficient of correlation for all 8 pairs during the tone was not significantly different from zero, and the same pattern was maintained in the after-period. In the visual cortex, only one of nine recorded pairs of neurons fired synchronously before application of the tone; dur-ing and after the end of its action this pair of neurons continued to fire synchronously. The same relationships were found also in the auditory area. During background activity, one of the eight re-corded pairs of neurons fired synchronously in this area and it continued to do so during and after the action of the tone. The re-maining pairs continued to discharge asynchronously.

With a single flash applied irregularly, at intervals of 2–3 min, or with flashes applied regularly at 1/sec, the following re-lationships were observed. Of 21 pairs of neurons in the visual cortex, three pairs fired synchronously during stimulation by a single flash. Synchronism was absent in the background activity from one of these pairs and it appeared only during the flash, while a second pair was silent in the absence of stimulation. Un-fortunately, no background recording could be obtained for the third pair.

In this series of experiments 11 of 21 pairs of neurons were examined for a long period (30–60 min). During this period their coefficient of cross-correlation was determined from three to nine times. It is interesting to note that stable relationships were maintained between these pairs of neurons, i.e., if they discharged synchronously, this synchronism persisted throughout the period of investigation of this pair. The same principle was observed in the case of asynchronously discharging pairs of neurons.

Application of flashes of identical brightness, with a frequency of 5/sec, and acting for 30–40 sec, showed that of two pairs of neurons in the visual cortex firing asynchronously, one pair began to fire synchronously during stimulation, while the other fired reciprocally.

In the sensorimotor cortex, of five pairs of neurons firing asynchronously in the control interval, during the action of regular flashes three pairs began to fire synchronously. On stopping the stimulation, this synchronism was not maintained.

The action of electrical stimulation of the skin was investigated on five pairs of neurons in the sensorimotor cortex, of which two pairs fired synchronously in the absence of stimulation. During stimulation their synchronism was maintained; in addition, one other pair of neurons began to discharge synchronously.

Next the correlation between neurons was investigated after administration of morphine, bromide, and caffeine to the rabbits in the doses given above. The action of bromide and morphine was to increase the percentage of synchronously firing neurons. This was seen most clearly in the sensorimotor cortex, where five of the 27 pairs (18.5%) of recorded neurons fired synchronously after administration of morphine, and five of 21 observed pairs (23.8%) fired synchronously during the action of bromide. After injection of caffeine, five pairs of neurons were recorded and in all cases their activity was asynchronous.

Because of the small number of recorded neurons in the visual cortex in this series of investigations, it was difficult to detect any particular tendency.

How can these results be explained and how do they agree with those obtained previously by other investigators? It first had to be determined unequivocally that, when using correlation analysis, only the synchronism in the discharge of two neurons was determined and not their relationship in general. By obtaining the mean spike activity for 100 msec for calculating the coefficient of cross–correlation, the frequency transmission band was determined, using the formula:

$$f(v) = \frac{\sin \pi \cdot v \Delta t}{\pi v \Delta t} \Delta t = 0.1 \text{ sec}$$

where v is the frequency in cycles per second.

Hence, at the level $(e - 1)/e$ the frequency transmission band ranged from 0 to 5.2 cps.

Proceeding with the analysis of the results, only a small percentage of neurons in all cortical areas investigated was found to discharge with a significant degree of synchronism during background activity. No essential difference was found between the various areas. These results agree with those obtained by other investigators using visual analysis of traces of unit spike activity or comparison of the mean frequency of impulse activity and show that the spontaneous activity of two or more neurons in the sensorimotor cortex of the cat and rabbit is mainly not correlated (Amassian et al., 1959; Li, 1959; Livanov, 1965). However, the study of the distribution of a large number of neuron pairs (170 pairs) taken from all cortical areas by the coefficient of correlation of their background activity showed that the mean coefficient of correlation was 0.1, with limits of 0.15 and 0.05, and with a probability of 0.95. It can thus be concluded that the degree of synchronization in the background activity of closely situated cortical neurons differs significantly from zero but is nevertheless low.

As mentioned above, this type of investigation could not be carried out for different forms of stimulation because of inadequate data for each individual series.

Correlation between discharge of two, three, and four neurons in the cortex and thalamic nuclei of the cat and monkey has been demonstrated by recording the spike activity with several electrodes placed close together (Verzeano, 1956, 1962; Verzeano and Negishi, 1960; Verzeano et al., 1965). However, besides the temporal parameter the spatial parameter in the discharge of the neurons was considered in these investigations. These workers demonstrated the circulation of spike activity in closed circuits. Clearly the correlation method is inadequate for this type of relationship.

We next showed that the relationships established between neurons are maintained steadily for a long period of time. This applies both to synchronously and asynchronously working pairs of neurons. Presentation of a pure tone or flash (single or repeated at 1/sec) does not alter these relationships essentially. They can be changed to some degree only by application of flashes at 5/sec or electrical stimulation of the skin.

The fact that definite relationships are maintained between neurons is not unexpected bearing in mind the strict organization of the cortical analyzers (Hubel and Wiesel, 1962). Closely situated neurons probably constitute functional groups with fixed relationships, and only relatively strong stimuli can modify them to some extent. In our case this was observed during the action of repeated flashes or electrical stimulation of the skin. Earlier investigators also observed the appearance of clear correlation (direct or reciprocal) in the activity of neurons during active or passive movements or electrical stimulation of an afferent nerve (Mountcastle, 1957; Amassian et al., 1959; Li, 1959; Evarts, 1965).

In a hitherto unpublished investigation, one of us, (M. N. Zhadin) noted that the coefficient of correlation between potentials of two points in the rabbit's cortex is stable over a period of time. On the assumption that the electroencephalogram reflects, albeit indirectly, neuronal activity, this phenomenon is in good agreement with the stability of the coefficient of cross-correlation between two closely situated neurons in time revealed in the present investigation.

Why is it that only a few neurons fire synchronously during background activity? On the basis of the work of Hubel and Wiesel (1962) and of Mountcastle (1957), who demonstrated the strictly columnar structure of the cortical analyzers and the similar functional characteristics of neurons located in the same column and close together, a higher percentage of synchronously active neurons should be expected, at least during the action of stimuli. Its absence must probably be explained by the very large number of influences to which a given neuron is exposed at each moment. Possibly with a decrease in the variety of influences and retention of only those which are vitally essential, the synchronism between the neurons might become clearer. This took place at least in the sensorimotor cortex after administration of morphine and bromide, under the influence of which, as previous investigations (Melekhova, 1965) showed, the number of spontaneously active neurons in the rabbit's cortex decreases. This hypothesis is confirmed by other investigators who have found that the .cooperative or antagonistic activity of neurons in the cat's cortex is much more clearly demonstrable against the background of marked hypoglycemia, in the stage of appearance of high-amplitude slow waves (Creutzfeldt and Meisch, 1963), in the encéphale isolé or in a drowsy animal compared with an unanesthetized animal (Murata et al., 1966), and in the transition from waking to sleep in the period of appearance of large slow waves in the cortex and thalamic nuclei of the cat and monkey (Verzeano and Calma, 1954; Verzeano, 1956, 1962; Verzeano and Negishi, 1960; Verzeano et al., 1965). Indirect evidence in support of this hypothesis is given by the high degree of correlation (reciprocity) in the activity of the ommatidia during stimulation by light from a point source (Hartline et al., 1956).

The results of the present investigation show that only a few pairs of neurons work "synchronously." Although, as mentioned above, the degree of this synchronism cannot be assessed in each concrete case, a valid general assessment of its magnitude can be made. Even among the "synchronously" firing pairs of neurons a high value of the coefficient of cross-correlation was observed in only one of 213 pairs, for which the coefficient of cross-correlation during acoustic stimulation was 0.9 ($n = 223$). The coefficient of cross-correlation for the other "synchronously" discharging neurons did not exceed 0.6 (for different values of n). For most of the other, "asynchronous" pairs of neurons, the coefficient of cross-correlation was of the order of 10^{-2}. This fact, i.e., the absence of high cross-correlation between the discharge of the neurons both during background activity and during the action of different stimuli, apparently contradicts the well-known phenomenon of rhythmic, synchronized slow activity which can be recorded constantly from the surface and the deep structures of the brain.

However, in a model investigation of the phenomenon of synchronization of cortical potentials undertaken by M. N. Zhadin (unpublished data), it was found that with strict orientation of the individual dipole generators even minimal cross-correlation of discharge provides a high level of synchronism on the surface. This fact may remove the apparent contradiction between the absence of perceptible synchronization in activity of individual pairs of neurons and the presence of synchronization of slow potentials.

SUMMARY

1. A method of cross-correlation analysis was used to study the activity of 213 pairs of cortical neurons of unanesthetized, nonimmobilized rabbits exposed to stimuli of different modalities (photic, acoustic stimulation, and electrical stimulation of the skin) and of pharmacological agents (bromide, morphine, and caffeine).

2. In the absence of any stimulation a significant degree of correlation is found between the activity of a few pairs of neurons: up to 9.6% in the sensorimotor cortex and 15% in the visual cortex.

A study of the distribution of the total number of pairs of neurons by coefficient of correlation demonstrates the existence of a significant difference from zero in the background activity, but only a low or moderately high level of synchronization.

3. The relationships established in terms of the degree of synchronism between most neurons (whether discharging synchronously or asynchronously) are maintained unchanged in time, whether during background activity or during weak stimuli.

4. An increase in the number of pairs of synchronously discharging neurons is observed during rhythmic photic stimulation and also after injection of pharmacological agents with sedative action (bromide and morphine).

Spatial Synchronization of Cortical Potentials in Patients with Disturbances of Association[*]

N. A. Gavrilova

Institute of Higher Nervous Activity and Neurophysiology
Academy of Sciences of the USSR
Moscow, USSR

Previous work has shown that during performance of intellectual operations the relationships between potentials in different parts of the healthy human cortex are definitely modified, in accordance with specific rules (Livanov, 1962b; Livanov et al., 1964). The idea thus developed that this fact could be used to study certain psychopathological phenomena associated with disturbances of intellectual activity.

In the clinical picture of the psychoses, the disturbance of intellectual activity is one of the most serious symptoms disabling the patient. Intellectual activity shows pathological changes in many diseases, and these disturbances may take different forms. One of the severest forms of its disturbance is delirium. This is found in nearly all mental diseases. The overwhelming majority of patients with delirium require hospitalization, and they constitute the permanent inmates of psychiatric hospitals.

The urgent importance of the study of this form of mental disturbance becomes clear when an attempt is made to elucidate its physiological mechanisms. So far two groups of patients with delirium have been investigated. The first group consisted of patients with the delirious form of schizophrenia, the second of patients with epilepsy with a syndrome of delirium in the clinical picture of the disease.

---

[*] Pages 54-64 in the Russian edition.

Potentials were recorded simultaneously by means of the electroencephaloscope from 50 points of the cortex on film moving at 24 frames/sec. The data thus obtained were analyzed with the M-20 or Dnepr electronic computer.

Correlation between pairs of potentials from all 50 cortical points was investigated. The criterion of correlation was the number of changes in potentials taking place in the same direction from one moment to another. The epoch of analysis was 1.5 sec (Livanov, 1962b, 1962c; Glivenko et al., 1962a).

In this paper a high degree of positive correlation between potentials was denoted if the potentials were in the same direction for more than 75% of the period of investigation.

Analysis of the material obtained by studying correlation between potentials in different parts of the cortex of patients with the paranoid form of schizophrenia revealed a number of specific abnormalities. These results were fully described in earlier papers (Gavrilova, 1965), and only the two principal conclusions will be mentioned here.

One conclusion is that in the cortex of schizophrenic patients with delirium there is a background of a larger than normal number of points whose potentials show a high degree of correlation. The number of pairs of "interconnected" points is particularly large in cases of severest delirium.

In the healthy person at rest a high degree of correlation is shown on the average by 5-15 pairs of points, while in these patients their number at individual moments may reach hundreds or more (Fig. 1a and 1b).

Second, the smallest number of positive correlations is found as a rule in the prefrontal areas of the cortex, and this is seen most clearly during the solution of mental problems, especially oral multiplication of two- or three-figure numbers. The frontal areas of the cortex, in which high correlations between potentials are most clearly defined in the healthy person during mental arithmetic, are considerably depleted in patients with the paranoid form of schizophrenia (Fig. 1c).

The second disease to be studied was epilepsy. Of the wide range of its clinical forms, epileptic psychoses with the presence of delirium were chosen.

Electroencephaloscopic investigations were carried out on 25 patients; in six of them the correlations between cortical potentials were studied. All six patients had a severe form of epilepsy. Besides major epileptic fits, observed not less than once every 5-20 days, their clinical picture included personality changes and also psychotic disturbances in the form of paranoid experiences of varied intensity — from the degree of unstable ideas of association to severe, systematized delirium.

The total number of high correlations between potentials in different parts of the cortex of these patients was higher than normal and was unstable, varying both in the course of one investigation and from one investigation to another.

The inertia or monotony which, judging from the literature and clinical picture, might be expected in such patients was not observed in the high correlations between their potentials. This is probably due to their possession of a delirious pathodynamic structure constantly fluctuating and differing in intensity among patients. This is confirmed by the fact that if paranoid experiences were absent or had subsided in these patients, the electrical relationships in the cortex were more stable and the number of correlations diminished.

The changes in high correlations between potentials in patients with epilepsy in the presence of delusions were as follows.

In each investigation there were periods of time when the number of high correlations was considerably above normal. The correlated points were situated in the prefrontal areas of the cortex. The correlation relationships in such periods differed from those in schizophrenia (where, as a rule, the frontal areas are depleted in correlations) and resembled the pattern observed in the healthy person during mental effort or during exacerbation of the phobia in a patient with an obsessive neurosis (see the chapter by Aslanov in this volume).

However, by the next moment the number of high correlations, although still quantitatively increased, had changed its localization. A pattern developed with the majority of correlations now in the posterior areas of the cortex. The pattern of synchronization was now similar to that observed in the paranoid form of schizophrenia. Subsequently, at one moment the total number of correlations could fall considerably, and then rise again equally

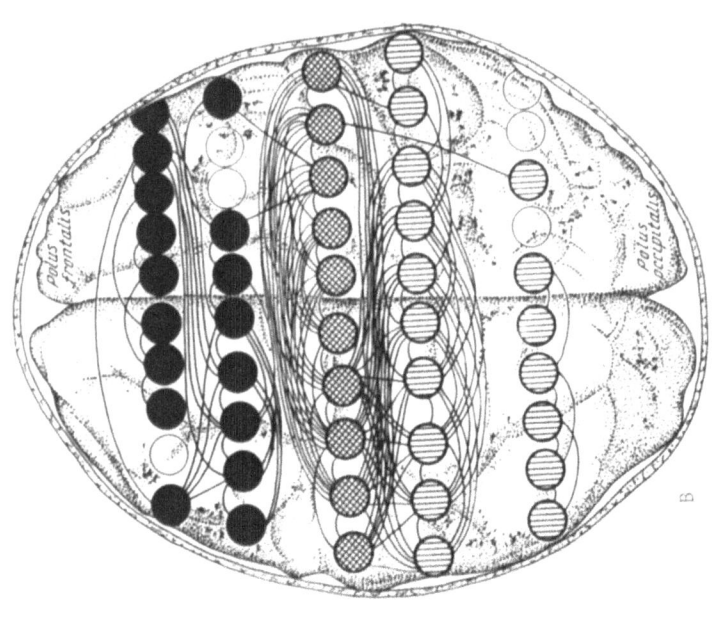

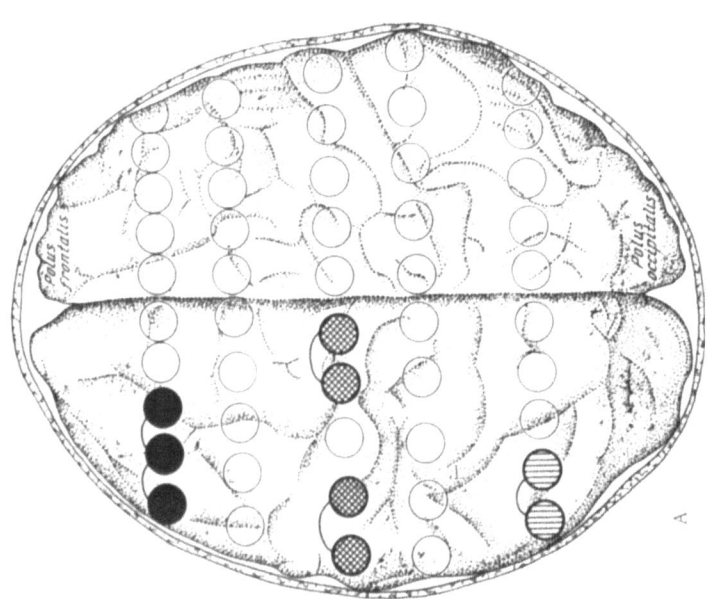

Fig. 1. High correlations between potentials in various points of the cortex of a healthy person at rest (A), of a patient with schizophrenia at rest (B), and during the solution of an arithmetical problem (C). Circles denote locations of recording electrodes. Points whose potentials give a high coefficient of correlation are joined by lines. Black circles denote points in the prefrontal area; crosshatched circles points in the motor area; vertically shaded circles points in posterior areas.

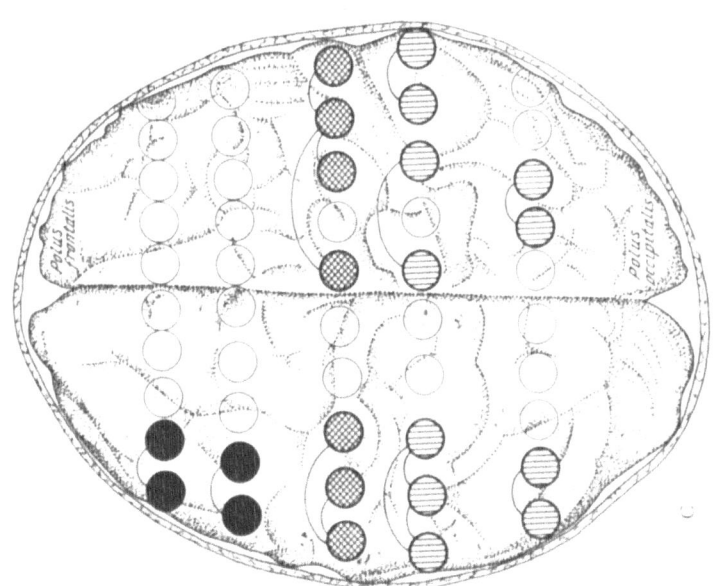

sharply. Their greatest number was observed sometimes in the anterior, sometimes in the posterior areas (whereas in schizophrenia usually no such change in the localization of high correlations was observed).

The dynamics of mutual influences as described above is characteristic of patients with epilepsy with well-defined paranoid structures.

In cases without persistent delusions, but with paranoid interpretations sometimes appearing in the clinical picture, sometimes subsiding, the spatial synchronization of potentials could be close to normal on days when the condition was relatively satisfactory.

This happened in two patients: in patient Priv. during the first and fourth examinations (Table 1) and in patient Al. during the second (Table 2). On these days, separated by a considerable length of time from fits, the patients' state was relatively satisfactory. Meanwhile, during the seventh examination of patient Priv. (Table 1), the clinical condition was satisfactory, no fits were observed, while the correlàtions between cortical potentials were higher than normal. They were most numerous in the posterior cortical areas. This fact is consistent with Deglin's (1960) findings, showing that in patients with epilepsy disturbances of conditioned-reflex activity may be observed independently of fits.

Electroencephalographic studies of patients in the absence of delirium, in agreement with the results of electroencephaloscopy, demonstrated the hypersynchronized activity previously described by many workers, in the form of more or less regular, slow waves (3-5/sec) or paroxysmal complexes: slow wave — pointed wave. No information appears in the literature on what changes take place in the EEG of the epileptic patient under the influence of delirium, and my encephalographic material is insufficient to allow conclusions to be drawn.

Six patients were investigated. All the points described are illustrated by two of them.

Patient Vas., a woman aged 23 years who had been ill since childhood. She had often been admitted to psychiatric hospitals because of frequent epileptic fits, with a diagnosis of epilepsy. Some 2-3 years ago she began to express paranoid ideas of a religious

TABLE 1. Dynamics of High Correlations during Various Examinations of Patients with Epilepsy (under Background Conditions and during Multiplication of Two Numbers)

Patient	Examination No.	Serial No. of 1.5-sec intervals	Background					Solution of arithmetical problem (multiplication of two-digit numbers)				
			frontal region		motor area	parietal region	occipital region	frontal region		motor area	parietal region	occipital region
			first pair	second pair				first pair	second pair			
			of electrodes					of electrodes				
Vas.	2	1	11	10	11	7	4	5	44	32	47	16
	3	1	43	13	6	13	13	2	1	0	3	5
		2	1	0	7	68	25	7	1	3	5	18
Priv.	1	1	8	12	4	1	0	17	14	8	1	7
	2	1	117	71	44	18	2	2	1	7	9	4
	3	1	21	48	33	8	6	27	4	31	22	14
	4	1	4	12	9	5	10	21	13	12	2	9
	7	1	18	11	29	38	41	15	24	16	18	28
Healthy subject	Mean data		2	4	4.47	5	3	20	45	31	24.3	15.9

The table gives results obtained with each patient on different days of the examination, when presented with problems of similar difficulty. The normal is shown as the mean of the data obtained with six healthy subjects.

nature, went away from home, and became a vagrant. In 1964, from paranoid motives, she threw herself out of a window. She was treated at the Sklifosovskii Institute for multiple fractures of the bones of the lower limbs, the chest, and the lumbar spine. After discharge she found home life intolerable; she was malicious, dysphoric, and aggressive.

During the period of investigation in the hospital, no organic disturbances were found in the internal organs or central nervous system.

Mental State. She lay motionless with eyes closed, without speaking, and refusing to eat. She thought that she was buried in the ground, surrounded by dead bodies. Several times a day she cried out spontaneously: "Dig me out." She had major epileptic fits in groups of two or three every 5-12 days. After the fit, for 2-3 days she was selectively communicative, stating that she had been "dug out" but she still remained negativistic and malicious. She then again became motionless, silent, and unwilling to eat.

In a period of worsening of the clinical condition, the correlations between potentials differed at different moments of the same examination (Table 1, Examination No. 3). At some moments the number of high correlations was greater than normal, but most of them were in the frontal areas of the cortex (in the first interval of the third examination, for example). At the next moment of time, the number of high correlations in the anterior areas of the cortex was sharply reduced, but they were very numerous in the posterior areas (the second interval of the same examination). Hence, the pattern of high correlations changed considerably during the course of one examination (Fig. 2). In the second examination, when the patient's condition was relatively satisfactory, there were some time intervals in which the number and distribution of high correlation were almost normal.

Patient Priv., a woman aged 23 years. She had suffered from epilepsy since the age of 15 years and had repeatedly been admitted to psychiatric hospitals because of frequent fits and severe dysphoria. Since March of the previous year she had thought that she was a witch. She was able only to cast wicked spells and consequently people hated her and she hated everybody else. Since the last exacerbation of her delusions in May, 1965, she had been admitted to a psychiatric hospital. No organic disturbances were found in the internal organs and central nervous system.

Mental State. She was malicious, dysphoric, and expressed ideas of witchcraft. She repeated that she was casting spells on people. She was very aggressive. In the ward she was constantly brawling and quarreling with the other patients and staff. After a few days her dysphoria weakened, she no longer felt that she was a witch, she became gentler and selectively communicative. It was in this state that she was examined. The last fit had taken place two days before the examination. Almost normal correlations between potentials were found (Table 1, Examination No. 1). She had no further fits for 15 days. Her paranoid expressions then reappeared, her dysphoria and aggressiveness

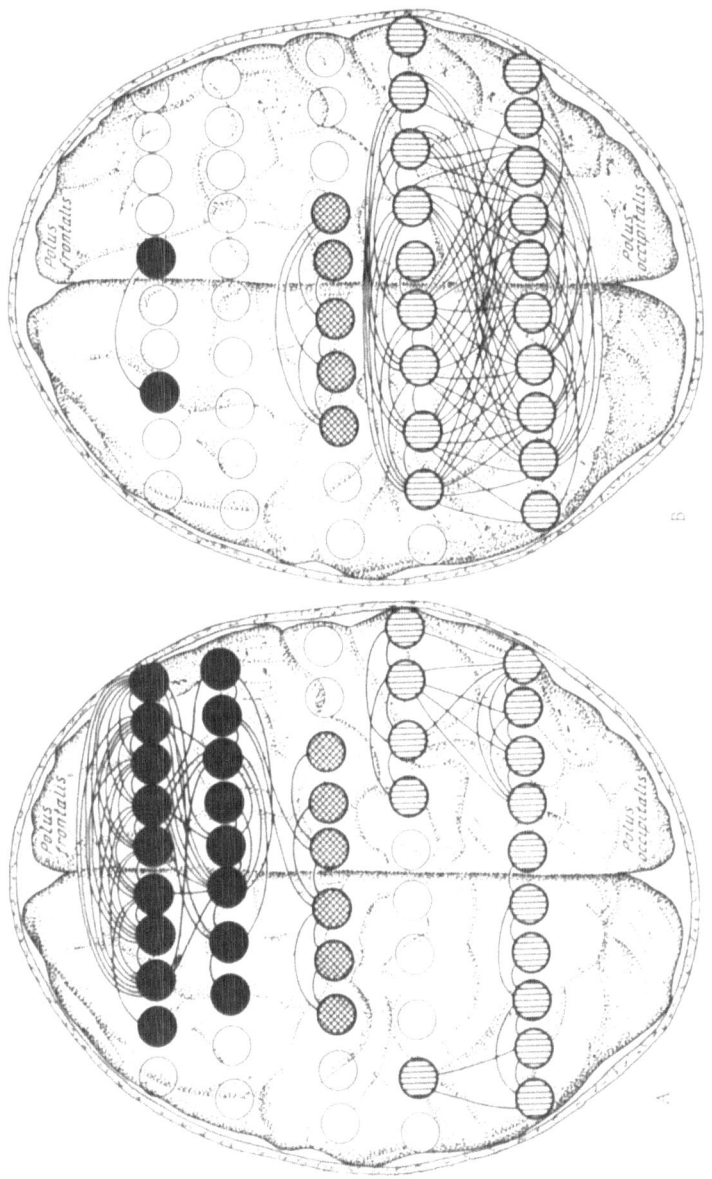

Fig. 2. High correlations between potentials in cortex of patient with epilepsy at rest, in different (successive) time intervals. A) Interval 1, Examination No. 3; B) interval 2, Examination No. 3. Remainder of legend as in Fig. 1.

increased. She was examined 2 h before a major fit. The number of high correlations between potentials was much higher than normal (Table 1, Examination No. 2). Later her condition fluctuated; the high correlations between potentials were unstable from one examination to another.

The changes in correlations between potentials during function tests, notably when the patients solved arithmetical problems, will now be considered. The results obtained were inconsistent. The interactions between points of the cortex differed in the same patient on different days when performing mental work.

The use of problem solving as a test revealed the functional capabilities of the cortex particularly well on each experimental day. All the patients who were investigated said that they had difficulty in remembering the original or the intermediate numbers. They calculated with difficulty and for a long time (2-5 min), while the healthy subjects finished problems of the same difficulty in 40 sec -1.5 min. The only exception was patient Rev., who did her calculations rapidly and correctly.

On days when the patient's clinical condition was poor (preceding a major epileptic fit, with exacerbation of psychotic symptoms or with marked dysphoria), correlations between potentials during function tests were abnormal. When solving the problem, for instance, no clearly defined increase in the number of high correlations was observed and, what is more important, they were not predominantly localized in the anterior parts of the cortex.

In some examinations, giving a mental problem to solve actually reduced the number of correlations, including the frontal areas. The impression was gained that the problem to be solved had become the dominant feature of cortical activity, while the existing pathodynamic structure (the paranoid structure, in particular) had disappeared under negative induction and was inhibited during the period of action of this dominant. Patient Vas. (both intervals of the third examination, Table 1) and the second examination of patient Priv. were examples of this. In the background, the number of correlations reached 88 in Vas. in the first interval and 101 in the second interval, while during the tests (multiplying 112 × 17) it fell to 11 and 34 respectively. In the second examination of patient Priv. the number of high correlations in the background was 252, falling to 23 during mental arithmetic (26 × 3 + 21) (Table 1).

TABLE 2. Number of High Correlations in Various Parts of the Cortex of All Investigated Patients with Epilepsy in the Background and during Multiplication of Two Numbers

Patient	Background					Mental work				
	frontal region		motor area	parietal region	occipital region	frontal region		motor area	parietal region	occipital region
	first pair	second pair				first pair	second pair			
Priv.	117	71	44	18	2	2	1	7	9	4
Vas.	43	13	6	13	13	7	1	3	5	18
Al.	0	5	7	6	1	9	8	32	30	12
Rev.	1	12	11	25	16	43	51	25	11	8
Aks.	9	8	55	33	5	Refusal				
Sav.	1	12	12	26	16	16	22	61	30	11

In other patients (for example, Al., with major epileptic fits without psychotic symptoms but with marked personality changes and frequent dysphoria) mental activity apparently "decompensated" the cortex. The correlation relationships between the potentials during control periods were almost normal (Table 2), while during the multiplication of numbers they differed considerably from the characteristic pattern of the healthy person. The number of high correlations increased in this case not in the prefrontal areas of the cortex, but predominantly in the posterior and (more so than normally) in the motor area (Fig. 3A). The problem was solved with great difficulty after 3 min and the answer was incorrect.

Finally, mental work could restore the correlations between potentials to normal. In the case of patient Rev. (with major epileptic fits repeated once a month, with unstable ideas of association, and dysphoria), for instance, during the second examination (Table 2) on a day when her condition was relatively satisfactory, during the arithmetical test the correlations between potentials were perfectly normal, despite the fact that in the control intervals they differed — the number of positive correlations being then increased (Fig. 3B).

It may be assumed that if the functional capacity of the cortex on a particular experimental day was adequate, during the test the influences between different points were just as in the healthy human cortex. Even if the correlation relationships in the background activity were quantitatively changed, mental work lowered

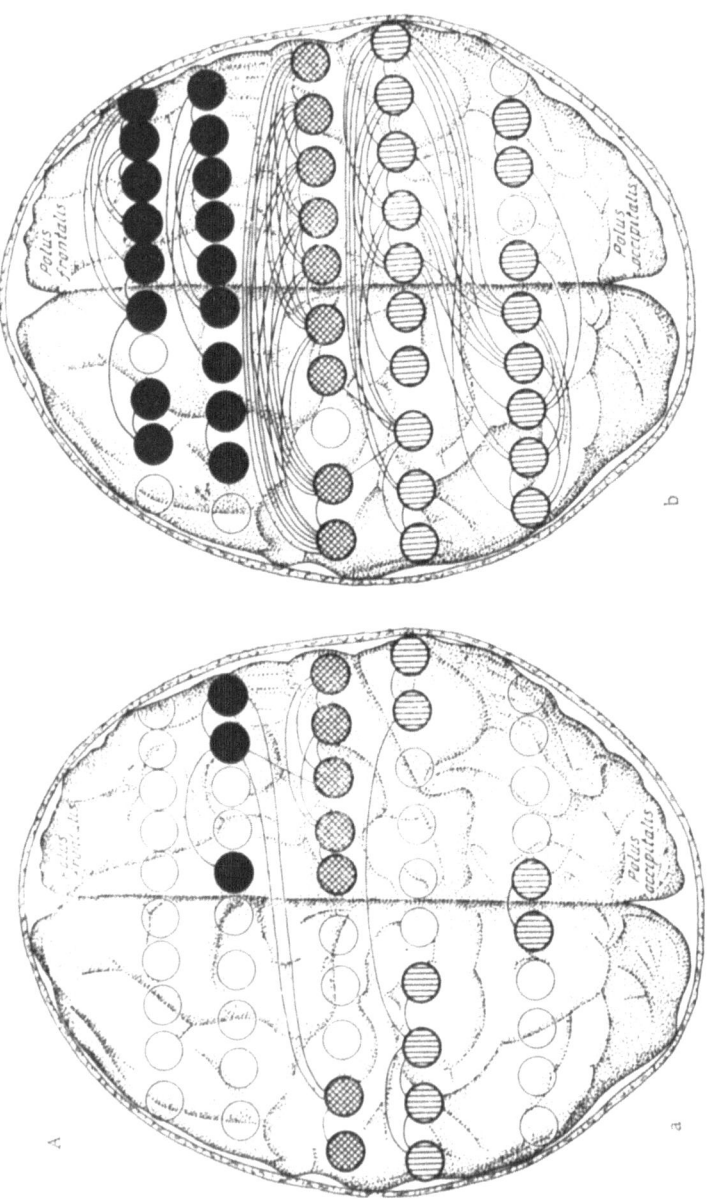

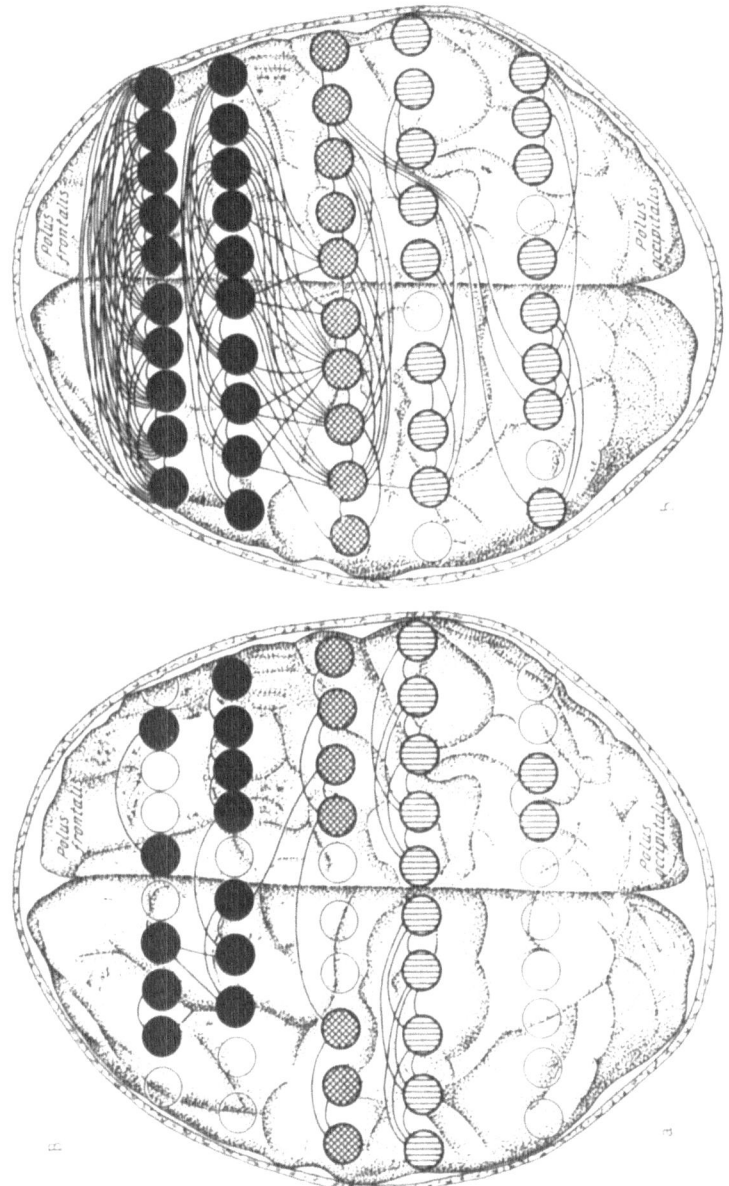

Fig. 3. Changes in high correlations between potentials in cortex of patients with epilepsy in background (a) and when solving an arithmetical problem (b). A) Patient Al.; B) patient Rev. Legend as in Fig. 1.

them to the values observed in the cortex of the healthy person
when solving similar problems. If the functional capacity of the
cortex was reduced, this did not happen. On the contrary, having
demonstrated its normal spatial synchronization, the cortex was
decompensated by the additional load and the correlation relationships between potentials were distorted.

Hence, spatial synchronization of potentials in the cortex of
patients with epilepsy and a paranoid syndrome is modified. Abnormalities are recorded both in the background electrical activity and during function tests. The character and degree of disturbance depend on the clinical condition. If the patient's mental state
deteriorates, especially during exacerbation of delusions, the normal number and distribution of correlations between potentials in
different parts of the cortex are distorted.

The similarity between the pattern of correlation of potentials during delusions in patients with schizophrenia and epilepsy
is evidence that they are associated with the functional state of the
cortex which, in all probability, is identical in these patients because their psychopathological syndrome is the same.

Obviously high correlation between potentials does not reflect the nosological specificity of the disease, but the functional
states existing in the cortex at the time of examination.

What is the physiological significance of high correlations
between potentials? It can be assumed that if electrical processes
at two remote points of the cortex are synchronized, some form of
mutual influence exists between them.

I consider that high correlations characterize an active state
of the cortex. They arise in a state when the irradiation of neural
processes is facilitated. It can be assumed that excitation is
slightly dominant over inhibition.

From my point of view proof of this is given by the fact that
the number of high correlations increases when static excitation,
lying at the basis of delusions and obsessive states, exists in the
cortex. Furthermore, in epileptics (even if free from paranoid
structures), in whose cortex excitation is frequently increased and
not balanced by inhibition, the number of high correlations is increased at such times. From results obtained with implanted electrodes (Vvedenskaya et al., 1964), spatial synchronization in the

cortex and in certain subcortical structures of patients with epilepsy may be intensified during a subclinical fit. Synchronization
(complete co-phasing) of the potentials of the frontal and occipital
poles was observed by Melik-Pashayan (1965) during an epileptic
fit. I have also observed a sharp increase in sychronization during a minor epileptic fit.

The facts described above appear to support this hypothesis.
At the same time, some forms of increase in spatial synchronization ("overflows," for example) probably reflect fluctuations in
the interaction between excitation and inhibition with, evidently,
the facilitation in such cases of irradiation of inhibition. An example of this may be the increase in "overflows" on going to sleep
(Monakhov, 1960), or the onset of interhemispheric overflows in
the cortex when an epileptic focus is present in the subcortex,
when induced cortical inhibition may be postulated (Nudel', 1962).
The investigation of correlation relationships between potentials
is a promising method of study of certain psychopathological syn-
dromes, notably states connected with disturbance of association.

Temporal Organization of the Auditory System*

G. V. Gershuni

Pavlov Institute of Physiology
Academy of Sciences of the USSR
Leningrad, USSR

The problem of the means by which information concerning an external stimulus is transformed at the receptor surface of the cochlea and the higher divisions of the auditory system is fundamental to the study of the mechanisms of auditory function.

The fact that frequency is reflected three-dimensionally in various parts of the auditory system has been demonstrated by many familiar investigations. However, the fact that frequency of the signal (spectrum) is reflected by spatial or other means (periodicity) by itself inadequately describes the transformations taking place in the auditory system (as is seen particularly clearly in examining the functions of the auditory cortex), primarily because of the absence of data for the temporal course of the transformations. Because of this an investigation (Gershuni, 1963; Gershuni, 1965) was carried out with the aim of determining the temporal characteristics of activity of the auditory system by a combination of psychophysical, behavioral, and electrophysiological methods.

The following dynamic characteristics were studied: The absolute thresholds and differential frequency thresholds as a function of duration of acoustic stimuli (tones and noise); latent periods as a function of signal intensity; and characteristics of single unit discharges for stimuli of different durations within different ranges of intensity.

---

* Pages 65-70 in the Russian edition.

By measuring thresholds as a function of stimulus duration in behavioral experiments on animals (dogs) after extirpations of various parts of the auditory cortex (Fig. 1), it was shown that disturbances of absolute threshold and differential frequency thresholds can be observed only in response to sounds lasting on the average under 10 msec (Baru, 1964).

For sounds whose duration exceeds 10-20 msec, no disturbance of discrimination criteria could be found, in agreement with published data indicating that extirpation of the auditory cortex has no effect on sound discrimination (Neff, 1961; Mering, 1959; Khananashvili, 1964). Facts demonstrating the importance of cortical auditory projection only for sounds of very restricted duration were also obtained on patients with injuries of the temporal cortex (Baru et al., 1964).

Electrophysiological investigations of various parts of the auditory system in animals (cats and rats) were carried out using both integral (Gershuni et al., 1964) and intracellular recording from single units. The basic principle of investigation of the phenomena as functions of time remained as before. The regions investigated were the cochlear nuclei, inferior colliculi, and the primary auditory cortex.

The results described in this paper are based mainly on those obtained by single unit recordings from these parts of the auditory system.

Acoustic stimuli (white noise and tones) of different duration, mainly between 2 and 100-150 msec were used. Potentials were recorded by tungsten microelectrodes, not more than 1 μ in diameter, moved by means of a hydraulic micromanipulator. The experiments were carried out on cats and rats anesthetized with urethane and chloralose.

The results obtained are based on investigation of more than 500 single units. The results of investigation of each part of the auditory system are described in separate communications (Radionova, 1966; Radionova and Popov, 1965; Maruseva, 1966; Vardapetyan, 1966).

A general survey of the results shows, to begin with, that the characteristics studied are interconnected and that types of neuronal activity can be distinguished in all parts of the auditory

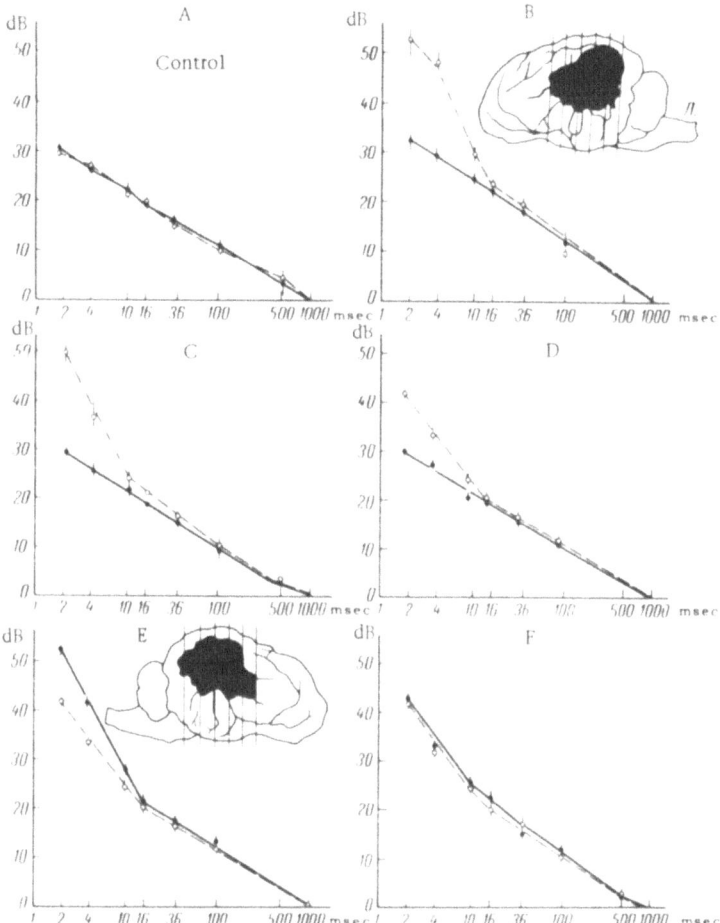

Fig. 1. Changes in threshold intensity of sounds of different duration in dogs after extirpation of auditory cortex (by motor-defensive methods). Abscissa: duration of stimulus in msec; ordinate: thresholds in dB relative to zero level (thresholds for duration of 1000 msec). Continuous line represents left ear, broken line right ear. Vertical strokes indicate scatter (standard deviation). A) Control (before operation); B) 20-60 days after extirpation of auditory cortex on the left (measured at frequency of 1000 cps); C) the same, measured for white noise; D) the same, 90-180 days after; E) 20-45 days after operation on the right (measured for white noise); F) the same, 60-120 days after. Curves show that thresholds after extirpation of cortex are increased only for short durations (up to 10 msec) of stimuli acting on ear contralateral to cortical injury (from A. V. Baru).

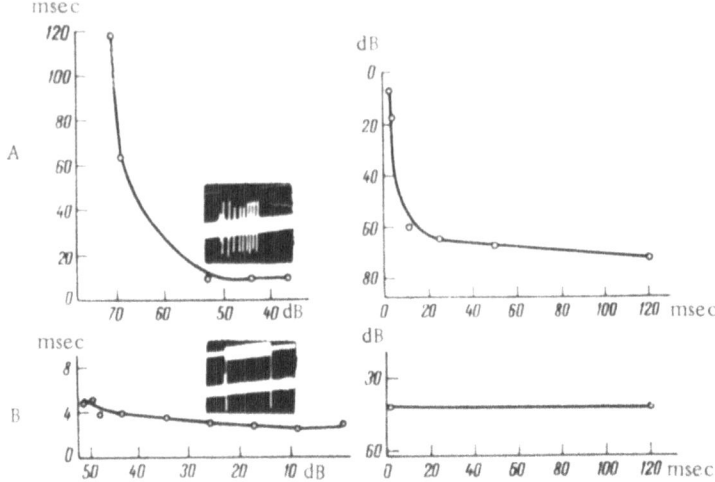

Fig. 2. Characteristics of two types of unit response from the
cochlear nucleus. A) Neuron No. 267, summating slowly, with
tonic discharge; B) neuron No. 263, summating rapidly, with on/off
response. On the left: changes in latent period as a function of
stimulus intensity. On the right: changes in threshold intensity
as functions of stimulus duration. Scale of intensity in dB of attenuation, acoustic tone bursts of optimal frequency (for A, 21 kc/sec, for
B, 6 kc/sec. Duration 100 msec). It can be seen that latent period
for neuron No. 267 falls from 118 to 6 msec with an increase in sound
intensity. Range of temporal summation more than 50 dB. Latent
period for neuron No. 263 falls from 5 to 3 msec; temporal summation absent (from E. A. Radionova).

system. This applies, above all, to summation established relative to the degree of changes in thresholds with a change in stimulus duration and to the degree of changes in latent periods over
the range from minimal to maximal stimulus intensity. It applies
also to the definite character of the changes during unit impulse
activity between 100 and 200 msec from the beginning of action of
the stimulus.

The discovery of sharp differences in temporal characteristics of the unit response as a function of its spectral sensitivity is
highly significant: temporal summation and the duration of the
unit discharge are much greater for an optimal than for a nonopti-
mal frequency.

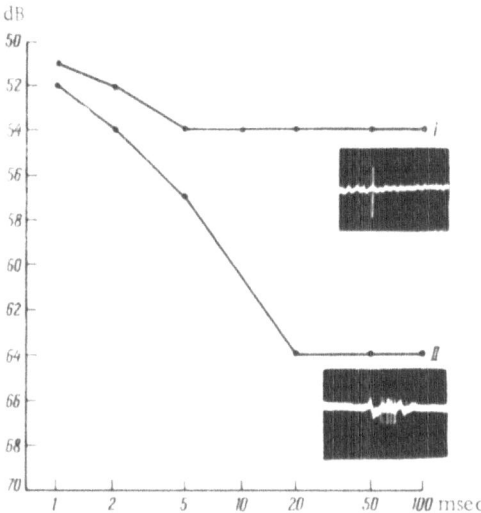

Fig. 3. Characteristics of two types of neuron
from inferior colliculus. Changes in threshold
intensity as functions of stimulus duration.
I) Phasic response, temporal summation ill defined; II) tonic response with marked temporal
summation (from A. M. Maruseva).

Assessment of differences between the activity of individual
units in relation to these characteristics revealed two extreme
types of responses possessing the following properties.

One type possesses well-marked signs of temporal summation — a difference between thresholds for short (2 msec) and long
(from 20 to 30 msec or more) stimuli amounting to 15-50 dB; with
clear changes in latent periods within a dynamic range of intensities of from 3 to 20 times; with a discharge continuing throughout
the period of stimulation (100-200 msec), and without marked
signs of adaptation (Fig. 2A).

The second type of response was characterized by ill-
defined phenomena of temporal summation (no difference between
thresholds for short and long stimuli, or difference not exceeding
1-4 dB); the latent periods changed by not more than 1.5-1.8
times; discharges are found only at the beginning (and in some
cases, at the end) of action of the stimulus and consist of few

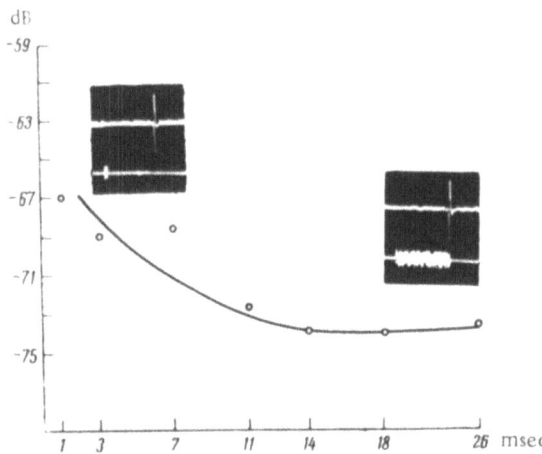

Fig. 4. Curve of temporal summation (change in threshold intensity as a function of stimulus duration) for units of the cat's auditory cortex with stable responses (from G. A. Vardapetyan).

(from 1 to 2-5) impulses throughout the dynamic range of intensity (Fig. 2B).

Differentiation between the unit responses using temporal summation as criterion justifies describing these two types of response as slowly summating and rapidly summating. I prefer this subdivision to the very frequently used classification of tonic and phasic responses. Nevertheless, it follows from what has been said that slow summation is more commonly associated with a response of tonic character and rapid summation with a phasic response (Fig. 3).

Examination of the distribution of types of response in different parts of the auditory system reveals the following. In the cochlear nuclei slowly summating types of response with a prolonged tonic discharge are found much more often than rapidly summating types with a phasic discharge (40% of the former, 20% of the latter, and 40% of intermediate types of response).

In the inferior colliculus 40% of the responses are of the rapidly summating type and 50% of the slowly summating type.

Although the character of the unit response in the auditory cortex is much more difficult to classify, unquestionably many of the neurons have the property of rapid summation, and quantitatively speaking their response closely resembles the primary evoked response of the auditory cortex (Fig. 4).

In the examination of these data the extremely important question of the relationship between a particular type of response and the character of stimulation arises. The results obtained show that both in the cochlear nuclei and in the inferior colliculus, neurons are present which show very clearly defined phenomena of temporal summation and discharges of a tonic character in response to stimulation of optimal frequency, but ill-defined temporal summation and discharges of phasic character in response to frequencies above or below the optimal value.

The frequency-threshold curve for such neurons shows lowering of the thresholds at the optimal frequency with a change from sounds of short duration to long, i.e., they show temporal frequency peaking. A spontaneous rhythmic activity is found only in the slowly summating neurons with a tonic type of discharge and is not observed in the rapidly summating phasic neurons.

The results described above show that the slowly and rapidly summating responses must have actions which differ in many respects during the transmission of information concerning an acoustic stimulus. In the first place, these actions develop completely differently along the time axis. For instance, the rapidly summating neurons* (bearing in mind their short latent period in response not only to strong, but also to weak stimuli, the small number of discharges, and the short summation time) must develop their action within a short time interval; the slowly summating neurons must develop their activity over the whole time of action of the stimulus and must find the optimum of accumulation of information concerning frequency to correspond to the critical time, which may reach up to 100 msec (especially with weak sounds).

Comparison of these results with those of psychophysical measurements mentioned earlier shows that the mechanism of

---

* All the results described above suggest that it is terminologically more correct to speak of rapidly and slowly summating working programs of neurons rather than of rapidly and slowly summating neurons.

cortical detection and discrimination of short stimuli can function only by the activity of rapidly summating neurons. It is these neurons or, more precisely, the activity of neurons in a rapidly summating (phasic) regime, which can provide for the mechanism of projection in the primary auditory cortex.

The auditory projection cortex thus behaves as the projection of neurons responsible for only a certain initial segment of the time axis of the stimulus. This phenomenon provides definite evidence of demarcation between processes taking place in different segments of the time axis in spatially different divisions of this auditory system.

During the testing of this hypothesis the question arises of the connection between this temporal-spatial organization and the receptor surface of the cochlea. On the basis of data obtained recently concerning the special sensitivity of the inner hair cells to transitional phenomena in a stimulus (Neubert, 1960), and also of the well-known data concerning the point innervation of the inner hair cells, equipped with radial fibers, it may be postulated that the rapidly summating type of organization of the auditory system, reflected in the cortical projection, is connected at the periphery predominantly with the inner hair cells.

In this hypothesis, subdivision of analysis of information concerning the stimulus along the time axis begins with the receptor elements of the cochlea. The rapidly summating type of organization may be described as a type of organization intended for the transmission of information about transitional phenomena. This type of organization possesses a cortical projection and the basic attributes of high discriminative sensitivity.

The problem of the spatial projection of longer sounds, which must be discriminated by the slowly summating neurons, whose activity is characterized by a temporal peak of discrimination, is not yet clear.

All that can be said is that this mechanism is not disturbed by extirpation of the auditory cortex, and it must therefore be associated to a greater degree with subcortical structures. Its study will be the subject of a future investigation. It can only be postulated that the sharp differentiation between the character of activity of the auditory system, revealed when this activity is examined at different time intervals, may be connected with certain mechanisms of perception for different acoustic structures, possibly of the sounds of speech and music, the organization of whose central mechanisms has been found to be surprisingly different.

Electrical Correlates of the Direction of Movement Elicited by Subcortical Stimulation*

E. Grastyán and L. Angyan

Institute of Physiology
University Medical School
Pecs, Hungary

INTRODUCTION

In the course of a study of motivating effects elicited in free-ly moving cats by stimulating different subcortical loci, it was a consistent finding that the stimulated and the contralateral sides of the hippocampus showed characteristically different patterns of electrical activity. More concretely, while in subsequent trials stimulation intensity was gradually increased, at a definite value a desynchronization in the ipsilateral hippocampus and theta waves in the contralateral hippocampus appeared simultaneously. Because this electrical asymmetry regularly coincided with contra-lateral turning of the animal, it was suggestive of a correlation between the direction of locomotion and the hemispheric locali-zation of fast- and slow-wave electrical patterns.

This correlation, if valid, would have provided an excellent approach to the intimate neural organization of motivation, and its verification therefore seemed important. The most advanta-geous target for this study seemed to be the nonspecific system of the thalamus; considerable insight has been gained recently into its intrinsic synaptic organization by analyzing the recruiting as well as the spike and wave complexes (Purpura and Cohen, 1962; Purpura and Shofer, 1963; Pollen et al., 1964a; Pollen, 1964; Poll-en and Sie, 1964; Jasper and Stefanis, 1965).

* Pages 71-78 in the Russian edition.

METHODS

The experiments were conducted on 12 freely moving adult cats bearing stereotaxically implanted stimulating as well as recording electrodes. Generally three pairs of side-by-side steel needle electrodes destined for stimulation were implanted in the anterior, dorsomedial, and posterior parts of the thalamic diffuse projecting system (involving in some cases the pretectal region). Bipolar electrodes of a similar type were located in the dorsal hippocampus, and pairs of silver ball electrodes (with the poles 2-4 mm apart) over the frontal (pre- and postcruciate gyri), parietal (suprasylvian or anterior ectosylvian gyri), and occipital (gyrus marginalis) regions for recording.

A square-wave stimulator with independently variable parameters was used for stimulation and an 8-channel Schwarzer EEG machine and a Disa oscilloscope for recording electrical activity.

After recovery from surgery (generally 4-7 days) the animals were stimulated in an empty cage of 1 m × 1 m size. The movements of the animal were recorded cinematographically simultaneously with the recording of the electrical manifestations. In the film record the trace of a small electric bulb attached to the animal's head and lit during the stimulation period helped to identify the corresponding periods of EEG and movement records. After the animals had been sacrificed the stimulated loci were identified histologically with a modification of the Klüver and Barrera (1953) combined method, staining both cells and fibers (Szabó, 1965).

RESULTS

General Comments

Since the early studies of electrical brain stimulation it has been known that stimulation of the same locus with different parameters may result in different or even opposite effects. With this fact in mind, care was taken at the beginning of this experiment to study the effects of different stimulus parameters (frequency, duration, intensity) in all possible combinations. However, it soon turned out that some of the parameters, e.g., low frequencies, despite their marked electrical effects (recruiting complex), were inappropriate for the purpose of the present study because of the paucity of the accompanying overt manifestations. For instance, in most cases low-frequency stimulation induced the well-known arrest reaction first described by Hunter and Jasper (1949) in freely moving cats. (Functional evaluation of the electric effects of low-frequency stimulation was successfully accomplished in a conditional-reflex situation. These findings, not being directly related to the present problem, are disregarded here.)

Difficulties of the opposite kind could be expected in the case of high-frequency stimulation, namely a monotonous picture of desynchronization accompanied by complex motor patterns. Fortunately this expectation proved to be false. A characteristic electrical pattern, spindles composed of mainly surface-negative, high-amplitude slow (4-7/sec) waves appeared in different neocortical and subcortical regions quite regularly during high-frequency (100-200/sec) stimulation. Since to our knowledge this peculiar electrical pattern had not previously been recorded as a stimulation effect in the freely moving animal, the possibility of its artifactual origin (muscle activity, movements in general) arose. This could, however, to all probability be excluded by stimulating under the paralyzing effect of curare, when the spindles could be identified in almost unchanged form. Spindles elicited by single pulses applied to the caudate nucleus (Buchwald et al., 1961) or the median thalamic structures have different frequency characteristics; moreover, their anatomical location is different. On the basis of the data available it is difficult to decide whether the two phenomena represent the same mechanism or not.

With the help of the slow-wave spindles two different regions, showing characteristic differences of organization, could be separated in the nonspecific thalamic system. One of these includes the anterior group of nuclei (including the n. reuniens) and the dorsomedial nucleus, the other the region of centrum medianum and the pretectal region.

The Anterior Thalamic Region

A fairly constant motor pattern was elicited from this region by a wide range of different intensities of high-frequency stimulation. Essentially the effect of a continuous stimulation of several seconds consisted of a circling movement in a direction

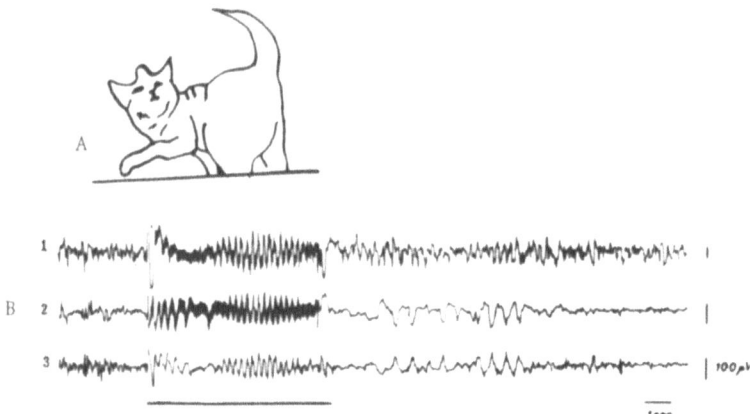

Fig. 1. A: Typical phase of the complex motor action (copied from film recording) elicited with high-frequency stimulation (100/sec, 0.3 msec, and 6 V) from the region of the anterior thalamic nuclei; B: the accompanying electrical manifestations of the contralateral hemisphere. Further details in text. 1: Right dorsal hippocampus; 2: right frontal; 3: right parietal cortex (suprasylvian gyrus), Monopolar leads. Stimulation: horizontal line beneath traces.

opposite to the side of stimulation, interrupted and also terminated by periods of arrest. During circling the trunk of the animal became curved with the convexity toward the side of stimulation and the homolateral side somewhat lowered toward the floor. In addition, the forelimb contralateral to stimulation was often lifted in the air. The whole picture resembled the normal defensive pattern of the cat (Fig. 1A).

Simultaneously with the active phases of circling, slow-wave spindles (4-7/sec) appeared predominantly or exclusively in different regions of the opposite hemisphere relative to stimulation (Figs. 1B and 2). A detailed analysis with the help of film records showed that the active phase of movement had been terminated when the potentials of the spindles reached maximum amplitude. The declining phase of the spindles and the desynchronized period coincided with a deceleration of movement and periods of arrest. Sometimes a single slow potential emerged in between the first two spindles and was accompanied by a quick lifting of the contralateral forelimb.

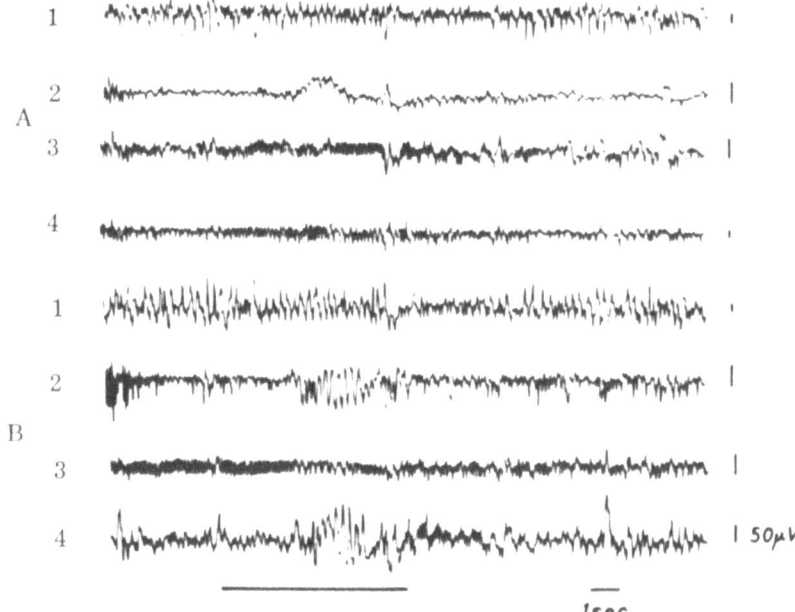

Fig. 2. If the motor effect of stimulation (left dorsomedial n.) consists in a pure lateral deviation in space, spindles appear exclusively on the hemisphere contralateral to stimulation. Activity recorded from symmetrical points of the two hemispheres. A: left, B: right hemisphere. 1: Hippocampus, 2: frontal, 3: parietal, and 4: occipital cortex. Monopolar leads.

The frequency of the waves of the first and the subsequent spindles showed differences, with the first frequency in most cases higher than the rest. In addition the polarity of the waves of the second spindle became reversed in several leads. A possible behavioral correlation corresponding to the frequency differences of the subsequent spindles was a difference in speed of the accompanying phasic movements, the first always being faster than the rest. This phenomenon also suggested that the effectiveness of stimulation progressively decreased as stimulation proceeded and this might be responsible for the reversal of polarity of the slow waves. This latter assumption seemed to be valid because a temporary increase of intensity at the time of the expected appearance of the second spindle elicited a spindle identical with the first one. Simultaneously the speed of locomotion was also restored by this procedure.

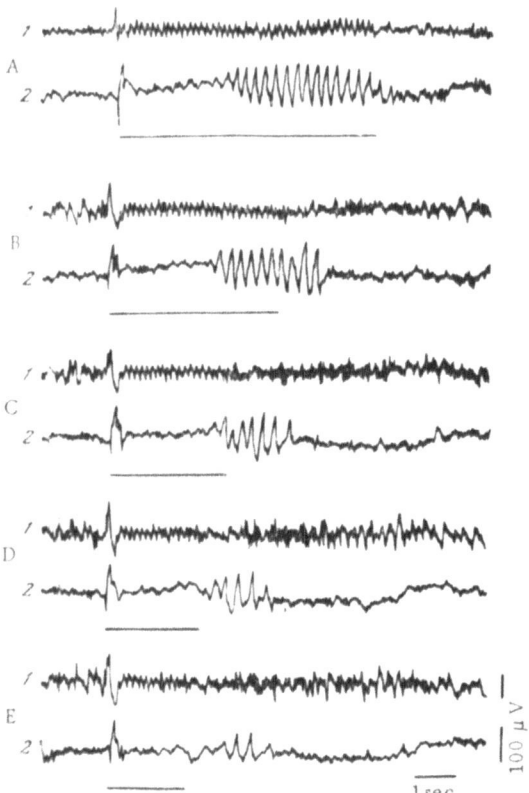

Fig. 3. Constancy of latency of the spindle appearing
on the contralateral (left) frontal cortex elicited by
stimulating the right dorsomedial thalamic nucleus.
A—E: duration of stimulation is progressively shortened
and at a certain length of stimulation the spindle appears as if it were a rebound effect. Phasic circling
movement toward the side of the spindle starts simultaneously with the appearance of the spindle; in C, D,
and E, it takes the shape of a rebound movement. If
the spindle appears as an after effect, high-amplitude
fast activity appears on the contralateral frontal cortex.
1: Right frontal cortex, 2: left frontal cortex.

Interruption of stimulation resulted in conspicuous and varied after-effects. Their character depended on the time of interruption. The general rule seemed to hold that if interruption of stimulation occurred during the build-up of the phasic turning movement (i.e., just before or during the ascending limb of the spindle), then this same movement continued to appear in a facilitated manner. This phenomenon therefore was called an "auto-rebound" reaction. If interruption of stimulation occurred during the slowing down of the phasic movement (i.e., during the decaying phase of the spindle or immediately afterwards), a movement opposite to the direct effect of stimulation, sometimes an arrest reaction, appeared. The phasic antagonistic after-effect was called a "hetero-rebound" reaction.

In good agreement with this, if interruption of stimulation occurred during the incremental phase of the spindle or immediately before its expected appearance, the spindle continued uninterruptedly or appeared without any preceding slow-wave component in the stimulation period, i.e., as a pure after-effect (Fig. 3).

The most important feature of the spindles, which deserves special emphasis in the present context, is their unilateral localization. On the basis of the findings described above, the general rule seemed to be that the direction of movement is always toward the slow waves. It is noteworthy, however, that in some cases discrepant observations were also made, especially in the occipital regions, for even with bipolar recordings spindles with amplitudes comparable to those of the contralateral hemisphere appeared ipsilaterally also. This discrepancy seemed partially resolved by the observation that in these cases the direction of the accompanying motor manifestation was often ambiguous, i.e., it was also contaminated by a postural or phasic component in the direction of the stimulated hemisphere.

As far as the occipital slow-wave spindles are concerned, an additional complicating factor emerged, namely the question of volume conduction. The frequency characteristics of the slow-waves spindles were in most cases identical with the theta rhythm of the hippocampus. The only difference between the two was that the slow waves of the hippocampus generally appeared in a continuous form and only exceptionally as spindles. In the minority of cases a clear-cut difference was established between the hippocampal and neocortical slow waves, at least in certain phases of spindling. Despite this latter finding, the probability that volume conduction caused the appearance of a hippocampal theta rhythm on the neocortex could not definitely be ruled out.

The Posterior Thalamic and Pretectal Region

A striking difference between the electrical and behavioral manifestations showed itself in response to stimulation in the region of the centrum medianum and in the pretectal region.

As far as the overt manifestations are concerned the character of the locomotion elicited from these regions resembled more those of a physiologically motivated animal, i.e., they did not show a forced character. The direction of movement was not as strictly determined by the side of stimulation as in the former case; moreover, the effect of stimulation could easily be influenced by environmental stimuli.

As to the electrical effects of stimulation, no peculiar manifestation except a monotonous desynchronization was observed. However, spindles with amplitude comparable to those elicited from the anterior region often appeared in the poststimulation period and quite unexpectedly in the ipsilateral hemisphere (Fig. 4). These rebound-like spindles appeared in cases where the direct behavioral effect of stimulation was a definite contralateral turning and where interruption of stimulation induced a "hetero-rebound" reaction, i.e., an immediate reversal of turning toward the opposite direction. This unexpected finding offered exceptionally good evidence in favor of our former statement that the direction of movement is always toward the side of slow-wave spindles.

DISCUSSION

Apart from minor discrepancies most of the facts described above strongly support the assumption that the direction of movement induced by subcortical stimulation is strictly correlated with unilateral appearance of slow-wave spindles. An obvious objection that electrical stimulation is artificial by its very nature and thus might create correlations not existing in natural circumstances reduces the importance of this statement, but this objec-

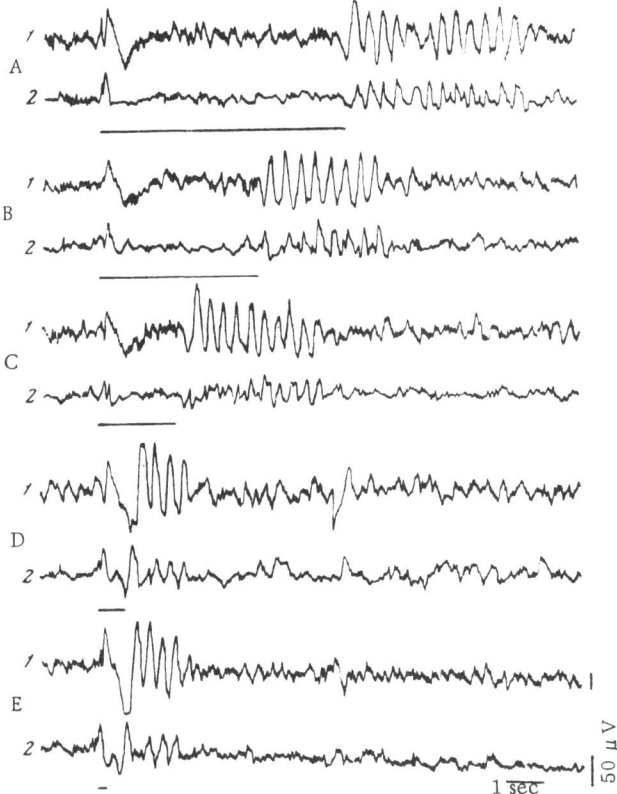

Fig. 4. Spindles appearing on the ipsilateral hemisphere after interruption of stimulation in the region of the centrum medianum. A—E: the fact that by progressive shortening of the duration of stimulation the latency of the spindle also decreases progressively shows that, in contrast to the spindles in Fig. 3, the effect here corresponds to a true rebound-like after effect. Direction of the rebound motor after effect coincides with the laterality of the spindle. 1: Left caudate nucleus, 2: left frontal cortex.

tion is somewhat weakened by the fact that the same correlation manifested itself also in the after-effects of stimulation. We admit that this is not enough evidence for our present thesis and any further generalization must await verification under the effect of physiological driving forces. The special purpose of the present thesis might be, however, that it urges us to search for the required electrical events when recorded under normal circumstances, because they may be present, unnoticed, in our previous records or have been discarded as artifacts.

We think, however, that the weakness of the present argument should not deter us from anticipating the final evidence with some conclusions concerning the functional significance of the electrical events described.

The first approach is offered by an important finding of Hassler and Hess (1954), who showed by means of small lesions that adversive patterns are reciprocally organized in the two hemispheres. More concretely, an adversive pattern elicited by electrical stimulation will be automatically replaced in a tonic manner by its exact opposite following a lesion of the stimulated site. This necessarily means that during the elicitation of a given posture or movement its reciprocal action represented by the opposite hemisphere comes under the influence of reciprocal inhibition. By applying this evidence to our case the slow-wave spindles appearing in different structures of the contralateral hemisphere may correspond to a reciprocal inhibitory process. This assumption is supported by the fact that the development of the spindles proceeds pari passu with the increase of speed and force of the elicited turning movement, moreover that the movement slows down during the decaying phase of the spindle and is eventually arrested in its absence.

In addition to this strict correlation there is some indirect electrical evidence supporting the inhibitory nature of the slow-wave spindles. By analyzing the late surface-negative wave of the spike and wave complex elicited from the median thalamic structures, Pollen et al. (1964) established that unit firing definitely ceases in the presence of these waves. On the basis of the suppressing influence exerted on conditional performance the present authors attribute an inhibitory nature to the recruiting potential (Ángyán and Grastyán, 1963; Ángyán et al., 1963) and recently more especially to the late surface negative wave which follows the recruiting potential. The weakness of this evidence lies in the fact that these slow waves are induced by low frequency stimulation. Although there are findings to show that under special circumstances recruiting-like potentials can be elicited also by higher frequency stimulation (Grastyán et al., 1955; Schlag and Chaillet, 1963), definite proof that the slow potentials composing the spin-

dles and those elicited by low-frequency stimulation represent
identical mechanisms is still lacking.

The inhibitory nature of the slow wave spindles, even if valid, would not settle all the problems raised by the present findings.
In particular, the periodic appearance of the elicited motor action,
i.e., its interruption by phases of arrest reactions, requires further explanation.

The fact that the motor effect elicited by stimulation progressively slows down and finally stops requires the assumption
of a second inhibitory action affecting the excitatory process itself. On purely logical grounds, this inhibition cannot originate in
the contralateral hemisphere because it begins to develop at a moment when the reciprocal inhibition has just reached its maximum.
Therefore it seems more probable that a kind of negative feedback
action, or we might call it "autogenous inhibition," is set into operation (at a certain level of excitement) simultaneously with the éx-
citatory action and begins to control the elicited motor action as it
proceeds. The progressive weakening of the excitatory process
on the other hand necessarily results in a subsequent release of
the reciprocally inhibited action.

Recent microelectrode findings offer several synaptic mechanisms which conform in principle to the requirements of this postulated negative feedback-like inhibition (Andersen and Eccles,
1962; Purpura and Cohen, 1962). As the simplest process the so-
called spike inactivation mechanism, which was shown to occur as
a result of higher intensity stimulation in the thalamus by Purpura
and Shofer (1963), has also to be considered in this context.

Autogenous inhibitory action would explain the otherwise in-
explicable finding that interruption of stimulation may, if appropriately timed, result in an "auto-rebound," i.e., in the facilitation of
the direct effect of stimulation. An additional unavoidable result
of this complex regulation, incorporating reciprocal and autogenous
inhibitory processes, is the arrest reaction. It is on account of
autogenous inhibition that a period of balance always appears, in
which the direct effect of stimulation is opposed by the motor activity of the symmetrical area, freed from inhibition. This freeing
from inhibition may be caused by postanodal exaltation, which is
known to take place at various levels of the central nervous system. The arrest reaction may be the consequence of conflicting

excitatory forces and not of generalized inhibition. This hypothesis also matches the general character of the arrest reaction.

SUMMARY

High-frequency (100-200/sec) stimulation of the anterior
portion of the nonspecific thalamic system evokes circling movements in freely moving cats, interrupted periodically by arrest
reactions. High-amplitude slow-wave spindles interrupted by periods of desynchronization appear simultaneously in the hemisphere
contralateral to stimulation.

High-frequency stimulation of the centrum medianum and
pretectal region evokes a variety of motor responses accompanied
by diffuse neocortical desynchronization. In these cases high-
amplitude slow spindles appear as a rebound effect in the ipsilateral hemisphere after the cessation of stimulation.

The functional significance of the correlation between direction of movement and localization of the slow-wave spindles and
the possible inhibitory nature of the slow waves is discussed on
the basis of recent microelectrophysiological evidence.

Conditioning of Single Unit Activity
by Intracerebral Stimulation*

Y. Hori, I. Toyohara, and N. Yoshii

Second Department of Physiology
Osaka University Medical School
Osaka, Japan

With the development of microtechnique, many papers were published on the neuronal mechanisms of temporary connections: Buchwald et al., 1965; Bureš and Burešová, 1965; Jasper et al., 1960; Hori and Yoshii, 1965; Kamikawa et al., 1964; Morrell, 1960; Olds, 1965; Yoshii and Ogura, 1960. In all of these experiments, the applied conditioning procedure was usually to present two stimuli in combination to the peripheral sensory receptors, though some experimenters tried to present one of the two stimuli to the central nervous system directly.

The purpose of our experiment is to study conditioning by combined presentation of two intracerebral stimuli as conditional and unconditional stimulus instead of peripheral stimulation, in order to delimit the excited areas in the brain. Conditioned effects were tested on thalamic, hippocampal, and cortical unit discharges.

METHODS

Thirty-eight cats were used. Twenty-five of them as encéphale isolé. The cats were placed in a stereotaxic instrument. Necessary operations were performed under ether anesthesia. After all pressure points and cut edges had been infiltrated by 2% procaine hydrochloride solution at the end of the operation, ether inhalation was discontinued. The experiment was conducted in an unanesthetized and immobilized state of the animal. For immobilization Flaxedil was administered intravenously. Cortical EEGs

* Pages 281-289 in the Russian edition.

and cortical evoked potentials were observed to check whether the animal was kept in good condition.

Stimulating electrodes were of concentric type, made of stainless steel wire 300 μ or less in diameter. They were inserted into the lateral geniculate nucleus (GL) for applying the conditional stimulation (CS). Another pair of electrodes for unconditional stimulation (US) were inserted into the midbrain reticular formation (RF), the nucleus centrum medianum (CM), or the nucleus ventralis anterior (VA) of the thalamus. High-frequency stimulation of RF or low-frequency stimulation of either CM or VA were used for this purpose.

Glass capillary microelectrodes filled with 3 M KCl were used to pick up unit discharges. Cortical EEG and unit discharges were displayed on the screen of a dual-beam cathode ray oscilloscope and photographed on running film by a kymograph camera. Mostly, the recordings were made once every two trials.

While picking up unitary discharges from any cortical or subcortical structure, it was seen that both high-frequency stimulation of RF (0.2 msec, 3 V, 100/sec) and low-frequency stimulation of CM or VA (0.5-5 msec, 5 V, 5-7/sec) evoked a clear change in the discharge frequency or the discharge pattern, and low-frequency stimulation of GL (0.2 msec, 2-10 V, 10-50/sec) induced almost no change or a slight effect opposite to that of the US.

The combination of CS and US was such that the CS was presented for 6 sec and its latter half was overlapped by the US, both ending at the same time. An adequate interval, 1 min at shortest, was set between two successive trials.

When an encountered unit had been found showing stable discharges, it was habituated to the CS before the reinforcement began.

To decide whether conditioned change was induced, the following analysis was done. Usually three or more serial records were sampled from the late period of reinforcement and extinction trials. With the record so sampled, mean discharge frequencies per second during 5 sec before the CS and during 3 sec after the CS were calculated. The results of calculation were compared with those of respective controls that were recorded before reinforcement. Since the discharges were often seen to occur forming a burst, it was necessary to treat them separately from the discharges occurring with a well spaced pattern. For this purpose, the term "burst index" was introduced, which is defined here as the ratio of the number of discharges forming bursts to the total number of discharges.

After the experiment the positions of the microelectrode tip were examined histologically in serial sections of the brain stained with thionin.

RESULTS

General Remarks

When tetanic stimulation of RF was applied, it induced an increase of discharge frequency in more than half of the tested units. Tetanic stimulation often decreased the burst index. Low-frequency stimulation of CM or VA used as the US also induced an increase in the discharge frequency and a decrease in the burst index. Enhancement of discharges by these stimuli reached 40/sec at the highest. Burst index during US presentation was usually less than 10%.

Unit discharges were classified according to whether or not they showed changes in discharge frequency per second and/or in burst index as the result of reinforcement (Table 1). Eight out of 58 (type A in Table 1) showed a clear conditioned effect. Five out of eight returned to the original pattern after extinction, and the remaining three units did not receive extinction. The conditioned effect seen in these units (type A) is such that the response to CS in terms of the discharge rate and/or the burst index increases or decreases as the reinforcement is repeated. To identify the type A change, progressive changes in the rate of background discharges seen during the intertrial intervals were strictly excluded. Seventeen units showed changes not only in the response to CS but also in the rate of the background discharges. This type of conditioned change is termed type B. Thirteen out of 17 showing the type B effect did not receive the extinction procedure. Group C includes units from which we failed to get a clear-cut effect. Among these not fully conditioned units, six survived after 45 trials of reinforcements. Discharges from 27 units disappeared before 45 trials with reinforcement could be applied.

TABLE 1. Classification of Examined Neurons

Group	A	B	C	Total
Number	8(5)	17 (4)	33 (5)	58

Group A shows the conditioned change only in response to CS.
Group B shows the conditioned change both in response to CS
and in background discharge of the intertrial intervals. Figures
in parenthesis of A and B show the number of neurons for which
extinction was tested. Group C shows no conditioned effect.
Figures in parenthesis in C show the number which had more
than 45 reinforcements.

Increase of burst index was not conditioned in this experiment, although sometimes low-frequency stimulation of CM or VA
showed a transient increase in burst activity.

Parallel to the fact that there were no great differences in
the discharge frequency and the burst index depending on the location of the units, we failed to find any marked differences in the
conditioned effects among units of different origins. Also as regards the ease with which the conditioned effect can be obtained at
the unitary level, there were no great differences between the
sites of recording.

Yoshii and Ogura (1960), Kamikawa et al. (1964), and Bureš
and Burešová (1965) reported a conditioned response which was
not similar to the unconditioned response, but it was not observed
in this experiment.

Some Typical Cases of Unitary Conditioning

Unit Showing Increased Discharge as Conditioned Response. The unit illustrated in Fig. 1 and summarized in Table 2 was the one recorded from the nucleus lateralis dorsalis of the thalamus. CS (10/sec stimulation of GL) initially decreased the discharge and US (100/sec stimulation of RF)
increased it slightly (Table 2, column 1). In the course of the
whole experiment, the frequency of the spontaneous discharge
in the intertrial intervals changed, but there was no significant
difference between the discharge frequency in the pre- and post-

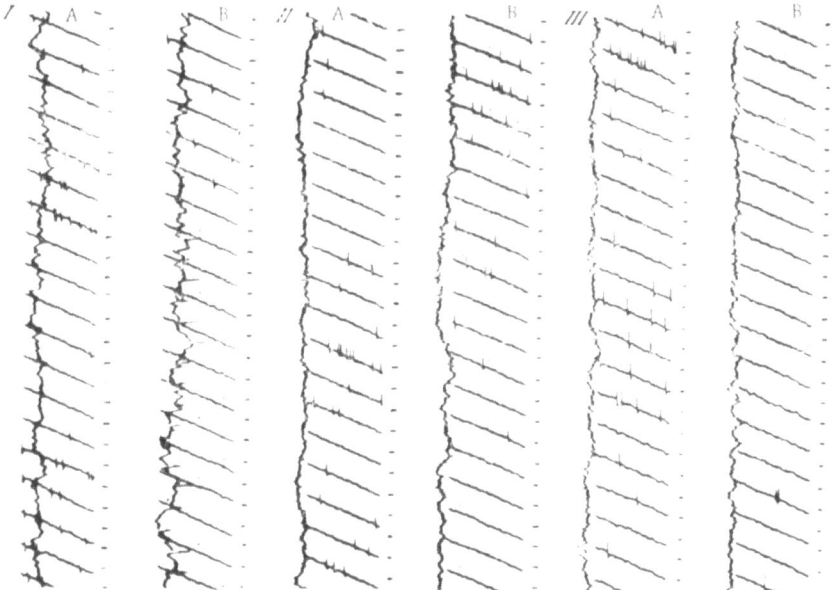

Fig. 1. Sustained discharge pattern of unit recorded from nucleus lateralis dorsalis.
I: Before conditioning. II: At 17th reinforcement. III: At the 14th extinction.
A: Background discharge in the intertrial intervals. B: Discharge during CS presentation. Abbreviations above are the same in subsequent figures. Stimulus pulses are
shown by dots. Continuous record is the cortical EEG. Note the increase in discharge
rate as the conditioned effect in the column II B.

reinforcements and extinction. The discharge frequency during
CS presentation increased a little in the early period of reinforcement, but this increase was not statistically significant. The mean
value of the discharge frequency during the late period of reinforcement increased significantly. This means that CS which had
a decreasing effect upon the discharge before reinforcement had
lost its original effect after reinforcement. Since the discharge
frequency during the intertrial interval was maintained at approximately the same level as that recorded at the prereinforcement period, the conditioned change seen in this unit was identified as type A.
Following 15 reinforcements, the CS was presented alone about 12
times (extinction), whereupon discharge frequency during the CS decreased and tended to return to the prereinforcement level.

Unit Showing Decreased Discharge Rate as
the Conditioned Response. The unit in Table 3 was recorded from the nucleus ventralis lateralis. The CS had no effect

TABLE 2. Conditioned Effect in Discharge Rate per Second

	Before rein-forcement	Late period of reinforcement	Late period of extinction
Interval	14.0	20.0	12.1
During CS presentation	2.9	15.0 *	7.7
During US presentation	15.8	30.0	—

This table contains data recorded from a unit in the nucleus lateralis dorsalis. Late period of reinforcement: 16th, 17th, 18th, and 19th trials. Late period of extinction: 13th, 14th, and 15th trials. The value marked by * is significantly different from the prereinforcement value. The same convention is followed in the subsequent tables.

TABLE 3. Conditioned Effect in Discharge Rate per Second

	Before rein-forcement	Late period of reinforcement	Extinction period	Late period of reconditioning
Interval	11.7	9.4	10.5	6.1
During CS presentation	10.1	5.3 *	12.3	3.7 *
During US presentation	2.2	5.6	—	3.2

Unit recorded from nucleus ventralis lateralis. Figures are mean discharge frequency per second. Late period of reinforcement: 11th and 12th trials. Extinction period: 3rd and 5th trials. Late period of reconditioning: 15th, 17th, 19th, and 21st trials.

initially upon this unit while the US (VA stimulation) significantly decreased the discharge frequency. After about 10 reinforcements, the mean discharge frequency during the CS decreased significantly (in 11th and 12th trials). Thus, the effect of the US was simulated by the CS to some extent. A decrease in discharge frequency in the intertrial intervals of the same period was, however, not significantly different from the control value. Therefore, the conditioned change seen in this was identified as type A. The conditioned effect disappeared after two applications of the extinction proced-

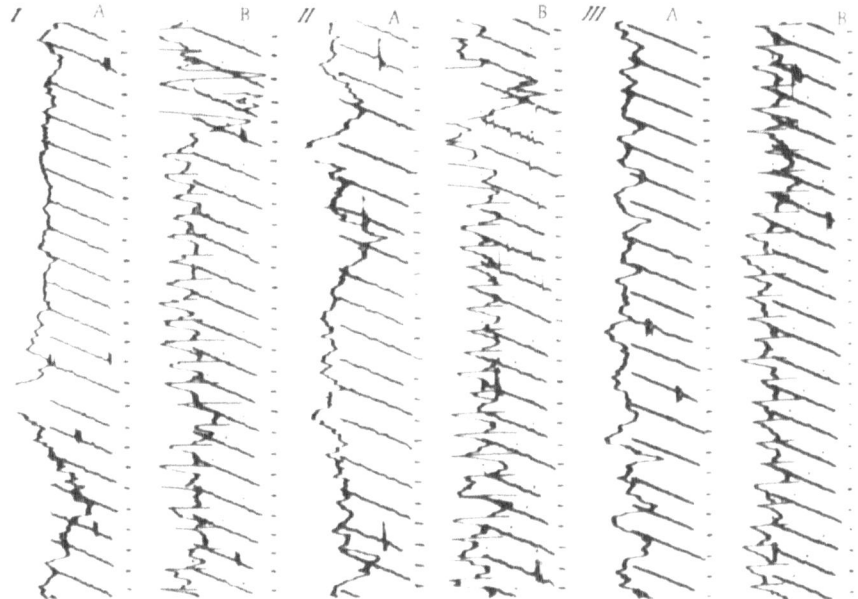

Fig. 2. Unit discharge recorded from the cingulate gyrus with burst activity. I: Before conditioning. II: At the 43rd trial of reinforcement. III: At the 13th extinction. Note the decrease of burst index in the column II B as the conditioned effect.

ure. After about 14 reconditioning combinations the conditioned effect was again induced.

Unit Changing Its Discharge Pattern but not Its Discharge Rate as the Result of Conditioning. The case in which the conditioned effect appeared only in the burst index is illustrated in Fig. 2 and Table 4. This unit was recorded from the cingulate gyrus. The burst index both during CS presentation and the intertrial intervals was as high as 90% before conditioning. At around the 50th trial of reinforcement, however, the burst index during CS presentation decreased to the minimal value of 32%. In the late stage of extinction, the level of the burst index returned to the level of 60%. This was still significantly small compared with the prereinforcement (93%), and was significantly different from that in the late period of reinforcement.

Unit Showing Conditioned Effect Both in Discharge Rate and Burst Index. The unit illustrated in

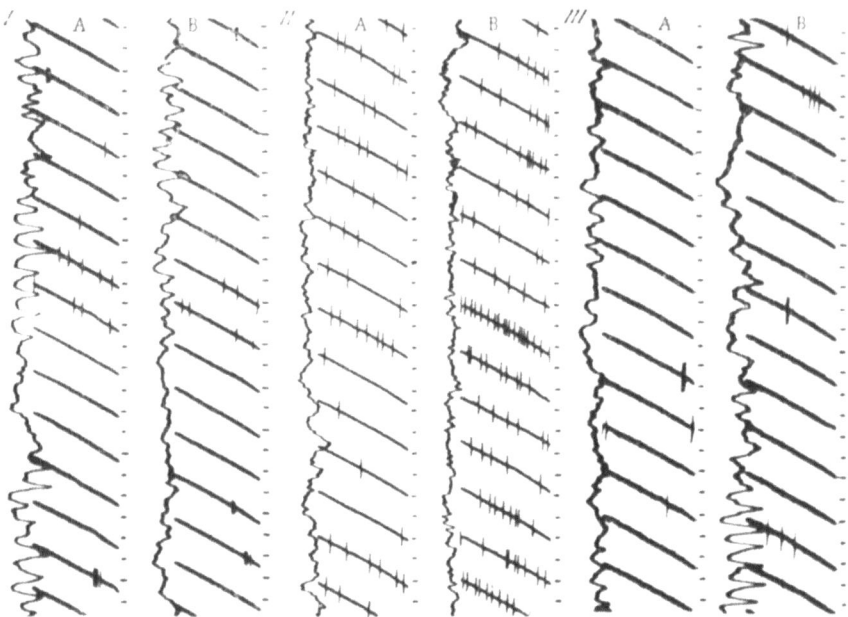

Fig. 3. Unit with burst activity recorded from the nucleus ventralis anterior. I: Before conditioning. II: At the 43rd reinforcement. III: At the 18th extinction. Note the conditioned effect in combination of the increase of discharge frequency and the decrease of burst activity in the column II B.

Fig. 3 and summarized in Table 5 was recorded from nucleus ventralis anterior. The CS was 50/sec stimulation to GL and the US 5/sec stimulation to CM. In the early period of reinforcement, the number of spikes in a burst during CS presentation significantly decreased. When reinforcement approached 40 trials, the discharge frequency during CS presentation rose with a lowering of the burst index, while during the intervals it was maintained at the prereinforcement level. The conditioned effect seen in this unit belongs to the A type, at least at the period described. But in the next stage of the experiment, namely in an early period of extinction, the discharge frequency and the burst index both during CS presentation and intertrial intervals showed values which were significantly higher than the respective controls. This means that at this period the conditioned effect turned into the B type. After repetitions of nonreinforcement, the conditioned effect was extinguished. During reconditioning, the neuron again showed the B-

TABLE 4. Conditioned Effect Appearing only in Burst Activity

		Before rein-forcement	Reinforcement Sub-period			Late period of extinction
			I	II	III	
Discharge frequency per second	Interval During CS presentation	2.8 2.8	2.9 5.2	3.0 3.1	2.9 3.1	2.8 3.3
Burst index	Interval During CS presentation	83.0 93.0	86.8 91.5	78.7 85.7	73.0 32.0*	87.0 60.0†

This table contains data recorded from a unit in the cingulate gyrus illus-trated in Fig. 2. Sub-period I of reinforcement: 6th, 8th, and 10th trials. Sub-period II: 33rd, 35th, and 37th trials. Sub-period III: 47th, 49th, 51st, and 53rd trials. Late period of extinction: 11th, 13th, 15th, and 16th trials. The value marked by † is significantly different both from the prereinforcement value and the one at sub-period III of reinforcement.

type conditioned change, which was as intense as that seen dur-ing the early period of extinction.

DISCUSSION

In the present experiment we were able to obtain conditioned effects by intracerebral stimulation and to find them at the level of unitary activity. This is, of course, an extension of previous work, which established that the classical CS and US can be applied as intracerebral stimulation and that conditioned effects can be found not only for peripheral reflex performance but also for central electrical activity even at the unitary level.

Referring to the previous work, we found that intracerebral stimulation as substitutes for CS and US yielded results not essen-tially different from those obtained by peripheral stimulation. For example, there is good agreement between our own and previous in-vestigations with regard to the probability of obtaining a conditioned effect on units. We found that 44% of 58 units exhibit a conditioned effect, though in some of them this was not followed by extinction. The corresponding figures reported by Kamikawa et al. and Yoshii and Ogura are 61% and 46% respectively, although Bureš and Bur-ešová reported a rather smaller figure.

TABLE 5. Conditioned Effect Appearing both in Discharge Frequency and Burst Activity

		Before reinforcement	Reinforcement Sub-period I	Reinforcement Sub-period II	Extinction Sub-period I	Extinction Sub-period II	Late period of reconditioning
Discharge frequency per second	Interval	6.6	6.3	8.9	32.4 *	4.1	28.7 *
	CS presentation	5.5	9.4	16.8 *	24.9 *	4.2	19.4 *
Burst index	Interval	18.8	14.9	14.6	3.1 *	24.4	10.1 *
	CS presentation	36.8	18.8	4.2 *	2.7 *	36.8	6.9 *

Summary of unit discharge recorded from nucleus ventralis anterior. Sub-period I of reinforcement: 21st, 22nd, and 24th trials. Sub-period II of reinforcement: 41st, 42nd, and 43rd trials. Sub-period I of extinction: 2nd, 3rd, and 4th trials. Sub-period II of extinction: 16th, 17th, and 18th trials. Late period of reconditioning: 7th and 10th trials.

However, there are certain points which distinguish conditioning by intracerebral stimulation from that by peripheral stimulation. We found that more than half of the neurons responded to the US with an increase in discharge frequency, while Kamikawa and co-workers reported that nonspecific thalamic neurons, which were the target of their conditioning experiments, decreased their discharge rate in response to sciatic shocks as US. Corresponding to this difference, we observed that more than half of the units were conditioned to increase their discharge frequency in response to the CS, while Kamikawa et al. showed a number of cases where the conditioned effect appeared as a decrease of the discharge frequency.

It was found that most of the neurons, when successfully conditioned, responded to the CS with a similar pattern as to the US. However, two neurons were rather exceptional. In these the unconditioned effect was seen both in discharge frequency and in burst activity. After they had received a sufficient number of reinforcements with the US, the CS caused changes only in burst activity, and not in discharge frequency. In other words, these units are said to be conditioned partially. It is notable that these exceptional units were obtained in an encéphale isolé preparation.

The type A conditioned change demonstrated in this experiment seems to be in good accordance with the conditioned change in the peripheral structures, in which the conditioned effect is latent when no CS is applied. Another group of type B neurons shows the conditioned effect not only during the CS but also during the background of intertrial intervals. Type B neurons are observed about twice as often as type A and this seems to be due to central conditioning and/or intracerebral stimulation.

SUMMARY

Unit discharges recorded from the specific and nonspecific thalamic nuclei, dorsal hippocampus, and cerebral cortex of unanesthetized and immobilized cats were investigated as to whether their discharge could be changed by combining presentation of two intracerebral stimulations, instead of peripheral ones. Among the 58 units examined, 25 were conditioned. Fewer than 40 trials were sufficient to induce conditioned changes. The conditioned changes were in discharge rate per second and in the burst index of discharge pattern. Some neurons among those successfully conditioned showed the conditioned effect only during the conditioned stimulus (CS) presentation, but more usually they did so not only during the CS but also during background discharges. Conditioned effects were extinguished by extinction procedures.

ACKNOWLEDGMENT

The authors are indebted to Prof. Iwama for his discussion and criticism.

Current Problems in Electrophysiological Studies of Memory[*]

E. R. John

Brain Research Laboratories
Department of Psychiatry
New York Medical College, New York

It gives my colleagues and me unusual pleasure to have the opportunity of joining in this tribute to Professor Livanov. For many years, we have studied the experimental work reported from his laboratory with great interest, because we have been engaged in the analysis of similar problems. Unfortunately, for some years after we began research on the nature of electrophysiological activity during conditioning we remained in ignorance of the contributions which Professor Livanov and his co-workers had made to this topic. This was due to the deplorable absence of communication between scientific workers in our two countries. When this situation was improved we found that some of the phenomena which we had observed in electrophysiological studies of conditioning using repetitive conditioned stimuli merely confirmed reports which Professor Livanov had published many years earlier, using almost identical techniques. On a number of occasions since that time, our two laboratories have independently reported new phenomena which were in agreement, and which were obtained using similar innovations in methodology. These similar, although independent, advances have impressed us with the fact that scientific work generates a kind of momentum which leads researchers in different countries to ask related questions in common ways at about the same time. At the same time, our increasing awareness of the many similarities between the researches of our laboratories makes us understand better the common intellectual bond which unites scientists of all countries. When I had the opportunity to

---

[*] Pages 84-95 in the Russian edition.

meet Professor Livanov some years ago, these feelings became augmented by my appreciation of his warm and friendly personality. I look forward to many more years of stimulating contributions to science by this imaginative and competent research worker.

This paper will focus upon some of the unsolved problems related to the brain mechanisms which store information acquired during the establishment of conditioned responses, and which make use of such information for the subsequent performance of conditioned responses. A great portion of our knowledge about these mechanisms comes from analysis of the changes in electrical activity of the brain during elaboration and performance of conditioned responses. Such analysis is particularly informative when intermittent conditioned stimuli with a characteristic repetition rate are used. In this case, it is possible to identify rhythmic electrical activity which possesses the same frequency as the stimulus repetition rate. The structures in which these frequency specific rhythms appear when the conditioned stimulus is presented can be considered to be involved in processing information about the stimulus. The size and distribution of these rhythmical brain responses change in a very complex way during conditioning, reflecting the development of new reactions and changed excitability in many regions of the brain. These changes will not be discussed in this paper. Some time ago, they were systematically reviewed and related to some of the theoretical formulations of Pavlov (John, 1961).

We will focus upon a set of phenomena which seem to provide special insight into some dynamic features of the brain mechanisms involved in the processing of information as new responses are learned and performed. Some of these phenomena reveal the ability of the brain to produce rhythmic electrical waves at the frequency of the intermittent conditioned stimulus during the interval between trials, while no stimulus is being presented. This property was first observed by Professor Livanov and his colleagues, who called it "assimilation of the rhythm" (Livanov and Polyakov, 1958). Since that time, assimilation has been observed in the rat, rabbit, cat, dog, monkey, and man, at frequencies ranging from 2 cps to 30 cps, in many different experimental situations. Clearly, assimilation reflects a brain mechanism of wide generality. It is interesting that studies with electrodes chroni-

cally implanted into numerous cortical and subcortical regions
show that assimilation arises earliest, is most pronounced, and
lasts longest in the nonsensory specific regions of the brain. It
appears early during conditioning, is most noticeable during the
middle period of conditioning when performance improves most
rapidly, and tends to disappear after the conditioned response becomes fully established, returning only if an error occurs and the
animal receives an unexpected consequence to his behavior. Assimilated rhythms appear when the animal is brought into the conditioning environment, but are not visible when the animal is in
the home cage (John, 1963).

When we observed these spontaneous rhythmic electrical
waves between conditioning trials, our first thought was that they
might be hippocampal theta rhythms related to orientation. A number of reasons made us abandon this interpretation. First of all,
they often appeared when no behavioral sign of orientation could
be observed. Second, they corresponded approximately to the frequency of the conditioning stimulus being used, although it was not
presented. These frequencies were often beyond the range of theta
rhythms. Third, they were often observed in regions like the mesencephalic reticular formation at times when hippocampal electrodes showed no rhythmic activity. Fourth, many times we observed that animals would spontaneously perform the conditioned
response shortly after the appearance of strong assimilated
rhythms. Other workers have observed a similar relationship
(John and Killam, 1959; Yoshii, 1962).

We next considered the possibility that such rhythms were
related to a reverberatory process representing the sustained
aftereffects of the preceding presentation of the rhythmic conditioned stimulus. This idea also had to be abandoned. Trained animals which had just been placed into the conditioned apparatus
displayed striking rhythmic activity of this sort while awaiting the
first presentation of the conditioned stimulus, although assimilated
rhythms were absent in the home cage. Similarly, when intertrial
intervals were made quite long relative to the usual spacing between stimulus presentations, such rhythmic waves would appear
spontaneously from a desynchronized resting activity. Therefore,
it was not plausible to suggest that they were merely persisting
reverberatory activity.

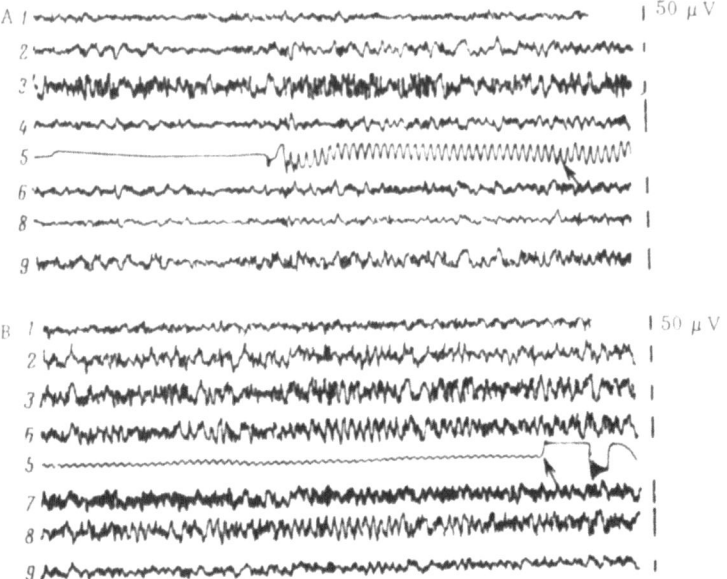

Fig. 1. Upper records: Effect of 10 cps flicker presentation to a cat which had learned to get milk whenever it pressed a lever. The arrow indicates when the cat pressed the lever. Flicker had no signal significance at this stage. Note the absence of labeled electrical rhythms at the flicker frequency. Bottom records: Effect of 10 cps flicker presentation after the cat had learned that milk could be obtained only if the lever were pressed during the flicker. The arrow indicates when the lever was pressed. Flicker was the conditioned stimulus at this stage. Note the marked labeled potentials at the signal frequency, greatly increased in comparison to the upper records. 1) medial geniculate; 2) visual cortex; 3) lateral geniculate; 4) auditory cortex; 5) flicker artifact; 6) fornix; 7) ventral hippocampus; 8) nucleus centralis lateralis; 9) medial suprasylvian cortex. Calibration: 50 μV.

Thus, we gradually came to the conclusion that assimilated rhythms arose from the release of characteristic patterns of neural activity representing the response of certain neural structures to the repeated presentations of the conditioned stimulus during the training procedure. What might be the functional significance of this phenomenon? We have come to the conclusion that these released temporal patterns of electrical activity are an electrophysiological reflection of activation of the neural mechanisms which store information about the repetitive conditioned stimulus. We reached this conclusion on the basis of two kinds of evidence.

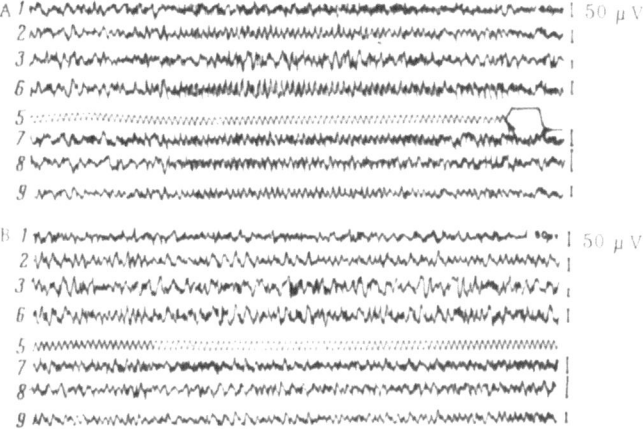

Fig. 2. Effect of 10 cps flicker presentation to same cat as illustrated in Figure 1, but after elaboration of frequency discrimination, in which 10 cps was positive stimulus and 6 cps was negative stimulus. Cat obtained milk if lever was pressed during 10 cps flicker, but was punished by long delay before next trial if lever was pressed during 6 cps flicker. Upper records: Labeled responses during correct response to 10 cps flicker. Lever was pressed at time indicated by the arrow. Notice that regularity and duration of frequency specific waves is much enhanced after differential training. Note correspondence of rhythmic potentials to signal frequency. Bottom records: Labeled response during error of omission, in which cat failed to press lever during positive 10 cps signal. Notice deterioration of frequency specific response. Note particularly the pronounced slow activity at about 6 cps, seen most clearly in nucleus centralis lateralis. Structures as in Fig. 1.

The first evidence in support of this interpretation of the functional significance of these assimilated rhythms came from comparison of the electrical activity observed during correct and incorrect behavioral responses to two different frequencies of flickering light which were the discriminative conditioned stimuli for performance of an alimentary response or an avoidance response in differentially conditioned animals (John and Killam, 1960).

The upper records in Fig. 1 show the effect of presentation of a strong 10 cps flicker to an animal which has learned that food can be obtained whenever a lever is pressed. The flicker is merely an environmental event of no significance for the acquisition of food by the animal. Notice the very minimal amount of frequency-specific electrical response to this stimulus. In contrast, the bottom records show the electrical activity elicited by the same flicker stimulus after the animal has learned that food will be provided only if the lever is pressed during the presentation of flicker. The flickering light is now the conditioned stimulus for the lever response. Notice that electrical waves at the flicker frequency now appear strongly in several regions of the brain.

The animal was then taught that food would only be obtained if the response was made during 10 cps flicker. Response during 6 cps flicker would not be rewarded with milk, but would result in a long delay before the next stimulus presentation. The sequence of stimuli at the two frequencies was random. The upper part of this figure shows electrical activity during correct response to the 10 cps flicker. Notice that the frequency specific waves are more widespread and more regular than in the previous figure. Compare these regular electrical rhythms with the records in the lower part of Fig. 2, obtained during an error of omission, in which the hungry animal failed to press the lever while the 10 cps positive stimulus was present. Notice the sharp decrease in frequency specific response, and the appearance of rhythmic activity at approximately the frequency of the 6 cps negative stimulus in some structures, most clearly in centralis lateralis of the thalamus.

Figure 3 permits the opposite comparison. During an error of commission, when the behavior appropriate to the 10 cps positive stimulus is elicited by presentation of the 6 cps negative stimulus, some regions of the brain show activity at frequencies which correspond to the positive stimulus. Although the animal illustrated in the second figure also displayed this phenomenon, Fig. 3 is taken from a different cat to show that this is not an observation seen only in one animal. Note the good correspondence between the stimulus and the electrical rhythms recorded in the upper part of the figure during correct inhibition of response, while the 6 cps negative stimulus is presented, and the clear appearance in some brain regions of faster activity approximating the frequency of the 10 cps positive stimulus, when an error of commission occurs, and the animal mistakenly presses the lever during the 6 cps negative signal. Additional data about this question have been presented elsewhere (John and Killam, 1960).

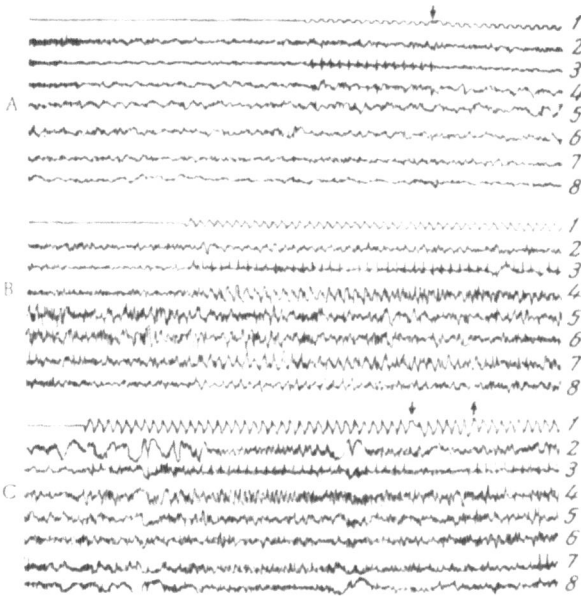

Fig. 3. Records obtained from a different cat subjected to similar training procedures as the animal illustrated in previous figures. U p p e r r e c o r d s: Effect of 6 cps flicker presentation to cat after it had learned that milk could be obtained whenever a lever was pressed. Flicker had no signal value at this stage. Notice how little labeled activity was elicited by the flicker, except in the lateral geniculate. Note disappearance of response in lateral geniculate due to internal inhibition, as cat presses lever and waits for milk. M i d d l e r e c o r d s: Effect of 6 cps flicker after elaboration of frequency discrimination, using 10 cps flicker as positive signal and 6 cps flicker as negative signal. These records were obtained during correct inhibition of lever press during 6 cps flicker. Note marked enhancement of labeled potentials at the stimulus frequency after differential training, and frequency specificity of the brain activity. B o t t o m r e c o r d s: Effect of 6 cps flicker during an error of commission, in which the cat pressed the lever during the inhibitory 6 cps signal. At stimulus onset the cat moves restlessly causing movement artifacts which can be seen in first part of the tracing. As the animal settles down, clear potentials at the 6 cps flicker frequency can be seen in the lateral geniculate. However, in other structures little activity corres - ponding to the signal frequency can be seen. In the visual cortex, mesencephalic reticular formation, and dorsal hippocampus, marked 10 cps electrical rhythms appear and are followed by performance of the lever pressing behavior which would be appropriate to a 10 cps signal. Note the disappearance of 6 cps activity in the lateral geniculate while the animal holds down the lever and looks in the dish for milk to appear. The 10 cps reticular activity continues during this interval. Finally, the animal releases the lever, 6 cps activity reappears in the lateral geniculate, and then the mesencephalic reticular formation begins to show activity corresponding to the actual stimulus frequency. This rhythm then appears in dorsal hippocampus and in the visual cortex. 1) stimulus artifact; 2) visual cortex; 3) lateral geniculate; 4) mesencephalic reticular formation; 5) nucleus centralis lateralis; 6) ventral hippocampus; 7) dorsal hippocampus; 8) posterior suprasylvian cortex.

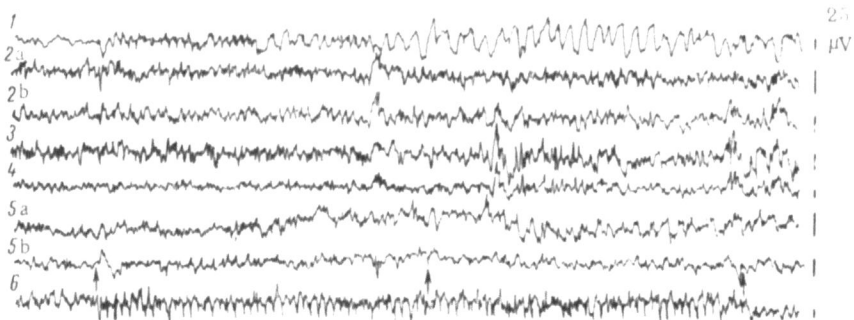

Fig. 4. This recording was obtained during generalization. It illustrates the effects of presentation of a 10 cps flicker to a cat after completion of avoidance training in which a 4 cps flicker was used as the conditioned stimulus. Upon presentation of the novel 10 cps signal, the lateral geniculate and visual cortex display clear frequency specific responses at the stimulus frequency. After several seconds, a slow wave at about 4 cps appears in the visual cortex and the animal displays a startle response at the second arrow. The animal then walked slowly across the cage and performed the conditioned avoidance response, pressing the lever at the third arrow. Throughout this interval, the visual cortex and reticular formation were dominated by slow activity at about the frequency of the 4 cps aversive stimulus, while the lateral geniculate continued to display a 10 cps discharge corresponding to the actual stimulus frequency. 1) visual cortex; 2) nucleus centralis lateralis; 3) nucleus center median; 4) nucleus medialis dorsalis; 5) mesencephalic reticular formation; 6) lateral geniculate. a) left side, b) right side. No stimulus artifact shown on this record.

This kind of evidence led us to propose that some of the electrical activity observed during the presentation of a conditioned stimulus was exogenous, evoked by the stimulus, while a portion of the activity was endogenous, arising in certain regions from the release of stored temporal patterns of response established during the conditioning procedure. That is, the electrical rhythms inappropriate to the actual stimulus which were observed during errors in behavioral performance were attributed to the release of information stored in the memory of the animal.

Additional support for this interpretation came from studies of generalization, in which an animal performed behavior previously established as a conditioned response to a stimulus of one frequency upon presentation of a novel stimulus of a different frequency.

Figure 4 illustrates electrical activity recorded from an animal trained to perform a conditioned avoidance response to a 4 cps flickering light. These records were obtained during generalization to a 10 cps flicker presented to this animal for the first time. At the onset of the novel stimulus, at the first arrow, evoked responses at the 10 cps flicker frequency can be seen clearly in the lateral geniculate body and the visual cortex. After a few seconds, a much slower rhythm appears in the visual cortex, at the time indicated by the second arrow. Shortly thereafter, the animal showed a behavioral startle response, stood up and walked across the apparatus performing the conditioned response at the time shown by the last arrow. During this period, slow waves at about the frequency of the 4 cps conditioned stimulus appeared in the visual cortex and the reticular formation, although the lateral geniculate continued to respond at the 10 cps stimulus frequency. These data show that some regions of the brain display electrical activity at the frequency of the conditioned stimulus actually used in training when the conditioned response is performed as a result of presentation of a new signal. The new signal does not directly cause these electrical rhythms; rather they seem to be released from a neural system established during the conditioning experience. Similar observations have been reported by Majkowski (1958).

Another example of this phenomenon, observed in a different animal, is shown in Fig. 5. The upper portion of this slide shows the electrical activity recorded during generalization to a novel 10 cps flicker after an avoidance response was established using a 4 cps flicker as the conditioned stimulus. Although the lateral geniculate shows a regular response at the 10 cps stimulus frequency, a slower rhythm can clearly be seen in several other structures, particularly the visual cortex. Inspection of the visual cortex tracings shows the presence of a faster component superimposed on the slow wave. The animal was then taught to differentiate between the two stimulus frequencies, receiving punishment if it performed the conditioned response to the 10 cps flicker. The bottom records were obtained following completion of differentiation, as the avoidance response is inhibited during presentation of the 10 cps signal. Notice that the dominant rhythm of the electrical activity in most structures now corresponds excellently to that in the lateral geniculate, which reflects the actual stimulus frequency.

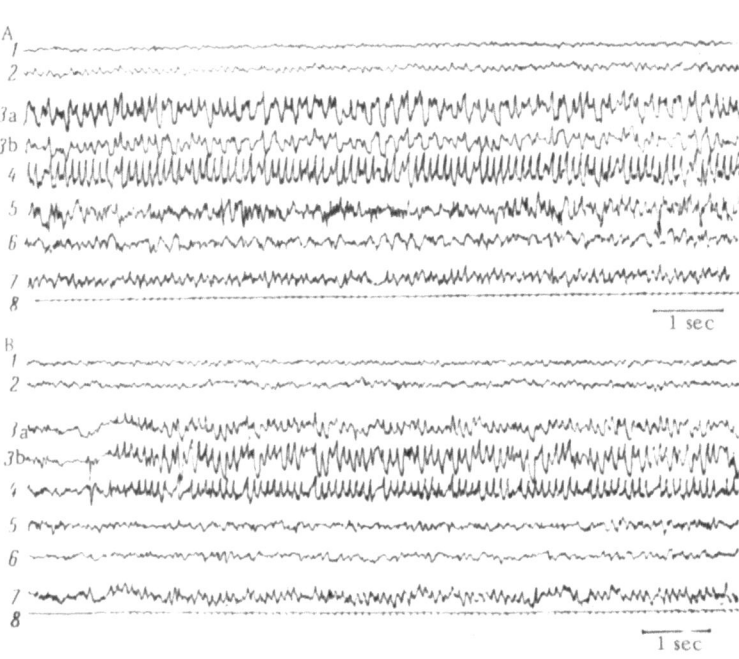

Fig. 5. Upper records: Another example of data obtained during
generalization from a different cat. These records show the electrical
activity elicited by a 10 cps novel stimulus presented to a cat after completion of avoidance training in which a 4 cps flicker was the conditioned
stimulus. Note the clear labeled potentials at the actual 10 cps stimulus
frequency in the lateral geniculate, while the visual cortex on both sides
and the reticular formation show marked slow activity at about the frequency of the 4 cps stimulus used in previous training. Bottom records: Records obtained from the same animal after completion of frequency discrimination training, in which the animal was punished if it
pressed the lever during 10 cps flicker. Note that the activity in the
visual leads and the reticular formation no longer contains the previous
slow rhythms, but corresponds well to the actual stimulus frequency.
1) nucleus reticularis; 2) medial suprasylvian cortex; 3) visual cortex;
4) lateral geniculate; 5) dorsal hippocampus; 6) mesencephalic reticular
formation; 7) nucleus center median; 8) stimulus artifact. a) right side,
b) left side.

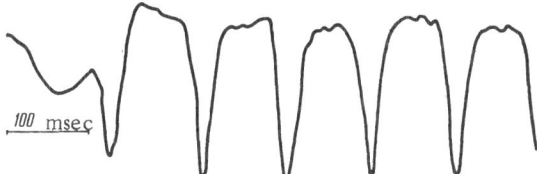

100 msec

Fig. 6. Average evoked response computed from lateral
geniculate body of cat illustrated in previous figure, during repeated performance of generalization on presentation
of novel 10 cps flicker. Computer was triggered at onset
of 100 msec dark period following every tenth flash. Analysis epoch was 625 msec. Computation is average of 100
stimulus sequences. These data show regular evoked potentials at 100 msec intervals, accurately reflecting the stimulus frequency.

Figure 6 shows the average response evoked in the lateral
geniculate body during generalization to the 10 cps flicker. The
evoked potential accurately corresponds to the stimulus frequency.

Figure 7 shows average response computations obtained
from the visual cortex under various conditions. The first waveshape (A) shows the average response of the visual cortex to the
4 cps conditioned stimulus after training. The second waveshape
(B) is the average response of the visual cortex during generalization to the new 10 cps flicker. Notice that this waveshape does
not possess a periodicity at 100 msec, although a component at the
stimulus frequency can be clearly seen. The third waveshape (C)
shows the average response evoked by the 10 cps signal after completion of differentiation. Notice that the waveshape now is periodic at 100 msec, corresponding accurately to the stimulus frequency. The slow component which was so marked in waveshape
B has now disappeared, as the informational difference between
10 and 4 cps flicker became established. These data show clearly
that when a flicker at one frequency is processed behaviorally as
though it possessed the same informational significance as a flicker at a second frequency, the electrical activity of the visual cortex resembles an interference pattern containing two components.
One component seems to represent the actual repetition rate of
the stimulus. The other seems to correspond to the frequency of
the familiar conditioned stimulus, presumably released from

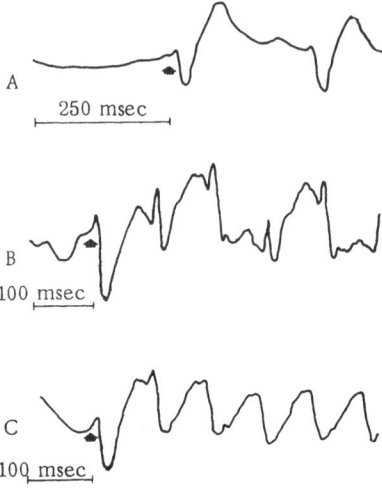

Fig. 7. These data illustrate average response
computations from the visual cortex of the
animal discussed in Figs. 5 and 6, under various
conditions. Again, each average was based on
100 repetitions, the analysis epoch was 625 msec,
and the computer was triggered at the onset of a
dark interval preceding stimulus delivery. U p p e r
r e c o r d : Average response of the visual cortex to
the 4 cps flicker actually used as the conditioned
stimulus during avoidance training. M i d d l e
r e c o r d : Average response of the visual cortex
to the novel 10 cps flicker during repeated performance of generalization. Note that the evoked
response waveshape is not periodic at 100 msec,
although a 10 cps spike is visible on the upper
contour of the waveshape, as if modulating a slower rhythm. B o t t o m r e c o r d : Average response
of the visual cortex to the 10 cps flicker after
completion of differential frequency training.
Note that the evoked response waveshape is now
regular and periodic at the actual stimulus frequency.

memory storage. After differentiation, this inappropriate release no longer occurs. Similar phenomena have been demonstrated to occur in the mesencephalic reticular formation (John,
1963).

The evidence which has been presented thus far shows that
rhythmic electrical activity, with a temporal pattern corresponding to the frequency of repetitive conditioned stimuli, appears in
the brain under certain conditions. Analysis of the characteristics of such activity during erroneous performance and generalization suggests that it may arise from the release of stored information which influences the behavior. In other words, these
electrical rhythms seem to reflect the activation of memories
about the stimuli with corresponding frequencies.

Perhaps the correlations observed between the frequency of
electrical rhythms and information about rhythmic stimuli represent only a special case of little general interest or relevance.
After all, informationally significant events in natural situations
are not characterized by particular frequencies of repetition. To
answer this question, it is useful to examine the actual waveshape
of the potential evoked by a stimulus. The evoked potential waveshape reflects the temporal pattern of activity in a large neuronal
population following the presentation of a particular stimulus.
This potential probably is related to the averaged membrane potentials of the cellular ensemble. Although the evoked potential
does not directly arise from summation of the discharges of single neural units, it may be considered to represent the probability
of coherent activity in the monitored population during that time.

We and other workers have noticed that the electrical responses of different brain structures seem to become similar as
the conditioned response is established (John and Killam, 1959;
Livanov, 1952). Using high-speed computers for more detailed
analysis we have shown quantitatively that as a stimulus acquires
informational significance during conditioning, the electrical activity evoked in many different brain regions by that stimulus comes
to contain markedly similar components (John et al., 1963). Such
common responses are seen in certain regions of cortex, rhinen-
cephalon, diencephalon, and mesencephalon — in other words, at all
levels of the nervous system. The same quantitative conclusions
have been reported by Professor Livanov and his colleagues (Livanov, 1962).

This anatomically extensive system displays similarity in evoked potential waveshapes as long as the conditioned stimulus elicits correct behavioral performance. However, if the animal displays erroneous behavior, certain regions of this system display idiosyncratic waveshapes, no longer corresponding to the responses observed in other areas. Like the previous data about the appearance of characteristic rhythms, this observation suggests that some of the components of the evoked potential waveshape in such regions may be released from storage rather than evoked by the direct action of the stimulus.

Evidence suggesting that this may be the case has been obtained from further studies of generalization (John et al., 1965). In such studies, we have observed that the evoked potential in many brain regions displays a pronounced second component when a novel stimulus elicits generalization, which is absent when the behavioral response is not performed.

Figure 8 compares the average response waveshapes which appear in the lateral geniculate and nucleus reticularis under various conditions. The upper waveshapes show the evoked potentials elicited in these structures during correct response to the 10 cps conditioned stimulus (flicker) which was used in training. The middle waveshapes were obtained during generalization to a 7.7 cps test stimulus. The bottom records were obtained when presentation of that test stimulus failed to elicit a conditioned response. Two components are clearly evident in the top records, and have been labeled I and II. Component II appears in some structures during the process of learning, and has also been observed by other workers (Asratyan, 1965; Killam and Hance, 1965).

First of all, note the striking correspondence in waveshape between the upper and middle records. During generalization, the test stimulus causes an electrical response which corresponds very closely in form to the potentials evoked by the conditioned stimulus itself. Secondly, comparison of the middle and bottom records shows that the waveshape is radically different when generalization fails to occur. Component II is essentially absent in the bottom record. Statistical evaluations of the difference between these two samples of data have been carried out, and show that the results are significant at better than the 0.01 level (Ruchkin and John, 1966).

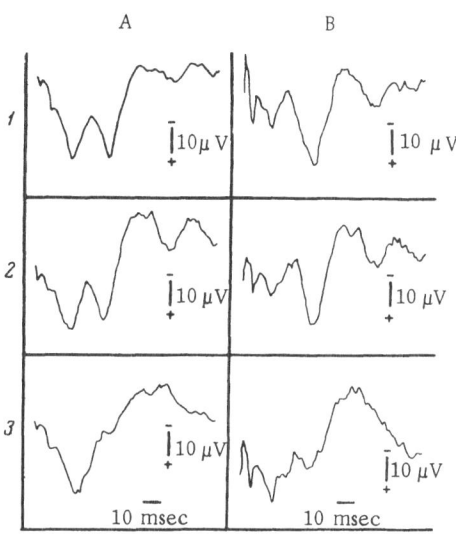

Fig. 8. Average response computations obtained from the lateral geniculate nucleus and nucleus reticularis of another cat, under various conditions during the same experimental session. All averages in this illustration are based upon 42 stimulus repetitions from a number of behavioral trials. Analysis epoch was 62.5 msec. Upper records: Average responses evoked in structures by the 10 cps conditioned stimulus (flicker) actually used in training, during repeated correct behavioral performances. Middle records: Average responses evoked by a novel 7.7 cps flicker, during repeated generalization behavior. Test trials with the 7.7 cps stimulus were interspersed among trials with the actual 10 cps conditioned stimulus, and were never reinforced. Bottom records: Average responses evoked by the 7.7 cps flicker on presentations when no generalization behavior was elicited. Note the similarity of the waveshape elicited by the actual conditioned stimulus to the response evoked by the novel stimulus during generalization. Notice the absence of the second positive component in the evoked potential when generalization failed to occur.

We have suggested that component I reflects the registration of afferent input upon the neural structure, and component II is related to the readout of information stored in the region and released in response to the stimulus. It is of interest to examine component II more closely. We can consider the bottom records of Fig. 8 as the effect of input alone, while the middle records show the characteristics of input followed by readout. If we subtract average evoked potentials caused by a novel stimulus when no generalization occurs from average evoked potentials produced during generalization, the difference waveshape provides a picture of the readout process. Figure 9 shows the results of this operation for a number of anatomical regions.

It is particularly interesting that the difference waveshape possesses extremely similar form and latency for regions of cortex, thalamus, and reticular formation. The latency differences are so small that it is unlikely that this electrical process arises in a particular one of these regions and propagates to the others. It is necessary to consider the possibility that the process arises independently in these various regions as something like a change of state.

Evidence exists that these released components are not merely related to the production of movements, or to similar unspecific factors. They are not correlated with the appearance of movement in generalization, as seen by study of trials with different response latency. Similar conclusions come from so-called conflict studies (Shimokochi and John, unpublished data). In such studies, we train animals to perform differentiated approach and avoidance responses to two different flicker frequencies, V_1 and V_2. The animals are then taught to perform the same two conditioned responses to auditory stimuli (clicks) at these two frequencies, A_1 and A_2. In conflict, the animal receives two contradictory signals simultaneously $(A_1 + V_2)$ or $(A_2 + V_1)$. Under these circumstances, it is possible to study the potentials evoked by the two different signals in the same neural regions at the same time. When a signal successfully controls the outcome of a conflict trial, the later components in the evoked potential of some regions are much more pronounced than when that signal fails to determine the conditioned response which is performed. Such findings seem to rule out unspecific origins to these components.

Fig. 9. "Difference" waveshapes constructed by subtraction of averaged responses evoked by 7.7 cps test stimulus during trials resulting in no behavioral performance from averaged responses evoked by the same stimulus when generalization occurred. Each of the original averages was based on 200 evoked potentials providing a sample from 5 behavioral trials. Analysis epoch was 62.5 msec. The onset and maximum of the difference wave has been marked by two arrows on each waveshape. The structures have been arranged from top to bottom in rank order with respect to latency of the difference wave. Note that the latency and shape of the initial component of the difference wave is extremely similar in the first four structures, and then appears progressively later in the remaining regions. 1) posterior marginal gyrus; 2) posterior suprasylvian gyrus; 3) mesencephalic reticular formation; 4) nucleus ventralis lateralis; 5) marginal gyrus; 6) dorsal hippocampus; 7) nucleus lateralis posterior; 8) nucleus reticularis; 9) anterior lateral geniculate. a) righ side; b) left side.

Thus, it appears that the change of potential during time which is recorded from a neuronal population may reflect processes of informational significance, and that particular wave shapes can be released from storage in local neural regions. The correlations between the temporal patterns of electrical activity and information are more general than merely the rhythms of electrical waves.

It is interesting that when a differentially trained animal performs two different conditioned responses to the same stimulus, one correctly and one erroneously, the same stimulus elicits two different evoked potential waveshapes in some regions. Such observations suggested that different waveshapes might be elicited by two stimuli which differed in meaning in a more natural way than by their frequency of repetition. For this reason we have

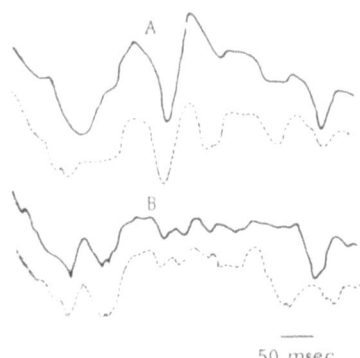

Fig. 10. Comparison of averaged responses evoked by weak flash illuminating a blank, white visual field or the same field containing a black square which was 64 in.² in area. Stimuli were presented in blocks of 25, arranged in a Latin square sequence so as to provide two replications for each sample. Each replication consisted of 50 repetitions of the stimulus. [In this and subsequent figures, the data were obtained from scalp recordings, monopolar derivation — inion versus earlobe reference. Analysis epoch was 500 msec.]

50 msec

turned to the study of the effect of the geometric form of visual patterns upon the evoked potential waveshapes (John et al., 1967). In all of the subsequent figures, the recordings were obtained from a scalp electrode located 3 cm above the inion in man, relative to a reference electrode on the ear lobe. Stimuli were presented in blocks, arranged in a Latin square design to provide replications of each observation. All potentials shown are based on averages of 50–200 stimulus presentations. The stimuli are mounted on the wall before the subject and are illuminated by a standardized light flash.

Figure 10 shows that the presence of a geometric figure in the visual field changes the evoked potential waveshape. Figure 11 shows that the waveshape elicited by a square figure is different from that elicited by a circle of the same area.

Figure 12 shows that the waveshape elicited by a large square is the same as that elicited by a small square. Figure 13 shows that the printed word "SQUARE" elicits a different wave-

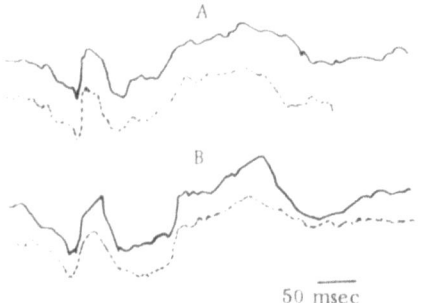

Fig. 11. Comparison of averaged responses evoked by weak flash illuminating a white visual field containing either a black diamond or a circle, equated for area. Details as in Fig. 10, but averages based on 100 repetitions each.

50 msec

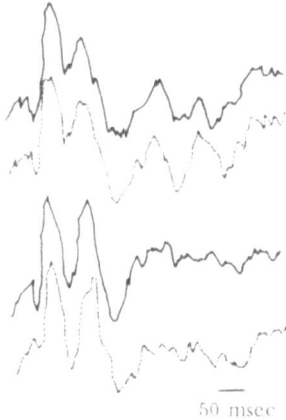

Fig. 12. Comparison of averaged responses evoked by weak flash illuminating a blank visual field containing either a square or a circle. Stimulus sequence was large circle, large square, large circle, small square. Note similarity of response to large and small squares. Each average based upon 200 repetitions.

shape than the word "CIRCLE," although the individual letters and total dimensions of the two have been carefully equated for area.

The foregoing data demonstrate that different waveshapes are evoked by presentation of psychophysically equated stimulus patterns which differ in meaning. These results demonstrate that the informational relevance of temporal patterns of electrical activity is not limited to the artificial laboratory case of frequency coded stimuli. Therefore, conclusions obtained from such studies may provide insights which are generalizable to the neural mechanisms which process, store, and retrieve information in more natural circumstances. Further research will be necessary to clarify the cellular processes which are responsible for the storage and release of these representational patterns in the brain.

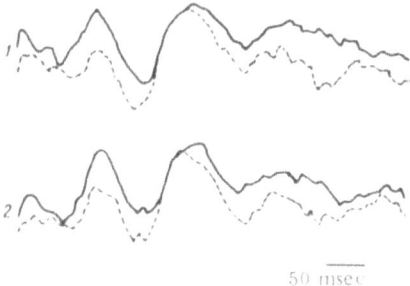

Fig. 13. Comparison of averaged responses evoked by weak flash illuminating the word "SQUARE" or "CIRCLE," printed in block letters. Total black area of both words was made equal. Note reproducibility of difference between responses to first samples, on top, and second samples, on bottom. Each average based upon 100 repetitions.

Electrophysiological Analysis of Responses of the Central Nervous System to Electromagnetic Fields [*]

Yu. A. Kholodov, S. N. Luk'yanova, and R. A. Chizhenkova

Institute of Higher Nervous Activity and Neurophysiology
Academy of Sciences of the USSR
Moscow, USSR

Throughout his working life M. N. Livanov has endeavoured to put scientific knowledge of brain activity to practical use. One example of this aspect of his activity is the study of responses of the nervous system to ionizing radiation. The view that the nervous system plays a considerable role in responses of the animal and human organism to ionizing radiation owes its development mainly to Soviet investigators, and an important contribution to this work was made by Livanov. The results of research conducted in his laboratory are summarized in the monograph "Some Aspects of the Action of Ionizing Radiation on the Nervous System" published in 1962.

However, Livanov's interest has not been confined to the short-wave part of the electromagnetic spectrum. His study (1960), carried out jointly with his collaborators, is one of the first important investigations of the influence of electromagnetic fields on the central nervous system, as many authorities have recognized (Presman, 1965; Fray, 1965; Kholodov, 1966, and others). Although the primary objective of this investigation was to study the effects of ionizing radiation (the EEG response developed when the heater to the tube of the x-ray apparatus was not powered, and only the high voltage was supplied), the effect of electromagnetic fields on the central nervous system emerged as an independent problem.

---

[*] Pages 273-280 in the Russian edition.

In this paper we summarize the results of a study of this problem. Of the wide range of electromagnetic fields attention has been concentrated mainly on fields of superhigh frequency (SHF) and ultrahigh frequency (UHF), and on static magnetic fields (SMF). The choice of these bands was made because of their extensive use in various technical devices, making the hygienic assessment of these factors an important practical problem.

The fundamental working hypothesis was that the electromagnetic fields listed above have a similar action on the central nervous system expressed as a synchronization response, developing after a long latent period. We were interested only in initial changes in the EEG, and for this reason the duration of action of the electromagnetic fields did not exceed 3 min, and most commonly was limited to 1 min.

METHOD

Experiments were carried out on rabbits. By means of nichrome electrodes implanted into the brain or imbedded in the skull the electrical activity of the sensorimotor and visual cortex and, in some cases, activity of certain subcortical structures, was recorded: The EEG was recorded by a monopolar method, the reference electrode being placed on the nasal bones. method, the reference electrode being placed on the nasal bones. Recordings were made on an "Ediswan" ink-writing electroencephalograph, and in some cases a Walter frequency analyzer was used.

The SHF* field was generated by means of an SHF-therapy apparatus (wavelength 12.5 cm), the UHF* field by means of a type UVCh-40-2M apparatus (wavelength 6.6 m) and the SMF by means of an electromagnet supplied by direct current from batteries. In every case the field was applied locally to the head region.

The unit of analysis in every case was a continuous EEG trace including a background segment, the EEG during exposure to this field, and the EEG after switching off the generator. In control experiments a similar continuous 3-min trace of the EEG was obtained: before the second minute of recording and at its end the manipulations of switching the generator on and off were carried out but no field was generated.

---

*Abbreviations follow Russian text. Wavelengths given, however, indicate that they should be UHF and VHF, respectively.

By visual comparison of the EEG recorded for 1 min in the background and the EEG recorded during exposure to the field, in each case absence or presence of change was observed, and if changes were present, the length of the latent period in seconds was recorded. The same indices were considered when the background minute of the EEG trace was compared with the minute of the trace after switching off the generator.

The response was assessed from the results of several tens of exposures. Each response was characterized by its stability (the ratio between the number of responses and the number of exposures, in percent, the mean latent period, and sometimes the intensity (the increase in number of spindles and slow waves during exposure over the background level, expressed in percent). The results were analyzed by statistical methods (Urbakh, 1963).

The EEG response to the field was compared with its response to weak (at the level of the threshold of human audibility) and strong acoustic stimulation (30 dB above the threshold of audibility). The duration of acoustic stimulation likewise was 1 min.

RESULTS

Neurophysiologists have become accustomed to the opinion that any stimulus may evoke a generalized response of EEG desynchronization, taking place after a latent period measured in fractions of a second. Our surprise will therefore be understood when we observed that application of the field usually evoked a generalized response of synchronization, with a latent period measured in seconds or even in tens of seconds. A detailed investigation of the form of the EEG changes during exposure to an SMF with an intensity of 400 Oe was made by Luk'yanova in experiments on 13 rabbits which were exposed 1270 times. The use of Student's criterion for relative variability showed that only the spindles and slow waves show statistically significant changes ($P \leq 0.05$) in the form of an increase during exposure. Neither the segments of desynchronization, nor the decrease or increase in amplitude without a change of frequency showed significant changes during exposure.

Similar conclusions may be drawn relative to the SHF and UHF fields, although during exposure to these more powerful

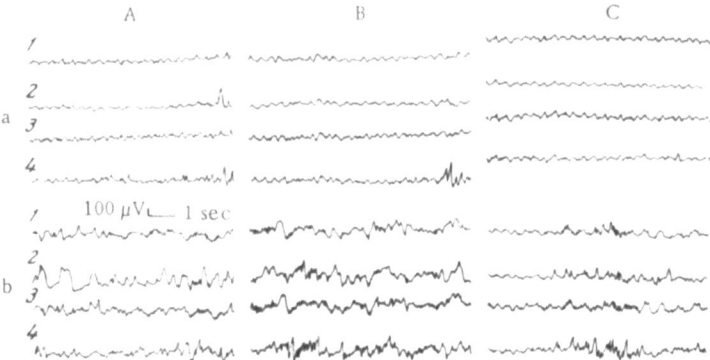

Fig. 1. Changes in EEG of intact rabbit after exposure to UHF (A), and
during exposure to SHF (B) field and to SMF (C). a) Before exposure;
b) during or after exposure. 1, 3) Sensorimotor cortex; 2, 4) visual cortex;
2, 1) right hemisphere; 3, 4) left hemisphere.

stimuli, responses were sometimes observed consisting of prolonged desynchronization, or even of the appearance of paroxysmal discharges, rather than an increase in the number of slow
waves and spindles (Zenina, 1964; Kholodov, 1966; Chizhenkova,
1966, and others). However, the only significant response of intact rabbits to the applied field is the synchronization response.

Even with visual analysis of the results, the detailed characteristics of the synchronization response were dependent both
on the character of the physical factor applied and on the physiological properties of the part of the brain whose electrical activity was recorded. For example, the UHF field mainly increased
the number of slow waves (Fig. 1a), the SHF field increased both
the number of slow waves and the number of spindles in the cortex (Fig. 1b), and exposure to the SMF led to an increase mainly
in the number of spindles (Fig. 1c).

On the other hand, the same SMF increased both the number
of spindles and the number of slow waves in the hypothalamus, the
nonspecific nucleus of the thalamus, and the parietal cortex. In
the mesencephalic reticular formation, the specific nuclei of the
thalamus, and the sensorimotor cortex only the number of spindles was increased, while in the hippocampus the increase affected
only the number of slow waves.

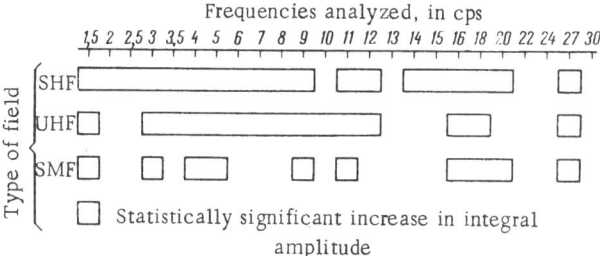

Fig. 2. Changes in integral amplitude of different EEG frequencies of an intact rabbit after exposure to SHF, UHF, and SMF fields.

Differences in the cortical electrical response to different types of field were still more marked when the method of automatic analysis of EEG frequencies was used. Chizhenkova showed that in control experiments when only the manipulation of switching on the field was carried out, no statistically significant changes were observed.

In experiments on the same rabbits, when the method of signs was used for evaluation, an increase in the content of different EEG frequencies was observed after the end of exposure to different types of field (Fig. 2).

Experiments using automatic frequency analysis not only confirmed the results of visual analysis of the EEG, but also gave additional information concerning the EEG response to a field. For example, an increase in the integral amplitude of fast rhythms was observed after discontinuing exposure to both SHF and UHF fields and to the SMF, which was not observed on visual analysis of the EEG. It is interesting to note that the SHF field increases the integral amplitude of nearly all frequencies in the sensorimotor cortex except 10, 13, 22, 24, and 30 cps. The UHF field has no effect on nine of the 24 analyzed frequencies, and the SMF on more than half (14 frequencies). Hence, the use of the frequency analyzer confirmed the existence of an EEG response to different types of field and demonstrated that these types may be arranged in the following descending order of strength: SHF field with power flux density of 40 mW/cm^2, UHF field with an intensity of 30 W/m, and SMF with an intensity of 460 Oe.

The reactivity of different parts of the brain has been studied in greatest detail by Luk'yanova in the experiments mentioned above. Although all the structures from which recordings were made responded with the same latent period (with an accuracy of up to 1 sec) and the same stability (diffuse responses predominated), as regards intensity of their response they may be arranged in the following descending order: hypothalamus, sensorimotor cortex, visual cortex, specific thalamus, nonspecific thalamus, hippocampus, and mesencephalic reticular formation. Similar results were obtained in experiments with the SHF field.

After administration of stimulants such as caffeine and adrenalin, the response to the field is increased (Kholodov and Zenina, 1964; Luk'yanova, 1965, and others) and the less reactive structures under these conditions predominate, as was the case with the mesencephalic reticular formation after injection of adrenalin. Inhibitors (chlorpromazine, Nembutal) as a rule reduced the EEG response to a field, although in each concrete case the effect depended on the dose of drug administered, the intensity of the physical factor, and the initial state of the central nervous system.

Experiments in which the electrical activity was recorded from many parts of the brain showed that to define the response of the central nervous system to a field it is sufficient to record the EEG response from one or two areas of the cortex. Later, therefore, we shall deal purely with the electrical response of the cortex, frequently confining our attention to the sensorimotor cortex where the response took the form of an increase in the number of spindles.

The dynamics of the spindles in the sensorimotor cortex of the same rabbits in control experiments and during exposure to an SHF and SMF field and to weak and strong acoustic stimulation is illustrated in Fig. 3. It will be clear that in the control experiments the number of spindles varied about a constant level. Strong acoustic stimulation produces the familiar desynchronization response, leading to a sharp decrease in the number of spindles during exposure. Weak acoustic stimulation and exposure to SHF and SMF fields increased the number of spindles with a long latent period, this increase persisting after discontinuing the stimulus.

The difference between penetrating stimuli (SHF and SMF fields) and the weak adequate acoustic stimulus in this particular case, disregarding the intensity of the response, is the appearance

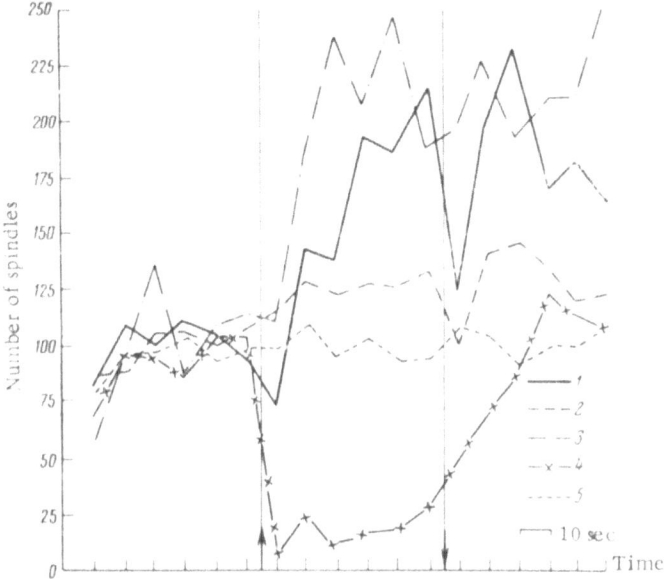

Fig. 3. Changes in number of spindles in experiments with VHF
(1) and SMF(2) fields; in experiments with weak (3) and strong (4)
acoustic stimulation; and in control experiments (5).

of a new increase in the number of spindles a few seconds after
turning off the generator. The simplest explanation of this increase would be to postulate that the end of stimulation causes an
interruption in the poststimulation response. But why does the
same phenomenon not take place after the end of weak acoustic
stimulation? Or, more important still, why does the increase in
the number of spindles sometimes take place only after the end of
exposure and not during exposure? It should be pointed out that
the same dynamics of the slow waves is observed during exposure
to the UHF field. We conclude from these facts that there is an
off-response to the field. A similar EEG off-response was observed during exposure to ionizing radiation (Grigor'ev, 1963).

Hence, the fields which we used are physiologically weak
stimuli, because they evoke a synchronization response similar to
weak acoustic stimulation in our experiments and weak electrical
stimulation of a cutaneous nerve in the cat (Pompeiano and Swett,
1962), although they differ from weak adequate stimulation by the
existence of an off-response.

Other facts which lead us to classify all the types of field used in this investigation as weak stimuli are the low stability of the response and the long latent period. For example, the response to fields of the intensities specified, producing changes of different character in the EEG frequencies, possessed stabilities of 41.04% for the SHF field, 39.76% for the UHF field, and 36.88% for the SMF. In other words, in this respect the fields were indistinguishable from each other and from the responses to weak acoustic stimulation, when the stability was 35.29%.

The latent period of the principal response to the SHF field was 29.45 ± 1.39 sec and to the SMF 24.41 ± 1.41 sec respectively. When a weak acoustic stimulus was applied, the response appeared after 21.45 ± 1.54 sec. The response to the UHF field (but in other rabbits) appeared on the average after about 30 sec.

Compared with the main response, the off-response was less stable and had a shorter latent period, but relative to these indices the resemblance between the responses for the different types of field was greater.

It is reasonable to suggest, and the facts confirm this, that the stability and latent period of the response to a field to some extent are interconnected and depend on individual characteristics of the central nervous system and on the intensity of the field. For example, an increase in the intensity of the SHF field to almost 300 mW/cm^2 led to an increase of stability to 90% and shortening of the latent period to 25 sec, while a decrease in intensity to 2 mW/cm^2 gave a stability of 20% and a mean latent period of 70 sec. Characteristically, whereas with respect to the stability of the response we can class the field with ordinary stimuli, increasing its intensity, with respect to latent period a very powerful field differs sharply from the stimuli customarily used in neurophysiology.

It is interesting to note that both weak and intense SHF fields produced EEG synchronization responses of similar form.

Could differences be found in the responses to weak and strong SHF fields? We obtained a positive answer to this question in experiments to study rhythm driving by flashes of increasing brightness in the EEG of the visual cortex of rabbits during exposure to different types of field. In Chizhenkova's experiments the

time of appearance of the rhythmic driving response to flashes was
shortened by exposure to the SHF and UHF fields and to weak and
intense acoustic stimuli, but was lengthened by exposure to the
SMF. Hence it follows that the outward similarity between the
EEG responses to different types of EMF may lead to different
functional changes in the central nervous system (the SMF causes
inhibition in the visual analyzer and the SHF field causes excitation, although both factors evoke a synchronization response in the
spontaneous EEG). On the other hand, different types of changes
in the spontaneous EEG (synchronization with weak acoustic stimulation and desynchronization with strong) may lead to similar
changes in the reactivity curve, increasing the excitability of the
visual analyzer equally.

Changes in the functional state of the central nervous system
depend not only on the quality of the stimuli but also on strength if
the same stimulus is used. For instance, an SHF field with an intensity of between 40 and 200 mW/cm^2 causes excitation of the visual analyzer detectable by the reactivity curve method, while the
same field with an intensity of about 300 mW/cm^2 causes inhibition.
This is a manifestation of phases in the action of stimuli observed
originally by Vvedenskii. From this point of view the SMF possibly causes inhibition, characterizing the first phase of parabiosis,
or preventive inhibition to use Simonov's (1962) term. An SHF
field of moderate intensity, a UHF field, and strong and weak
acoustic stimuli evoke the second phase of parabiosis, associated
with excitation, while a strong SHF field leads to secondary inhibition.

To conclude this discussion of responses to the field recorded by means of a test stimulus, it must be mentioned that when
the field plays the role of a correcting stimulus and not of a trigger
stimulus, stability of the response to it is approximately doubled,
in fact being increased by 90%.

Hence, the types of field studied may be classed in a special
group of stimuli evoking diffuse EEG responses of synchronization
with a long latent period and long after-effect, expressed as an off-
response. Although changes in the spontaneous EEG under the influence of the field were observed in about 40-50% of cases,
changes in the functional state of the central nervous system detectable by means of a test stimulus developed during nearly every

exposure. It may thus be concluded that recording the spontaneous EEG during exposure to a field gives too little information to
allow the study of responses of the central nervous system to electromagnetic fields to be continued by this method. This method
has played an important role in elucidating the mechanism of action of the field on the central nervous system.

When we tried to record the EEG response to a field in the
case of local exposure of any part of the body except the head,
nothing happened. It may be concluded from the results of these
experiments that the primary response to the field is formed
somewhere in the region of the head. It should be noted here that
throughout this paper we have been discussing only responses to
fields of the specified intensities and durations of exposure, for
an increase in any of these indices may give rise to a reflex EEG
response (Bychkov, 1957).

Many receptors are located in the head region which could
take over the additional function of reception of the field. However, destruction of the peripheral portions of the visual, auditory,
and olfactory analyzers did not prevent development of an EEG
response to the field. It was therefore concluded that there must
be certain hitherto unknown sense organs which are responsible
for the perception of electromagnetic fields.

Unilateral destruction of the mesencephalic reticular formation, the posterior ventrolateral nucleus of the thalamus, or the
posterior hypothalamus did not produce asymmetry of the EEG
response to a field.

The hypothesis was therefore put forward that the EMF has
a predominantly direct action on the central nervous system. This
hypothesis necessitated classification of the field with the penetrating stimuli, because the response to another penetrating stimulus
(ionizing radiation) was modified after the operation mentioned
above (Izosimov, 1961; Kondrat'eva, 1962; Lebedev, 1963). We
know that ionizing radiation possesses both a direct and a reflex
action on the central nervous system (Livanov, 1962a). It must
not be forgotten that in our comparison of the mechanisms of action of the different penetrating factors, we are not taking into account the duration of exposure, which may itself modify the character of the response.

The hypothesis was verified in experiments in which the EEG was recorded from an encéphale isolé preparation obtained by transection at the level of the mesencephalon, supplemented by division of the olfactory and optic nerves. Experiments of this type were carried out on 11 rabbits exposed to an SHF field, on 22 rabbits exposed to a UHF field, and on 11 rabbits exposed to an SMF.

It was found unexpectedly that the responses of neuronally isolated parts of the rabbit's forebrain possessed a longer mean latent period, and greater stability and intensity than the corresponding responses of the intact brain. The latent period was reduced by 50% during exposure to the SHF field, by 40% in the case of the UHF field, and by 30% in the case of the SMF. The stability of the response was increased by 40, 70, and 20% respectively.

The increase in the response to the field may be explained both by an increase in the sensitivity of denervated structures and by abolition of inhibitory influences from the receptor zones. It may be that both causes operate here, although the main question is to determine the role of the direct action of the field on the central nervous system in the intact organism. Is the increased sensitivity of neuronally isolated structures to a field found only in acute experiments? Our experiments involving local exposure of the head, with gradual denervation of individual parts suggest that the direct action of the field on the central nervous system is found in the intact organism.

We also carried out experiments to record the electrical response of a neuronally isolated strip of cortex to a field. An SHF field was used in experiments on 14 rabbits, and a UHF field and SMF in experiments on 23 and 7 rabbits respectively. The stability of the response after such an operation sometimes fell by 40% during exposure to the SHF field and by 6% during exposure to the SMF, while sometimes it rose slightly, by 8% during exposure to the UHF field. The latent period of the response of the neuronally isolated strip of cortex was much shorter by comparison with the intact brain. In an SHF field the latent period fell by 75%, in a UHF field by 50%, and in an SMF by 70%.

Hence, both large (the forebrain) and small (a strip of cortex) neuronally isolated brain structures have a much shorter latent period of their responses to all types of field used. An increase in the stability of the response was observed only in large isolated structures.

It is interesting to note that in neuronally isolated parts of the brain, the field off-response appeared qualitatively brighter, although quantitatively it was indistinguishable from the off-response of the intact brain. These results confirm the view that the principal response and the off-response differ in their genesis (Granit, 1957). Application of an acoustic stimulus did not evoke an electrical response in neuronally isolated structures, giving further evidence of the reflex nature of acoustic influences.

Hence, the types of field investigated constitute a group of physiologically weak stimuli, possessing a mainly direct action on the brain.

Since a neuronally isolated strip of cortex in the sensori-motor or visual areas responded equally to electromagnetic fields, while the responses of these intact regions were different, two conclusions may be drawn. First, each area of nerve tissue may respond independently to the field. This hypothesis has now been confirmed by Luk'yanova's experiments on the isolated ventral nerve cord of the crayfish, whose response to an SMF is independent of the humoral factors present in the experiments on isolated structures of the rabbit's brain. In addition, as the work of Aleksandrovskaya has shown, glial structures are involved initially in the response to a field, as revealed by morphological methods.

Second, in the intact brain the electrical response to a field is organized with the participation of nonspecific synchronizing structures, of which the hypothalamus plays an important role.

SUMMARY

The investigation begun by Livanov on the effect of electromagnetic fields on the central nervous system continues to develop. The fundamental idea expressed in his paper that the electromagnetic field acts directly on the central nervous system has repeatedly been confirmed. A detailed study of the mechanism of the effects described will not only help to explain the special features of new stimuli, but will also reveal new properties in the diverse activity of the central nervous system.

Spatial Synchronization of Cortical and Subcortical Potentials in Rabbits During Formation of Conditioned Reflexes*

I. N. Knipst

Institute of Higher Nervous Activity and Neurophysiology
Academy of Sciences of the USSR
Moscow, USSR

Among the various changes which recent investigations have revealed in cortical electrical activity during formation of conditioned reflexes, attention has been drawn to the regular changes in spatial synchronization of potentials in different parts of the cortex. The extensive spread of synchronized rhythms over the cortex observed by Livanov and Polyakov (1945) in the course of formation of a conditioned connection led to the suggestion that spatial spread of similar potentials may occur over the cerebral cortex. An increase in similarity between potentials under such conditions, even in the absence of synchronized rhythms, was found by Dumenko (1953a, 1955). Similar results were obtained by electrophysiological investigations of conditioned-reflex formation in dogs (Verzilova, 1958; Dumenko, 1960, 1965). Further analysis of this phenomenon (Knipst et al., 1959; Knipst, 1960) showed that the degree of spatial synchronization of cortical potentials varies with the phases of conditioning. Formation of a conditioned reflex in rabbits also led to the appearance of similarity between potentials at different levels of the cortex whose activity exhibits characteristic differences from animals in a resting state (Knipst, 1955, 1958).

Cases of the appearance of synchronization between potentials of the cortex and various subcortical structures have also been observed. Kolodnaya (1944) and Gershuni and Tonkikh

---

* Pages 127-137 in the Russian edition.

(1949), for instance, found changes of this type under superficial anesthesia, and Sokolova (1958) found them during the formation of a dominant. Adey (1961a) and L. G. Voronin et al. (1965) observed an increase in synchronization of potentials in the cortex and various subcortical structures during conditioning.

The use of the electroencephaloscopic technique, by means of which the electrical activity of 50 points of the cortex can be recorded simultaneously, made it possible to examine in detail the distribution of synchronously operating points of the cortex. For the objective analysis of distant synchronization of potentials, Glivenko et al. (1962a) developed a quantitative method. Their analysis of cortical electrical activity of rabbits at rest (in the absence of special stimulation), during application of various stimuli (Korolkova et al., 1966), and during formation of conditioned reflexes, revealed the characteristics of changes in the spatial correlation of brain potentials. For example, it confirmed the increase in synchronization of potentials at the investigated cortical points during conditioning. In addition, it was shown that the period of most marked spatial synchronization of cortical potentials correspond to the initial stage of formation of the conditioned connection. In the period of stabilization of the conditioned reflex, the number of cortical points with synchronized potentials decreases (Glivenko et al., 1962b).

Changes in spatial synchronization during conditioning and investigations of human cortical activity at rest and during solution of various problems (Livanov, 1962c; Gavrilova et al., 1964a; Livanov et al., 1964), and during performance of a skilled movement (Petrovich, 1965) show that the distribution of points with synchronized electrical activity over the cortical surface reflects definite functional relationships between them. Livanov (1962b, 1965) concluded from these findings, and also from Ukhtomskii's view on the role of isolability in the formation of connections between nervous centers, that synchronization between potentials, direct evidence of an identical course of processes in time, reflects the mutual adjustability of neural structures, permitting the formation of working connections.

Since the role of the subcortex in conditioning is not yet adequately clear, the use of an index such as synchronization of potentials in the cortex and certain regions of the subcortex may provide material for the further analysis of their role and interaction in the process of establishment of the conditioned connection.

Most investigators who have studied relationships between the cortex and lower centers have used electrical stimulation or extirpation of these structures. The results obtained, starting with the work of Moruzzi and Magoun (1949) have led to the belief that nonspecific influences of corresponding structures of the mesencephalon and diencephalon play a role of significance in cortical electrical activity and in animal behavior, and have also demonstrated reciprocal cortical influences on the activity both of the nonspecific thalamic projection system and of the nonspecific tegmental system (Narikashvili et al., 1960a; Narikashvili and Kadzhaya, 1963).

Thus, an attempt was made to investigate the cortical and subcortical areas (belonging mainly to nonspecific systems) when interacting under natural conditions, on the basis of a study of changes in correlation between their potentials, in order to obtain evidence of cortico-subcortical interaction during the formation of conditioned reflexes.

METHOD

A motor defensive reflex was formed in rabbits by a combination of rhythmic flashes and isorhythmic electrical stimulation of the skin of the animal's forelimb. The unconditional stimulus was applied 5 sec later, and they acted together for 3 sec.

Cortical electrical activity was investigated with the aid of 25 needle electrodes inserted into the preliminarily scalped cranial bone above the visual, parietal, and part of the motor areas of the cortex (see scheme in Fig. 4). Implanted electrodes made from nichrome wire $30-40\,\mu$ in diameter were used for recording the subcortical potentials. The implanted electrode was in fact a group of 6-8 electrodes glued together, their total diameter being $120-140\,\mu$. The recording ends of the electrodes were cut off at different levels, and the vertical distance between these ends usually varied from 0.5 to 5 mm (Knipst, 1961). Either five or six groups of electrodes were implanted into the animal. In this way, from 30 to 40 electrodes were present in the rabbits' subcortex. The positions of the burr holes through which the electrodes were inserted were determined from maps (Sawyer, Everett, and Green, 1954). After work with the animals had ended, the positions of the electrodes were verified morphologically. Electrical activity was recorded mainly by a monopolar technique, the indifferent electrode being applied to the nasal bone or to the animal's ear, for comparison of different methods of recording (Glivenko et al., 1965) showed that this method reflects most accurately the spatial distribution of synchronous cortical points. Some use was made of material obtained by an averaged recording method (recording of the potential of each point relative to the averaged potential of the other points). With simultaneous recording from cortical and subcortical areas whose potentials exhibit changes frequently independent or even opposite in phase (thereby excluding the possibility of synchronization of potentials at all investigated points), the disadvantages of this recording method have little effect on the results of analysis of spatial synchronization.

The potentials from the portions of the brain investigated were recorded by means of a 50-lead electroencephaloscope with motion pictures taken at 24 frames/sec. To analyze the material, the method of quantitative assessment of spatial correlation between potentials developed by Glivenko et al. (1962a) was used. To assess the degree of correlation between potentials of different

Location of Implanted Electrodes (Histologically Confirmed)

Rabbit No.	Adjacent subcortex	Diencephalon										Mesencephalon	
		thalamus						hypothalamus					
		medial and ventral nuclei	reticular nucleus	nuclei of midline	lateral nuclei	pulvinar	subthalamus	anterior portion	dorso- and ventro-medial regions	posterior portion	reticular formation	central gray
1	−	3	10	4	−	−	−	−	6	2	4	4
2	−	16	−	2	5	−	−	2	−	5	4	4
3	4	16	−	−	−	3	−	−	−	−	1	4
4	2	13	2	−	2	4	−	−	−	−	6	6

points of the brain, the first derivative of each lead was compared with all the rest and unidirectional changes of potential were chosen. The only difference was that during analysis of the amplitudes of the potentials, their repeated successive values were altered, depending on the tendency of the changes in amplitude in preceding and succeeding frames, in order to avoid the appearance of zero values (absence of changes from one frame to another). This method was justified by the fact that with the filming speed and frame enlargement used for the analysis, cases of this type formed a comparatively small percentage and it was more convenient to use only two values (positive and negative changes) when analyzing the material.

The amplitude measurements obtained were compared with the M-2 electronic computer.

RESULTS

To analyze changes in synchronization of potentials in different parts of the brain during conditioning, the results from four rabbits were used. The electrodes implanted in these animals were located mainly in various parts of the thalamus, hypothalamus, and mesencephalic reticular formation (Table 1).

Combinations of photic and electrodermal stimulation led to an increase in synchronization of potentials in various parts of the brain. Maximal spatial synchronization in intervals between combinations was observed in the initial stage of conditioning (after the first 5-10 combinations).

In the period of stabilization of the conditioned reflex, it gradually weakened. The dynamics of spatial synchronization (from the mean curves of distribution of correlations between pairs of points) in all the animals investigated shows that against the background of this weakening a second increase in synchronization of potentials is observed (Fig. 1A). This increase is seen particularly clearly on the graph reflecting the dynamics of the number of points with a high positive correlation only between the changes in potentials (Fig. 1B).

If these results are compared with the motor responses, the first increase coincides with the appearance of individual motor responses arising after the first combination, and evidently associated with revival of the orienting reaction as a result of the addition of a new component, namely of reinforcement. The second increase is observed after weakening or complete disappearance of these responses and it coincides with the appearance of stable motor responses to the conditional stimulus (Fig. 1A).

Analysis of the positive values of correlations between pairs of points shows that during conditioning the spatial synchronization of potentials is increased not only in the cortex, as previous investigations showed, but also in lower structures of the brain. The increase is most marked in the cerebral cortex, thalamus, and hypothalamus, and is much less marked in the mesencephalic reticular formation (Fig. 2).

Comparison of changes in spatial synchronization in various areas shows that the increases in its value do not coincide in time. Changes in spatial synchronization both in the cortex and in the lower structures are characterized either by two periods of increase, or (and this is more common in the subcortical regions) by a distinct increase at about the same time as the second increase in synchronization in the cortex (Fig. 2). The first period of increase in spatial synchronization in the cortex is usually of greater magnitude and longer duration than the second.

Hence, the first increase observed in Fig. 1 (from the general mean curves of distribution) is mainly determined by changes in cortical activity. The second increase is produced by an increase in spatial synchronization both in the cortex and subcortex, and it coincides in time with formation of the conditioned response.

To obtain information on cortico–subcortical relationships, the degree of correlation between potentials in the cortex and sub-cortex and lower structures was compared. These investigations were based on comparison between the number of pairs of leads in which a high percentage of synchronization of potentials (over 75% of the time of investigation) was observed, i.e., leads from points which could be regarded as connected with a reasonable degree of probability. This comparison showed that most cases of positive correlation in electrical activity of the investigated brain regions, with no stimuli acting on the animal, are usually found in cortical

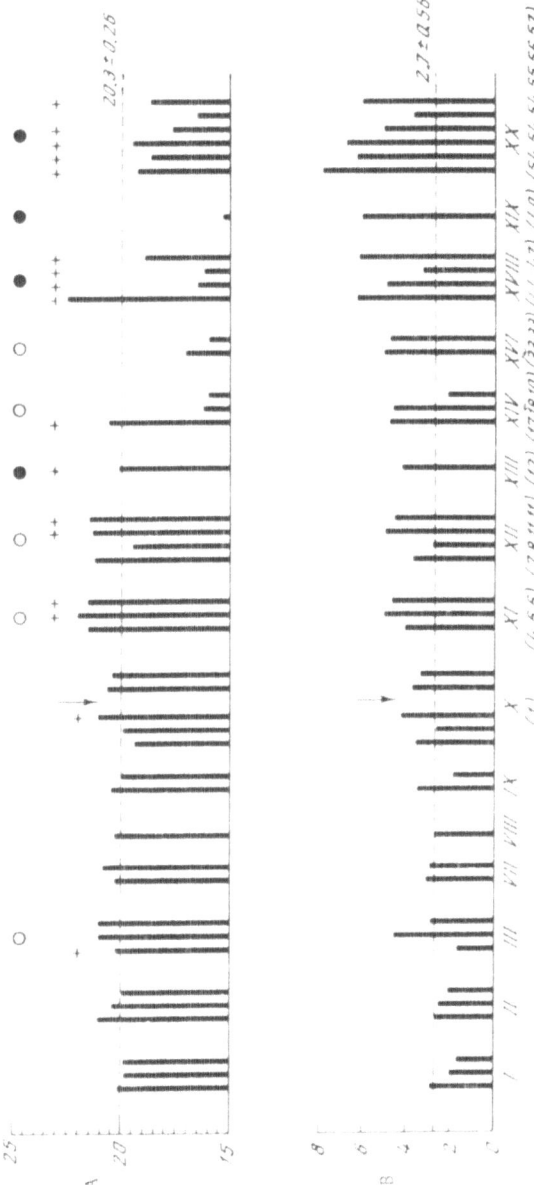

Fig. 1. Mean curves of distribution of positive correlation between potentials at pairs of points in different parts of the brain during conditioning (rabbit No. 4). Abscissa: serial number of experiments, number of preceding combinations in parentheses. Ordinate: values of mean distribution curves of correlation between pairs of points. Columns represent results of individual measurements during experiment (from background potentials). Broken line represents mean for data obtained before combinations. Arrow shows beginning of combinations; +, presence of motor response to photic (conditioned) stimulus in a particular record; O, less than 50% of motor responses to application of conditioned stimulus during experiment; ●, more than 50% of motor responses to application of conditional stimulus during experiment. A) For all correlations between pairs; B) for correlations above 70% between pairs.

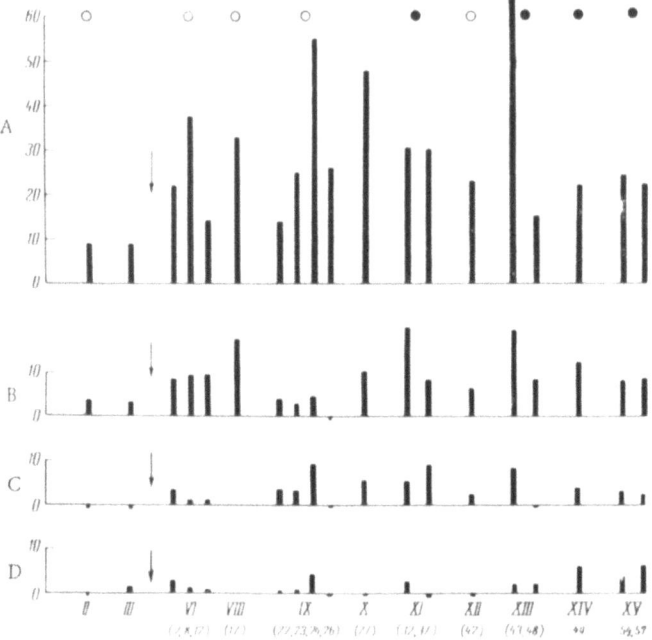

Fig. 2. Relative values of spatial synchronization in cerebral cortex
(A), thalamus (B), hypothalamus (C), and mesencephalic reticular
formation (D) during formation of a conditioned reflex (rabbit 1).
Abscissa: as in Fig. 1. Ordinate: relative units of spatial synchronization (for a high percentage of coincidence, above 70%). Remainder of legend as in Fig. 1.

leads. They are somewhat fewer in the diencephalon and mesencephalon, and fewest of all in cortico–subcortical comparisons.
Similar relationships were found in all animals (Fig. 3A) (changes
in these relationships were found only in isolated cases) and they
persisted during and after the action of photic stimuli.

The first combinations of stimuli, just as the single stimuli,
produce no visible changes in the relationship between the number
of pairs of leads with high correlation between potentials in the
cortex, in the subcortex, and between them. However, in the period of appearance of conditioned responses, a distinct increase is
observed in the coincidence between electrical changes in the cortex and lower structures in relation to the remaining two values
(the values of correlations between two points for the cortex and
subcortical structures) (Fig. 3B).

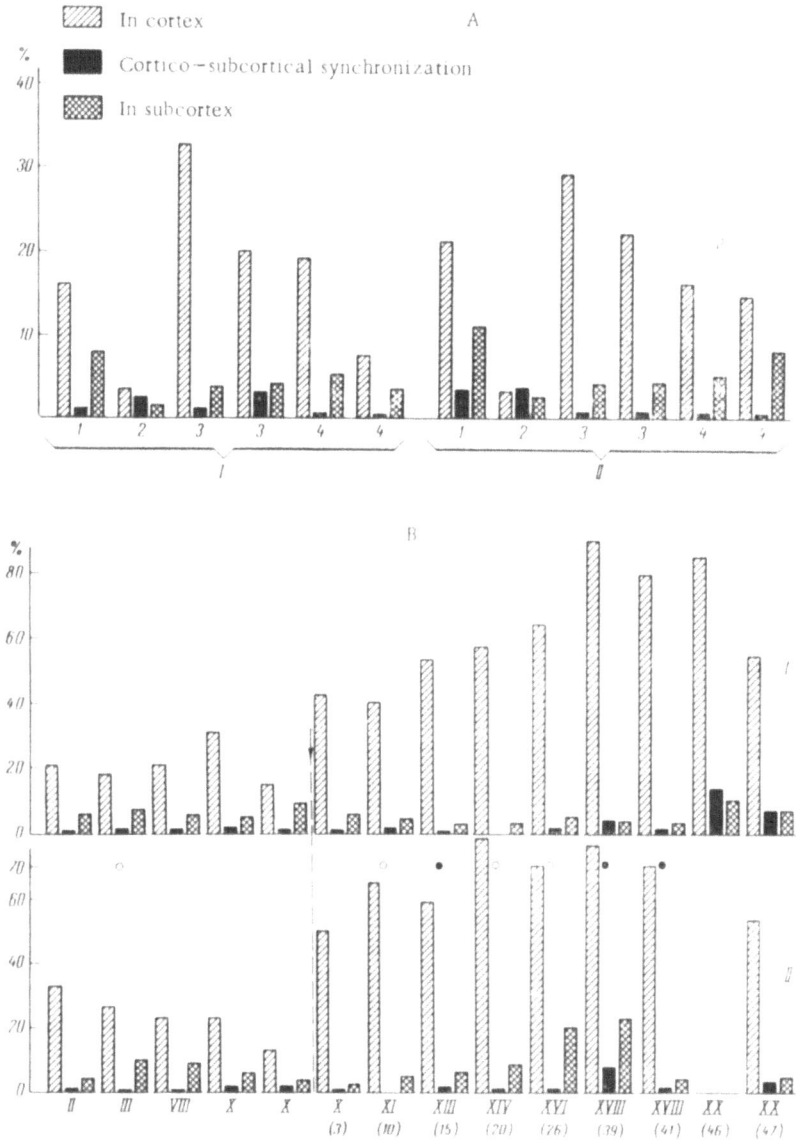

Fig. 3. Comparison of spatial synchronization of cortex and lower regions of the brain and coincidence between potentials in these structures before and during establishment of a conditioned reflex. A) Mean values for all experiments before conditioning for each rabbit. Abscissa: serial number of rabbits. Ordinate: number of pairs of leads with a high degree of synchronization (more than 75%) relative to the total number possible, in percent; B) results of individual experiments (rabbit No. 4) during conditioning. Abscissa: as in Fig. 1. Ordinate: as in A. I) Background before stimulation; II) during photic stimulation. Rest of legend as in Fig. 1.

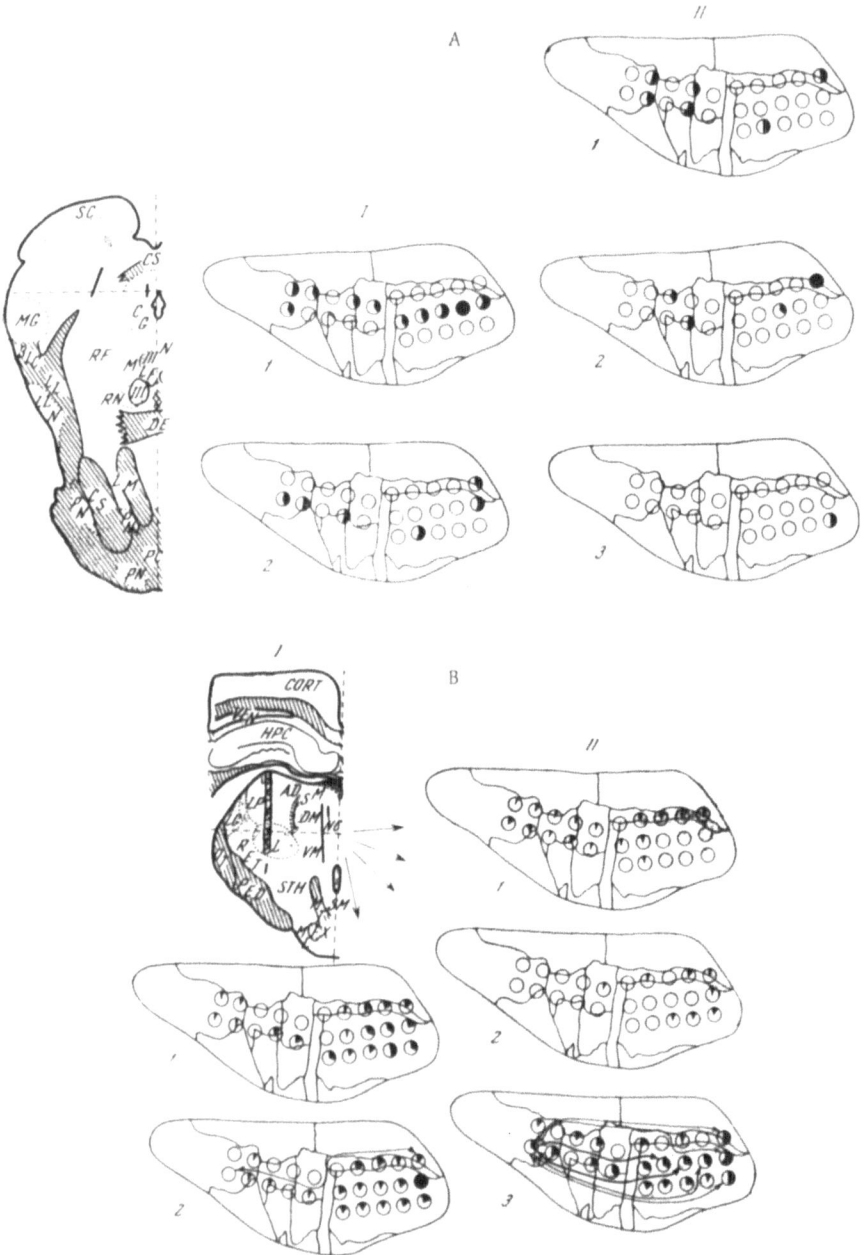

Fig. 4.

Fig. 4. Spatial distribution of areas in cerebral cortex whose fluctuations of potentials correlate mainly with electrical changes in lower regions (rabbit No. 4). A) For mesencephalic reticular formation; B) for medial and ventral nuclei of thalamus. I) Before combinations: 1) background; 2) during photic stimulation. II) During combinations: 1) background; 2) during photic stimulation (combinations 1-30); 3) during photic stimulation (combinations 31-52). Arrangement of electrodes in subcortical structures is shown on schemes of transverse brain section. 0, Arrangement of electrodes on skull surface. Shading of circles denotes relative degree of synchronization of that area with investigated areas of corresponding subcortical structures.

It is of interest to know whether areas exist in the cortex which are preferentially connected with subcortical structures, to discover where they are, and how the synchronization of potentials between these areas is modified.

For this purpose the results were summed for control periods and at different periods of conditioning.

In animals in a resting state, exposed to no external stimulation, no significant differences are observed in correlation between potentials in the subcortex and at all points of the cortex. Systematic application of photic stimuli (several times in the course of four or five experiments) changes this uniform distribution, and areas with potentials connected to a higher degree with those of lower structures than other parts of the cortex can be distinguished, mainly in the occipital region (Fig. 4).

During formation of a conditioned reflex to a combination of photic and electrodermal stimuli, these relationships between the cortex and lower structures are also modified by comparison with the initial background (Fig. 4). The development of the functional connections differs for different subcortical structures. For instance, besides the fact that potentials of the investigated parts of the mesencephalic reticular formation and medial portions of the diencephalon show positive correlation with various parts of the cortex, the degree of synchronization of these subcortical structures with the cortex differs at different stages of conditioning. Whereas in response to the conditioned stimulus it is more clearly visible in the reticular formation in the initial period (Fig. 4A), at a later stage (during stabilization of the conditioned-reflex response) the potentials of the investigated portions of the diencephalon are synchronized predominantly with cortical activity (Fig. 4B).

Hence, in the process of conditioning, areas are observed in the cortex whose electrical activity is synchronized to a greater degree than that of other areas with the potentials of lower brain structures. Comparison of the activity of these areas, located both in the occipital region and in the anterior portions of the cortex, also demonstrated an increase in correlation between fluctuations of their potentials (indicated by arrows in Fig. 4B).

It is not yet possible to discuss details of the dynamics of synchronization of potentials in different parts of the brain, but from the results so far obtained the formation of synchronously acting cortico–subcortical systems reflecting different stages of formation of the conditioned reflex can be postulated. The first combinations form a system with the participation of the mesencephalic reticular formation. Disturbance of these relationships in the course of subsequent combinations coincides with the findings of Naumova (1965), who also demonstrated disappearance of changes in electrical activity of the reticular formation observed after the first combinations at later stages of conditioning. The results also demonstrate that at the stage of formation of a specialized conditioned response, more highly specialized portions of the reticular formation (the medial portions of the diencephalon) are incorporated into the synchronously acting system.

Points of the cortex more clearly in functional connection with subcortical structures than the rest usually are not incorporated into a synchronously acting cortical system formed under the influence of combinations of conditional and reinforcing stimuli. This general spatial synchronization of cortical activity may be a preparatory stage, facilitating formation of a cortico–subcortical system which can take place only after an increase in spatial synchronization within separate morphological subcortical structures arising at a later stage. More probably these points correspond to areas of the cortex discovered by Luchkova (1965), at which the fluctuations of potentials were correlated during stabilization of the conditioned reflex as well.

The dynamics of the observed synchronously-acting "constellations" evidently reflects the formation of spatial systems incorporating various structures of the animal's brain, depending on the stage of conditioning. This conclusion corresponds to the much more commonly expressed hypothesis that the formation of

a conditioned connection is the formation of a system whose morphological structure is determined by the stage of conditioning.
This view was expressed by Livanov et al. (1951) on the basis of
their investigation of the localization of rhythms corresponding to
the frequency of combined stimuli at different stages of conditioning. It was confirmed by more detailed investigations (Yoshii and
Hockaday, 1958; John, 1961). The same conclusion was reached
by Naumova (1965b) on the basis of her study of frequency-ampli-
tude characteristics of the electrical activity of the cortex and
subcortical structures at different levels in dogs during conditioning. She also observed that the cortico−subcortical systems
formed as a result of combinations possess the property of a dominant. This may probably also account for the fact that the action
of separate stimuli is inadequate for the formation of a synchronously acting system during conditioning and for modifying its
structure; the systematic action of combinations of stimuli is
essential.

Hence, the results show that the relative independence of
the course of electrical processes in the cortex and subcortical
structures observed in an animal in a resting state has a tendency
to change, as a result of systematic action of combinations of
stimuli toward spatial synchronization, thus apparently facilitating
the development of the conditioned response.

SUMMARY

1. Mean curves of distribution of correlation between pairs
of points in the course of conditioning in rabbits showed two periods of increased synchronization of electrical activity: the first
was determined mainly by changes in cortical activity, the second
was created by an increase in spatial synchronization both in the
cortex and in subcortical structures, and coincided in time with
stabilization of the conditioned response.

2. Comparison between the cortical electrical activity and
that of lower structures (various parts of the diencephalon and the
mesencephalic reticular formation) showed that the largest number of pairs of points with a high percentage of coincidence of
electrical oscillations was given by cortical leads, a rather smaller
number by subcortical structures, but only a few by cortico−subcor-

tical comparisons. These values were modified by the action of stimuli, but their relative proportions remained unchanged.

3. Conditioning caused the following changes: (a) An increase in spatial synchronization in the cortex and all investigated subcortical structures except the mesencephalic reticular formation, where this increase was very slight or absent. Periods of maximal increase in spatial synchronization in individual areas did not coincide in time; (b) an increase in synchronization of cortical electrical activity with that of lower brain structures, appearing at the stage of formation of stable conditioned responses.

4. During conditioning, areas may be observed in the cortex whose functional connection with the subcortical structures is increased over the initial level, while an increase in synchronization in the potentials of these cortical areas also takes place.

Properties of Spike Activity of Neurons in Different Layers of the Cortex[*]

A. B. Kogan

Department of Physiology
University of Rostov-on-Don
Rostov-on-Don, USSR

Considerable progress has been made in the study of the mechanisms of neuronal activity and in the depth of its analysis as a result of the development of microelectrode techniques and investigation of processes at the neuronal level. However, as Livanov (1965) has rightly pointed out, these investigations have dealt mainly with the function of individual neurons and not with the principles governing their interaction. Yet there is no doubt that it is these principles governing interneuronal relationships which determine the onset and course of neural processes of various types and can act as the key to our understanding of their nature and properties. In contrast to the term "neurodynamics," denoting the course of and interaction between excitation and inhibition as integral acts of nervous activity, Livanov (1965) introduced the term "neurokinetics" to denote the interneuronal movement of neural processes and their integration. Among the main factors determining the character of this movement and its end results, in his opinion, are the chain character of the spread of excitation along neuronal systems and the limiting effect of inhibition. The role of inhibitory neuronal responses in the processes of cortical activity was the subject of a special examination (Livanov, 1963).

The problems of neurokinetics are intimately connected with the elucidation of the general principles governing the organization of neural mechanisms and the study of their construction from neuronal elements. It has been conclusively shown that central

---

[*] Pages 138-147 in the Russian edition.

neural processes are built up from the activity, not of individual
nerve cells, but of "constellations" (Ukhtomskii, 1945) or "ensembles" (Hebb, 1949) of them. Experimental morphophysiological evidence has been obtained of the absence of fixed interneuronal connections in these ensembles, and on this basis the hypothesis of the stochastic organization of neurons in functional systems
of the higher levels of the brain was put forward (Kogan, 1962).
Further investigations revealed the spatial characteristics of certain neuronal ensembles, within which the neurons reacted synergically to afferent stimulation, while neurons of different ensembles reacted antagonistically (Chorayan, 1965b).

To understand the structural basis of neuronal interaction
and the organization of neuronal ensembles it is important to know
how morphologically different nerve cells differ in the properties
of their activity. As a first approximation to this problem, it is
important to study the functional properties of neurons in different
layers of the cortex. It must not be forgotten, of course, that the
cell composition of each layer is heterogeneous. In this paper a
layer-by-layer comparison is made of the characteristics of background and evoked activity of cortical neurons of the rat (Sukhov,
1965), guinea pig (Klepach, 1965), and rabbit (Dikareva, 1965),
obtained in our laboratory.

METHOD

Spike activity of the neurons in sensorimotor cortex was recorded extracellularly by glass microelectrodes with a tip
0.5-1.0 μ in diameter. The rats used were immobilized with d-
tubocararine (1.5 mg/kg). The skin of the paw or the surface of
the cortex was stimulated with square pulses 0.1 msec in duration.
The guinea pigs and rabbits were investigated under semichronic
and chronic experimental conditions. A detachable device (Kogan,
1966) was used to permit simultaneous insertion of several microelectrodes into the sensorimotor and visual (rabbits) or only the
visual (guinea pigs) areas of the cortex. The rabbits received
photic and electrodermal stimulation, and also stimulation of the
sciatic nerve and of the skin and muscles separately. Guinea pigs
were stimulated with light from a 75 W lamp switched on for
4 sec.

TABLE 1. Distribution of Active Neurons among Cortical Layers
of Different Animals

Animals	Area of cortex	Total number of neurons with background activity	Number of neurons in layers					
			I	II	III	IV	V	VI
Rats	sensorimotor	51	–	2	2	5	23	9
Guinea pigs	visual	2144	168	210	402	570	475	319
Rabbits	sensorimotor	192	–	6	21		40	125
–	visual	286	–	11	96		107	72

Background Activity. The distribution of neurons
showing activity in the absence of spatial stimulation throughout
the layers of the cortex is shown in Table 1.

As Table 1 shows, neurons with background activity were
most common in layers IV-VI of the cortex. The number of active
neurons was somewhat higher in the deeper layers (V and VI) of
the sensorimotor and the less deep layers (IV and V) of the visual
cortex. However, it is difficult to say to what extent the distribution given in this table reflects the relative proportions of active
and inactive cells. It is probably determined by the number of
neurons lying in the path of insertion of the microelectrode, the
density of their distribution, their shape, size, and so on. More
definite relationships are discovered if differences in the type of
background activity are examined layer by layer (Fig. 1).

If an arrhythmic, continuous succession of impulses is described as the first type, and grouped discharges as the second
type of background activity, it can be seen that guinea pigs and
rabbits share the same predominant tendency. The proportion of
neurons giving discharges of the first type diminishes as deeper
layers are reached. Correspondingly, the relative proportion of
neurons producing grouped discharges increases. In rats this
tendency was found in layer V. However, the small number of recorded neurons in layer VI does not reveal the fate of this tendency in the direction of the deepest neurons.

However, if only one type of background activity is analyzed,
considerable differences in the character of unit activity may still
be seen as regards both the frequency and order of impulses.
Some idea of these differences is given by the interspike interval

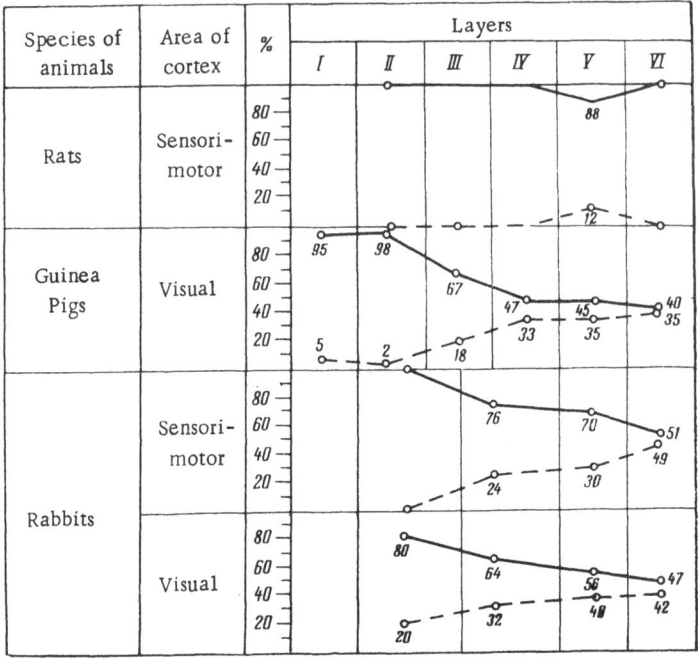

Fig. 1. Percentage of neurons with background activity of first (continuous line) and second (broken line) types in cortical layers of different animals. The balance of neurons making 100% possess an intermediate type of activity.

histograms (Fig. 2). This figure gives examples of recordings of discharges from two neurons showing the same type of continuous succession of impulses. However, the interval histograms of these traces show a marked difference between their frequency characteristics. Whereas the discharge of one neuron has a maximum in the region of 25-30 msec, the distribution of intervals for the second neuron has a maximum in the region 70-75 msec.

In all animals, many cortical neurons with background activity did not respond to stimulation of the corresponding analyzers. The mean number of neurons in the sensorimotor cortex of the rat changing their background activity during electrodermal stimulation, for instance, was only 36%, compared with 44% in the visual cortex of the guinea pig during photic stimulation, 61% in the sensorimotor cortex of the rabbit during stimulation of the skin and muscles, and 57% in the visual cortex of the rabbit during photic stimulation. Comparison of the results of experiments in

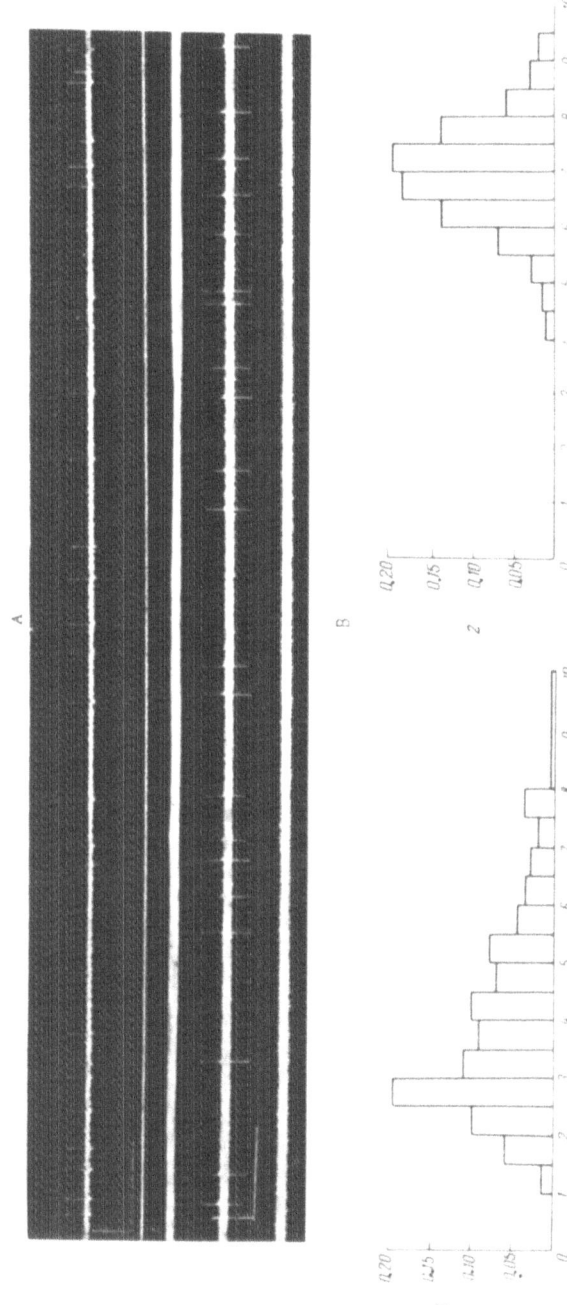

Fig. 2. Background activity of similar type (A) and stochastic interval histograms (B) of two neurons from layers II (1) and IV (2) of visual cortex of a guinea pig, recorded in one experiment on the same animal. A) Calibration 1 mV, time 0.2 sec; B) abscissa, time in tenths of a millisecond; ordinate, probability of encountering intervals of different durations between impulses (experiments of G.S. Klepach).

TABLE 2. Distribution of Neurons Responding to Afferent Stimuli
in Various Layers

Animals	Area of cortex and type of stimulation	Total number of neurons responding to stimulation	Absolute number of neurons and percentage of total number of neurons of this layer having background activity					
			I	II	III	IV	V	VI
Rats	sensorimotor cortex, electrodermal stimulation	18 (36%)	–	–	–	3	13 (40%)	2
Guinea pigs	visual cortex, photic stimulation	950 (44%)	70 (42%)	88 (42%)	181 (45%)	280 (49%)	204 (43%)	127 (40%)
Rabbits	somatosensory cortex, cutaneous and muscular stimulation	119 (61%)	–	5	13 (62%)		26 (65%)	75 (60%)
"	visual cortex, photic stimulation	160 (57%)	–	8	56 (60%)		57 (55%)	37 (51%)

which the sciatic nerve and its cutaneous or muscular branches
were studied separately showed that if a large number of mixed
afferent pathways from the limb are excited simultaneously, fewer
neurons are excited than with stimulation of cutaneous or muscular afferents only. Because of differences in experimental techniques, which undoubtedly influenced the cortical function of the
experimental animals, it is difficult to assess any differences in
the responses of cortical neurons between animals of different
species.

Participation of neurons of the various cortical layers in
the response to a specific stimulus is illustrated in Table 2.

Considering not the absolute number of responding neurons
but their number expressed as a percentage of the number of cells
with background activity tested by stimulation, a tendency can be
seen for the proportion of responding neurons to increase in lay
ers III and IV. This is particularly clear in the figures obtained
in experiments on guinea pigs which are numerous enough for
statistical analysis. Comparison of the results shows that in the
sensorimotor cortex of the rabbit the number of responding neu-

TABLE 3. Distribution of Neurons Responding to Stimulation by Excitatory or Inhibitory Response in Layers of the Cortex, and Percentages of Each Type

Animals	Area of cortex and type of stimulation	Type of neuronal response	Total number of responding neurons	Layers of cortex					
				I	II	III	IV	V	VI
Rats	sensorimotor cortex, electrodermal stimulation	excitatory	12 (67%)	−	−	−	2	8	2
		inhibitory	6 (33%)	−	−	−	1	5	−
Guinea pigs	visual cortex, photic stimulation	excitatory	668 (70%)	50 (71%)	64 (73%)	136 (75%)	187 (67%)	146 (71%)	85 (67%)
		inhibitory	282 (30%)	20 (29%)	24 (27%)	45 (25%)	93 (33%)	58 (29%)	42 (33%)
Rabbits	somatosensory cortex, cutaneous and muscular stimulation	excitatory	52 (63%)	−		1	8	12 (60%)	31 (65%)
		inhibitory	30 (37%)	−	3	2	8 (40%)	17 (35%)	
	visual cortex, photic stimulation	excitatory	84 (66%)	−	4	51 (74%)	16 (64%)	13 (48%)	
		inhibitory	44 (34%)	−	3	18 (26%)	9 (36%)	14 (52%)	

rons reaches a maximum in the deeper layer V, while in the visual cortex of the rabbit the maximum lies in the middle layers III and IV.

Neurons of the cortical analyzer zones are known to respond differently to stimulation, quickening or slowing their firing rate when the stimulation is switched on and off, and also during its action. Depending on the character of the initial response of the neuron, most of them can be divided into two categories: those reacting by excitation and those reacting by inhibition. Their distribution in the layers of the cortex is shown in Table 3.

The percentage of neurons responding by excitation and by inhibition in the various animals (rats, guinea pigs, and rabbits) varies within narrow limits. As a rule there is one inhibited

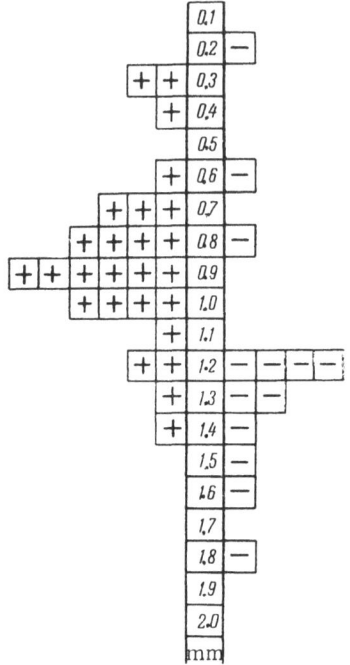

Fig. 3. Distribution by
depth of neurons partici-
pating in epileptiform activity by an increase (+)
or decrease (-) in firing
rate. Vertical row of numbers represents depth of
microelectrode in mm.
Seizures evoked by a series
of square pulses (0.1 msec,
10 cps, 15-20 V), acting
for 5-6 sec (experiments
of A. G. Sukhov).

neuron to two excited neurons. As regards the number of neurons responding by excitation or inhibition, in our material no
significant differences were found in recordings made from different analyzers such as the sensorimotor and visual cortex.

Because of the small number of neurons in each separate
layer, no conclusions can yet be drawn regarding the principles
governing distribution of neurons with different types of response
in the cortical layers of rats. However, the large number of observations made on guinea pigs shows that the ratio between neurons responding by excitation to neurons responding by inhibition
in the various layers of the cortex shows a tendency for predominance of excitatory responses in the superficial layers (II-III) of
the visual area. If the figures relating to the few responding neurons in the superficial cortical layers are disregarded, a tendency for the relative proportion of excitatory neurons to increase in
the more superficial layers of the visual cortex is also seen in
rabbits, whereas in the sensorimotor cortex the number of excitatory responses reaches a maximum in layer VI.

It is extremely striking that even with maximal excitation in the cortex, as in a paroxysmal response, the ratio between the numbers of excited and inhibited neurons may remain the same as during responses to adequate afferent stimuli. Direct electrical stimulation of the rat sensorimotor cortex, producing a convulsion, brought 66% of the recorded neurons of this zone into an excited, and 34% into an inhibited state. The same two to one ratio was found here as during adequate electrodermal stimulation.

The depth distribution of neurons participating in epileptiform activity by excitation or inhibition is shown in Fig. 3. Most neurons revealing changes in background activity during a convulsion lie in layer V (650-1300 μ). Most of these neurons responding by excitation were at the upper levels of this layer, and most responding by inhibition at its lower levels.

During simultaneous recording of activity of two or more neurons by several microelectrodes, the coefficient of correlation of intervals between successive spikes as a rule was less than 0.1, and there were no grounds for suggesting any regular connection between their discharges. However, during stimulation, the coefficient of correlation rises sharply to reach significant figures. Both positive (i.e., a synergic response of the neurons) and negative correlation (i.e., antagonistic relationships) between them were recorded (Fig. 4).

With distances of the order of millimeters between the recorded neurons, both synergic and antagonistic relationships were found without any marked predominance of one over the other. For example, of 28 neurons recorded in pairs in layer VI of the rabbit's sensorimotor cortex, 12 responded synergically and 16 antagonistically. If the distances between the recorded neurons were small (down to 100-200 μ), their relationships as a rule were synergic. The responses to stimulation of some neurons recorded simultaneously by one microelectrode were therefore always synergic. Table 4 shows the distribution of these pairs of recorded neurons in the layers of the guinea pig's cortex depending on the character of their responses.

The first feature to draw attention is that cases in which two neurons were recorded by one microelectrode occurred mainly in the upper and middle layers of the cortex. Although in layers IV

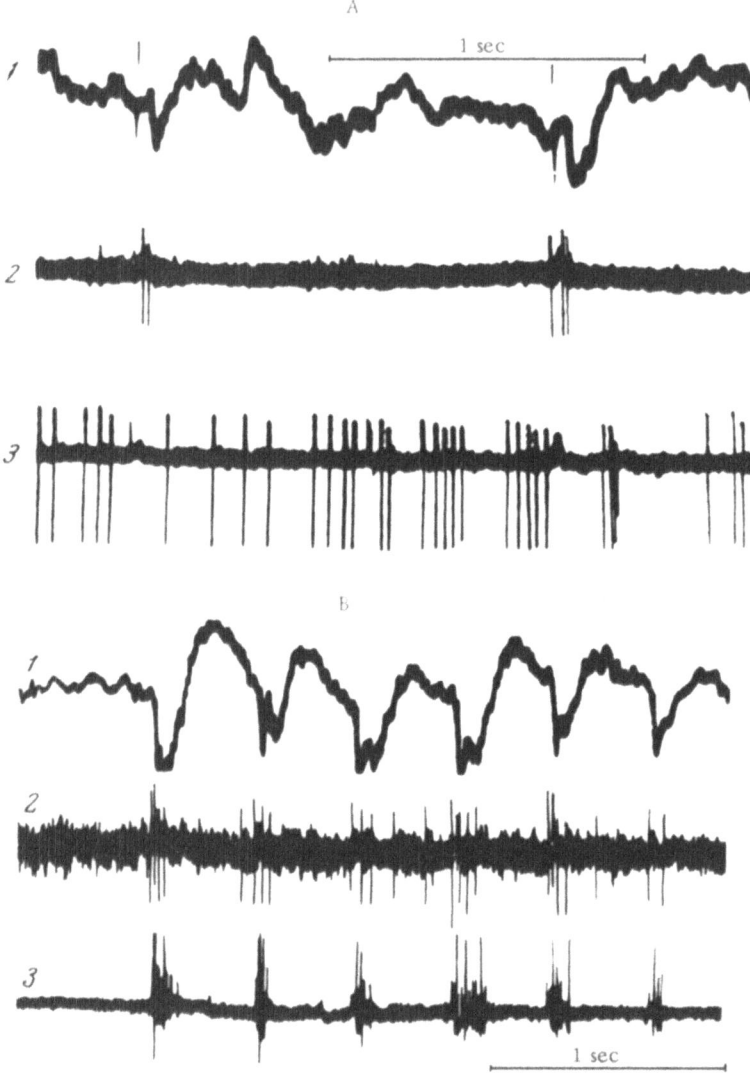

Fig. 4. Responses to electrodermal stimulation with single square pulses (short vertical lines above) from two neurons of the rat sensorimotor cortex recorded simultaneously. A) Antagonistic responses: 1) evoked potentials of electrocorticogram; 2) neuron at lower level of layer IV; 3) neuron of layer V;
B) synergic responses: 1) evoked potentials of electrocorticogram; 2) neuron at upper level of layer V; 3) neuron at lower level of layer V (experiments of Sukhov).

TABLE 4. Distribution of Neurons Recorded in
Pairs among Layers of Guinea Pig's Cortex
(Experiments of G.S. Klepach)

Layer of cortex	Total number of neurons recorded in pairs	Number of neurons responding to photic stimulation	Character of synergic response	
			Excitation	Inhibition
I	8	2	2	—
II	30	14	8	6
III	50	26	14	12
IV	14	4	4	—
V	—	—	—	—
VI	—	—	—	—

and V, more neurons with background activity were recorded than in layers I, II, and III taken together (Table 1), as will be clear from Table 4, in layers IV and V the microelectrode never once entered a zone of overlapping of the electric fields of neighboring neurons. At the same time, comparison of the total number of neurons recorded in pairs in this layer with the number responding to photic stimulation in layers where the figures are more suited to statistical analysis (II and III) shows the same ratio as before, varying within limits of 40-50%.

So far as the character of the response is concerned, in all layers, cases of excitation of both simultaneously recorded neurons predominate. However, the ratio between the number of pairs responding by excitation to the number responding by inhibition is rather different from that given in Table 3. Among neighboring neurons responding synergically to photic stimulation, the number of pairs with inhibitory responses is closer to the number with excitatory responses.

DISCUSSION

The differences discovered in the numbers of neurons with background activity in the various cortical layers may be attributed to the fact that in general there are few cells in the superficial layers (Vasilevskii, 1965) and they are inhibited by trauma during the operation (Melekhova and Shul'gina, 1963). In fact, no background rhythm was recorded in any of the three uppermost layers of the sensorimotor cortex of the rat (Sukhov, 1965) or in layer I of the sensorimotor (Dikareva, 1965) and visual cortex of the rabbit when the animals were immobilized, but it was regularly recorded in all the layers, starting from layer I, in guinea pigs allowed to move freely (Klepach, 1965). However, the fact that in the same animals (rabbits) under identical experimental conditions, the number of neurons with background activity reaches a maximum at different levels in the visual (layer V) and sensorimotor (layer VI) cortex is evidence of the probable functional significance of this distribution. It perhaps reflects relative predominance of activity of afferent elements at the middle levels in the visual analyzer and of the deeper effector elements in the motor analyzer. Whatever the case, a purely morphological explanation is inadequate here, because in the relatively undifferentiated cortex of rodents there are no corresponding differences in neuronal structure of the visual and somatosensory areas (Kukuev, 1953).

What is the possible physiological significance of the well-defined increase in relative number of neurons with a grouped type of discharge starting with layers III-IV? Since their incidence increases toward the deeper layers, the possibility of the traumatic nature of this type of background activity is ruled out. It may be supposed that these periods of comparatively high-frequency discharges reflect the activity either of intermediate receptive cells transforming single afferent impulses into discharges (Li et al., 1956), or of stellate neurons performing a specific sensory function (Beritov, 1961), or analogs of the Renshaw cells inhibiting other neurons with their high-frequency discharge (Burns, 1958), or finally, they may reflect effector discharges of pyramidal neurons, the high-frequency and grouped character of which was demonstrated long ago in corticofugal pathways (Adrian and Bronk, 1929). It must be remembered that even within the bounds of continuous or grouped types of activity there are definite differences in the character of the impulses, and as the results of the histogram analysis showed, these may form the basis of a more detailed classification.

The fact that the percentage of neurons responding to stimulation in relation to the total number showing background activity in a given cortical layer varies only very slightly on the average from layer to layer, strengthens the view that neurons may be interconnected in "columns" from the surface to the deepest layers of the cortex (Mountcastle, 1957). However, just as in relation to background activity, some tendency can also be seen here toward greater reactivity of neurons at the middle levels in the visual cortex and at deeper levels in the somatosensory cortex.

The mean ratio between the neurons responding by excitation or inhibition (2 to 1), which is surprisingly similar for rats, guinea pigs, and rabbits, is modified slightly if the cells of the visual and motor analyzers are examined separately by layers. For instance, the greatest preponderance of responses of excitatory type in the visual cortex is observed at higher levels than in the somatosensory area. This may be compared with the results of measurements of latent periods of responses of individual neurons belonging to different cortical layers to afferent stimulation (Patton and Amassian, 1960; Livanov, 1965). An even more striking fact is that the proportion of 2 to 1 between excited and inhibited neurons is preserved even in epileptiform responses to direct stimulation of the cortex (Sukhov, 1965), which apparently ought to bring all the neurons into an excited state. It may be postulated on the basis of all these facts that this ratio between excited and inhibited neurons reflects a general principle of the dynamic structure of cortical mechanisms.

The differences in neuronal responses in different cortical layers explain the contradiction arising between the widespread view that desynchronization is an expression of cortical activation and the observed facts that its excitability, as determined from thresholds of direct stimulation, is lowered under these conditions (Kogan and Nikolaeva, 1958). Layer-by-layer analysis of neuronal activity confirms the earlier hypothesis that desynchronization is not solid excitation of the cortex, but it conceals a complex process embracing excitation of some and inhibition of other neurons (Kogan, 1958). Neuronal interaction is evidently modified, with a shift of the fundamental ratio between excited and inhibited neurons discussed above in favor of the former for the system of afferent neurons of the middle layers of the cortex and in favor of the latter for the system of effector neurons of its deeper layers.

Layer-by-layer analysis of paired and multiple correlation between activity of the cortical neurons reveals the extreme com-

plexity of their system of interaction. Some relationships discovered between the responses of neurons oriented differently relative to the source of incoming impulses (Sukhov, 1965) and with
increasing distances between neurons may help to determine the
structural organization of neuronal ensembles. In particular, the
pattern discovered in the tectum mesencephali of the frog (Chorayan, 1965) has also been found in the cortex of rats, guinea
pigs, and rabbits. Adjacent neurons, evidently belonging to the
same ensemble, react synergically; cells lying up to 0.2-0.6 mm
apart (probably belonging to neighboring ensembles) react reciprocally, while neurons lying still further apart react independently.

Cyclic Changes in Cortical Neuronal Activity after Brief Stimuli*

I. N. Kondrat'eva

Institute of Higher Nervous Activity and Neurophysiology
Academy of Sciences of the USSR
Moscow, USSR

Stimuli of short duration (of the order of 1 msec) produce prolonged changes in the cortex measured in tens of milliseconds,
consisting of waves of evoked potentials, cyclic changes in excitability to subsequent stimuli, and phasic responses of the cortical
neurons. These changes are evidently relevant to formation of the
response to a short stimulus and to analysis of the information
concerning its properties. These changes in cortical excitability,
in evoked potentials, and in cortical neuronal responses are possibly interconnected and the mechanism of their genesis may be
common. This problem is analyzed herein, paying attention mainly to the visual cortex.

Phasic Responses of Cortical Neurons
to Short-Duration Stimuli

During application of flashes certain neurons of the visual
cortex, which are in a state of background activity, do not respond
(A-neurons). The remaining neurons may be classed, depending
on their type of response, as responding initially by excitation
(C-neurons) or by inhibition (D-neurons).

The main distinguishing feature of the responses of these
types of neurons is their phasic character: with the C-neurons an
initial increase in spike frequency (in some cells this is preceded
by pre-excitatory inhibition) is followed by an inhibitory pause and

---

*Pages 148-159 in the Russian edition.

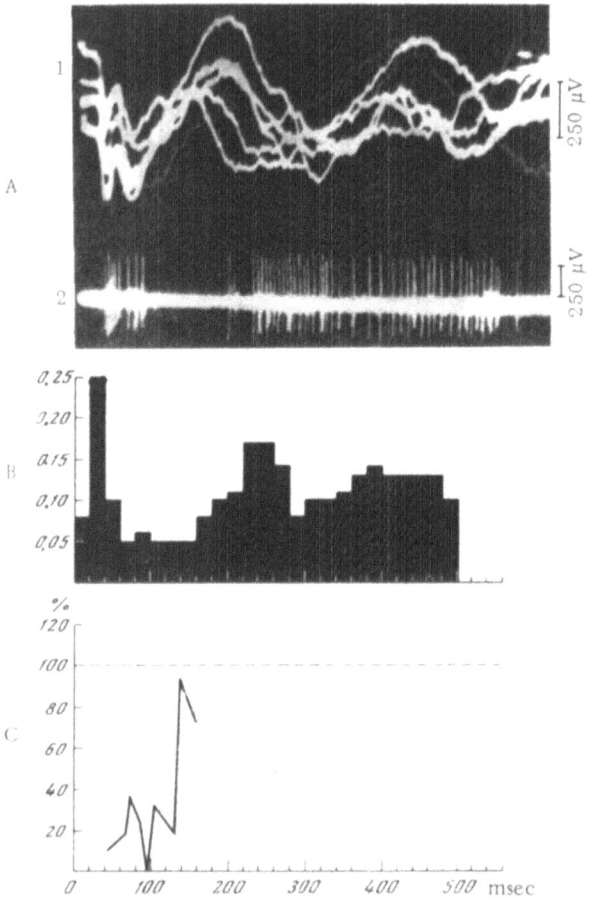

Fig. 1. Comparison of evoked response with spike responses
of neurons and their recovery cycle in visual cortex after
flashes. A) Surface ECoG (1) and spike activity of neuron
at a depth of 1350 μ (2). Flash starts oscilloscope beam.
Downward deflection inidicates positivity beneath active
electrode; B) poststimulation histogram of mean spike frequency of 36 neurons with different types of response to
flashes (1345 flashes). Ordinate: mean number of spikes
during interval of 20 msec; C) mean curve of recovery of
responses of 19 neurons to second flash applied at various
intervals after first. Ordinate: magnitude of response, response to flash 1 taken as 100%; abscissa: time between
flashes. All curves given on same time scale, indicated
along abscissa of graph C in msec.

a second phase of activation, which in turn is followed by a second phase of inhibition, accompanied by a third of activation (Baumgartner, 1955; Grüsser and Grützner, 1958; Grützner et al., 1958; Kondrat'eva, 1964; Polyanskii, 1965a, 1965b; Tolkunov, 1965; Fromm and Bond, 1965, and others). The first phase of activation is absent in neurons responding initially by inhibition. With some cells an increase in spike frequency is observed during the inhibitory pause of most other neurons (Grüsser and Grützner, 1958; Kondrat'eva, 1964). These cells are possibly internuncial inhibitory neurons, responsible for genesis of the inhibitory pause in spike activity (see below).

The cortical neurons respond to flashes as a series of neuronal systems, for identical phases of the response of most neurons occur at identical times after application of the stimulus (the latent period of phase I of excitation varies from 20 to 50 msec, of phase II of activation from 120 to 200 msec, and of inhibitory phase I from 25 to 80 msec). Because phases of the response of different neurons coincide in time, the mean poststimulation histogram of many neurons reflects all phases of the response precisely (Fig. 1B). For the same neuron, genesis of inhibitory pause I is independent of the latent period of the activation phase. For example, a decrease in the latent period of activation phase I could be observed with no change in latency of the inhibitory pause. Statistical analysis of the small fluctuation in latent periods of phases of the responses of the same neuron to application of successive stimuli showed, in accordance with the χ^2 criterion ($P = 0.05$), that changes in latent periods of excitation phase I and inhibitory pause I are independent (Kondrat'eva, 1964).

The phases of neuronal responses in the visual cortex are more clearly visible in response to flashes of relatively high intensity (Polyanskii, 1965a; Kondrat'eva and Bakaeva, 1966).

Responses of the same type as during the action of flashes are found in the cortex during electrical stimulation of the optic nerve. Grüsser and Grützner (1958) subdivided the visual cortical neurons of encéphale isolé preparations of the cat when stimulated in this way into the following types: 1) nonreacting units (corresponding to A-neurons); 2) responding initially by excitation with a latent period of between 1.7 and 17 msec (corresponding to C-neurons); 3) neurons becoming activated later, after

25 msec (possibly corresponding to internuncial inhibitory neurons); and 4) neurons whose response begins with inhibition, followed after 100–200 msec by activation (corresponding to D-neurons). The responses of neurons of the second and fourth types are phasic in character: phases of activation are followed by inhibitory pauses.

Electrical stimulation of the lateral geniculate body (LGB) in waking rabbits not immobilized by pharmacological agents, also leads to the appearance of phasic responses (Kondrat'eva, 1964). Comparison of mean poststimulation histograms for visual cortical neurons to stimulation of LGB and to flashes shows that all phases of the responses coincide in time, with the exception that during stimulation of LGB, excitation phase I begins and ends sooner than during photic stimulation.

Li et al. (1960) recorded intracellular potentials from visual cortical neurons of cats which were curarized and anesthetized with pentobarbital and found that stimulation of LGB gives rise either to one prolonged hyperpolarization potential, or to such a potential preceded by depolarization, leading to a spike discharge. Sometimes they found that the initial depolarization is apparently cut short by the subsequent hyperpolarization, so that the spike may not develop. They postulated on the basis of these observations that stimulation of LGB activates both excitatory and inhibitory fibers; the latter evoke hyperpolarization through internuncial neurons. Direct electrical stimulation of the visual cortex of cat encéphale isolé preparations produces the following successive phases in the neurons of this area: 1) inconstant primary activation; 2) a constant inhibitory pause of between 50 and 200 msec in duration; 3) a less constant secondary activation; 4) an inconstant later inhibition, sometimes followed by a third activation (Creutzfeldt et al., 1956). These workers report that electrical stimulation of varied strength caused cessation of spikes (the inhibitory pause) irrespective of whether a primary discharge was present or not. The threshold for evoking an inhibitory pause was lower in many neurons than that for primary excitation, while in others it could be much higher. Just as during stimulation by flashes, electrical stimulation of the cortex gave more stable responses at a frequency of 1–2/sec. These workers point out that all phases of the response are clearly dis-

tinguished on the integral poststimulation histogram of many
neurons.

Hence, during exposure to flashes and also during electrical
stimulation of the visual system at various levels, most neurons
of the visual cortex respond in phases, periods of activation alter-
nating with inhibitory pauses. A feature common to nearly all
neurons is the presence of a first, long inhibitory pause, appearing in most neurons at approximately the same time after action
of the stimulus, and during which, according to Tasaki et al.
(1945), Li et al. (1960), and Skrebitskii and L. L. Voronin (1966),
intracellular hyperpolarization is observed.

It is important to know where this periodicity of the response
to a short duration stimulus arises. Let us examine the responses
of neurons at a lower level in the visual analyzer.

In the LGB of rats anesthetized with pentobarbital, during electrical stimulation of the contralateral optic nerve or the ipsilateral
visual cortex, Sefton and Burke (1965) found two types of neurons.

The first type, described as "P" cells, accounts for 88% of
all neurons investigated. In response to stimulation they discharge by a short-latency phase of activation, followed by a period
of silence in spike activity, and then by a late discharge, after
which a whole series of grouped discharges may be observed,
separated from each other by inhibitory pauses. These neurons
give similar responses to orthodromic and antidromic stimulation.
During stimulation of the visual cortex, signs of antidromic conduction are found.

The second type, known as "I" cells, discharge in response
to the stimulus after a somewhat longer latent period and with a
larger number of spikes in the activation phase, and they were
classed as inhibitory internuncial neurons. The primary grouped
discharge of these neurons was also followed by bursts of activity
separated by pauses. During stimulation of the cortex the "I"
cells responded, but in this case there were no signs of antidromic invasion.

Phasic responses of LGB neurons during application of various stimuli to the visual analyzer have been observed by Tasaki
et al. (1954), Bishop and Davis (1960), and others.

Responses of the retina and ganglion cells have been carefully investigated by German workers (Grüsser and Grützner,
1958; Grüsser and Rabelo, 1958; Grüsser and Kapp, 1958). They
showed that retinal on-elements of cat encéphale isolé preparations respond to flashes by a first phase of activation with a latent
period of 8-50 msec, followed by an inhibitory pause lasting from
80 to 200 msec, and then by secondary activation. If the intensity
of the flash is increased, a second inhibitory pause and third activation may arise. The retinal off-neurons respond to flashes by
inhibition of background activity for between 50 and 500 msec, followed by a phase of activation. The pause in spike activity is
lengthened by increasing the intensity of the flash. The retinal on-
off cells also respond with a primary inhibitory pause followed by
a phase of activation. However, the pause in spike activity in this
case is short (30-80 msec), and is still further shortened by an
increase in intensity of the flash; a second inhibitory pause and
further activation then arise.

In the opinion of these workers, phases of activation and inhibition may be explained by the simultaneous application of synaptic excitatory and inhibitory impulses arriving from receptors
and internuncial neurons. They also consider that the first activation phase is due to the receptor potential, while the second activation indicates processing of excitation by retinal neurons.

Phasic responses of retinal neurons were found by Crapper
and Noell (1960) during electrical stimulation of the retina. An
increase in stimulus intensity likewise led to the appearance of
fresh phases of activation and inhibition. By dividing the optic
nerve, they showed that the phasic character was independent of
the generation of cycles of excitation between the retina and higher
structures of the nervous system.

In their investigations of retinal and cortical neuronal responses in cat encéphale isolé preparations, Grüsser and Grützner
(1958) found no differences in principle between the temporal relationships of excitation and inhibition in the cortex and retina,
apart from a strong dependence of cortical neuronal activity on
background activity and nonspecific influences. The number of
discharges in the primary and secondary activation phases of the
cortical neurons was modified by electrical stimulation of the in-
tralaminar thalamus, the mesencephalic reticular formation, and

the ipsilateral optic nerve. They accordingly postulated that retinal excitation is transmitted to the cortex, where it is repeated by cortical neurons, and cortical excitation is modulated by activation of nonspecific systems. However, no data appear in the literature concerning the temporal comparison of the phases of neuronal responses at different levels of the visual analyzer under more nearly physiological conditions, so that there is little point in discussing this question at the present time.

Nevertheless, it is difficult to concede that the retina is the only source of these phases, because stimulation of the optic nerve and LGB (see above) also evokes phasic responses in the cortex. According to Creutzfeldt and Struck (1962), the first phase of activation and the inhibitory pause arise in the isolated cortex of cats, but secondary activation is absent. Burns et al. (1957), however, observed the secondary phase of spike activity in the isolated cortex. I have also observed a second phase of activation in a strip of cortex in unanesthetized rabbits. It may be postulated on the basis of these findings that phasic responses arise at all levels of the visual analyzer (the mechanisms of their genesis will be discussed below).

Phasic responses of this type after stimuli of short duration are specific not only for the visual analyzer, but are observed in many other structures, namely:

1. They were found in neurons of the sensorimotor cortex by Creutzfeldt et al. (1956) upon direct electrical stimulation; by Mountcastle et al. (1957) upon tactile stimulation; by Suzuki and Tukahara (1963), and Stefanis and Jasper (1964a, 1964b) upon antidromic stimulation of Betz cells; by Lux and Nacimiento (1963) and Schlag and Balvin (1964) upon stimulation of the thalamus; and by Tolkunov (1965) upon stimulation of the sciatic nerve.

2. In the neurons of the orbital cortex by Cohen et al. (1957) upon stimulation of the chorda tympani and trigeminal branch of the lingual nerve, and also of the taste and tactile receptors of the tongue. They found phasic responses with depression of the same duration as in the visual cortex.

3. Phasic responses were found in the neurons of the thalamic nuclei by Andersen et al. (1964a, 1964b).

4. In neurons of the cerebellar cortex and hippocampus by Andersen et al. (1963a, 1963b, 1964d) and by Gloor et al. (1963).

Evoked Potentials of the Visual Cortex

and Their Comparison with Spike Activity

In the unanesthetized rabbit, under certain technical conditions (using an amplifier with long time constant) the surface-evoked potentials in the visual cortex consist mainly of the following components: a primary complex of a first positive and first negative wave, followed sometimes by a new, second positive wave, followed by a slow negative potential of 100–200 msec, and then by a third positive wave (Fig. 1A). Sometimes further slow negative waves are observed, ending with a positive wave (Bartley et al., 1937; Pearlman, 1963, and others).

Besides the primary response, the secondary slow negative wave and subsequent positive wave undergo inversion in the depths of the cortex (Shuranova, 1964; Kondrat'eva, 1965; Polyanskii, 1965a, 1965b).

When the mean changes in spike activity of many neurons ("mean poststimulation histogram") was compared with the surface-evoked potential, it was found that the first phase of increase in frequency of spike activity of the cortical neurons corresponds to the primary complex and the second positive wave. The pause in spike activity occurs at the time when the slow negative wave is recorded on the surface, and the second phase of activation in neuronal activity correlates with the third positive wave (Kondrat'eva, 1964, Fig. 1A, B).

During simultaneous recording of spike and slow activity from the surface or from different levels of the cortex by the same microelectrode, precise correlation between them may often be seen. Neuronal discharges are found mainly during the primary response and subsequent surface-positive and deep-negative waves. The slowing or complete stopping of spike activity occurs during the slow surface-negative and deep-positive wave (Kondrat'eva, 1965; Polyanskii, 1965a, 1965b; Fromm and Bond, 1965). More precise correlation was observed in neurons possessing all phases of the response, such as in Fig. 1A.

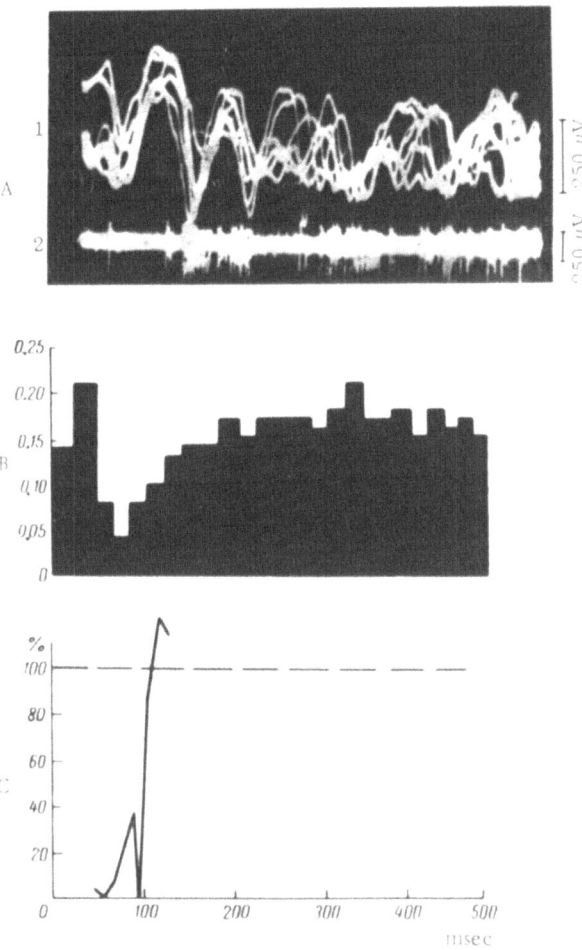

Fig. 2. Comparison of evoked response with spike responses of neurons and their recovery cycle for the visual cortex after flashes during Nembutal anesthesia (30-25 mg/kg). A) Focal activity (1) and spike activity (2) recorded by microelectrode at depth of 1080 μ. Flash starts oscilloscope beam. Upward deflection indicates positivity; B) the same as in Fig. 1B, 73 neurons, 2499 flashes. C) The same as in Fig. 1C. Curve obtained from 7 neurons.

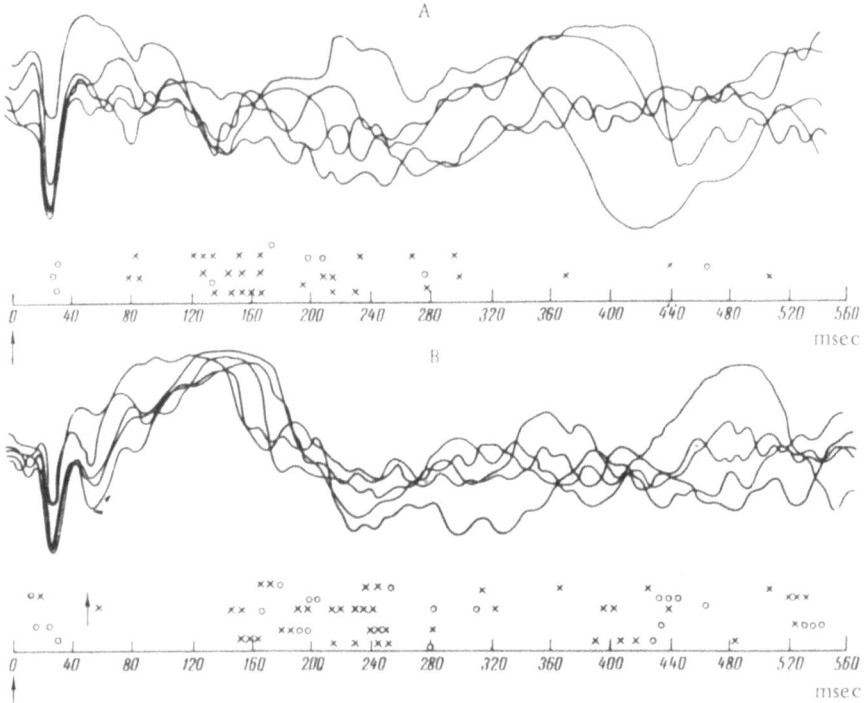

Fig. 3. Recovery of slow deep-positive wave and inhibitory pause by application of second flash 45 msec after first (1 hr 20 min after injection of Nembutal, 30 mg/kg). A) Single flash; B) two flashes denoted by arrow. 1) Curves represent focal activity from microelectrode; ×) spikes from 1 neuron; ○) spikes from another neuron recorded at the same depth of 1770 μ. Each sign represents one spike (five responses are superposed). Bottom line shows time in milliseconds.

Administration of small doses of Nembutal (25-30 mg/kg) to intact rabbits reduced the slow surface-negative and deep-positive waves in the evoked response, while the inhibitory pause in the spike activity of the neurons was shortened (compare Figs. 1 and 2). Sometimes after injection of Nembutal, the inhibitory pause in spike activity in response to a flash could be lengthened by applying the second flash a short time (30-60 msec) after the first. Meanwhile the surface-negative and deep-positive wave was increased (Fig. 3).

Correlation between the slow wave of potentials and inhibition in spike activity has been observed in other brain structures.

In the LGB, according to Vastola (1959) and Bishop and Davis (1960), the initial negative waves are followed by long positivity, and then by a late negative wave. All waves undergo inversion as the electrode moves from the geniculate bodies into the optic radiation. These workers found that negative waves in the LGB are associated with synchronous discharges of a large number of neurons, while the positive waves correlate with cessation of the discharges.

In the orbital cortex of cats during stimulation of the trigeminal nerves, Cohen et al. (1957) observed correlation between the negative waves in the depth of the cortex and neuronal impulses, and also a slow positive wave during the pause in spike activity.

In response to stimulation of a cutaneous nerve, Marshall (1941) obtained a high-frequency discharge from thalamic neurons, followed by a long positive potential associated with depression of synaptic transmission lasting 150 msec.

Andersen et al. (1964a, 1964b) stimulated the radial nerve and, by means of extracellular microelectrodes, recorded a negative potential with superposed spike discharges in the thalamus followed by a positive potential lasting 100 msec, during which spikes were absent. If intracellular recording was used, hyperpolarization of the membrane was observed at this time. Relative negativity was then restored (extracellularly) and discharges of the neurons appeared.

A precise correlation thus exists between the long inhibitory pause in spike activity and the slow waves of potentials arising after application of stimuli.

Recovery Cycles of Cortical Neuronal Responses

After the action of a stimulus (conditioning stimulus), primary excitation is followed by changes in excitability of the cortical neurons determined by application of a second hetero- or homosynaptic stimulus (testing stimulus).

Cyclic changes in excitability after stimulation in the visual cortex of rabbits were first described in 1933 by Bishop, and again in 1936 by Bartley. They showed that if the interval between

stimuli does not exceed 80 msec, there is no response (as judged
by the primary potential) to the second stimulus. If the interval
is increased further, the amplitude of the second response is increased, ultimately reaching the magnitude of the potential to the
first stimulus and sometimes exceeding it.

Since only limited investigations have been made of recovery cycles in the visual cortex, I shall describe the main principles governing this phenomenon, using data for other parts of the
cortex and subcortex, because the principles described above are
characteristic of them also.

The recovery cycle as a whole was divided later into the
period of absolute refractoriness (periods of areactivity), lasting
(for example in the rabbit's visual cortex) for up to 60 msec after
the first stimulus, a phase of relative refractoriness (period of
subnormality) in the rabbit's visual cortex, lasting up to 100-150
msec, and a period of facilitation followed by fluctuations of excitability.

The temporal parameters of the recovery cycles vary with
the strength of the stimuli and functional state of the animals. The
more intensive the conditioning stimulus, the stronger the depression, although its duration does not change greatly (Rosenzweig
and Rosenblith, 1953; Zhang, 1963; Merlis, 1965). The depression decreases with an increase in strength of the testing stimulus
(Evarts et al., 1960). According to Evarts et al. (1960) and Palestini et al. (1965), in the visual cortex of cats the phase of depression of the response to the second stimulus is deeper and longer
in the waking state than during sleep. In man in a state of stress,
as judged from the EEG, the magnitude of subnormality in the recovery cycle is less than if the a-rhythm is regular (Ciganek,
1964).

The recovery cycles are considerably modified by the action
of certain drugs. Most workers have shown that narcotics considerably prolong the depression phase in the recovery cycle, a
stronger action being shown by barbiturates (Heinbecker and Bartley, 1940; Marshall et al., 1941; Jarcho, 1949; Gastaut et al.,
1951; Morin et al., 1951; Schwartz and Shagass, 1962; Zhang,
1963; Berry and Hance, 1965; Greenbaum and Merlis, 1965, and
others). Some workers, however, assert that relatively small
doses of barbiturates initially reduce the phase of subnormality

both in duration and in depth (King et al., 1957; Schoolman and Evarts, 1959; Myslobodskii, 1966; Kondrat'eva, 1966). Substances causing convulsions (cardiazol and amphetamine), according to the observations of Gastaut et al. (1951) and Morin et al. (1951), shorten the time of recovery of the responses.

A number of workers have observed precise correlation between the depression phase in the cycle of excitability and slow waves. Vastola (1959) and Bishop and Davis (1960) found that depression in the LGB arises during a slow positive wave, corresponding, according to Sefton and Burke (1965) to an IPSP.

In the visual cortex the phase of subnormality appears simultaneously with the surface negative wave (Clare and Bishop, 1955; Pearlman, 1963) and the phase of depression of spike activity (Kondrat'eva, 1965; Polyanskii, 1965a, 1965b; Kondrat'eva and Volodin, 1966). However, the depression ends sooner than the slow negative wave (Fig. 1).

Precise correlation between changes in the surface-negative and deep-positive wave of the evoked potential, the pause in spike activity of the neurons, and duration of the phase of subnormality was found in the visual cortex of intact rabbits after injection of small doses of Nembutal (30 mg/kg), when the durations of all these parameters were simultaneously shortened (compare Figs. 1 and 2).

Andersen et al. (1964a, 1964c) established a close correlation for neurons of the ventro-basal thalamus between depression of spike activity after a stimulus, the slow extracellular positive wave, hyperpolarization of the cell membrane, and the depression phase in the cycle of excitability.

Blocking of the response to the testing stimulus in neurons of the sensorimotor cortex was parallel to the intracellularly recorded IPSP, but it ended before hyperpolarization returned to its initial level (Stefanis and Jasper, 1964b).

It has often been observed that the phase of subnormality in the recovery cycle ends with a period of facilitation, after which depression again arises, followed by exaltation, and so on (Morin et al., 1951; Chang, 1952; Rosenzweig and Rosenblith, 1953; King et al., 1957; Pearlman, 1963; Ciganek, 1964; Andersen et al.,

1964a, 1964c; Merlis, 1965; Greenbaum and Merlis, 1965; Polyanskii, 1965a; Kondrat'eva and Volodin, 1966).

According to Andersen et al. (1964a, 1964c), facilitation
followed by depression is observed only in the thalamus and cortex. At lower levels of the central nervous system (in the nucleus
cuneatus, for example), they are not found. Most workers ascribe
this facilitation to postanodal exaltation (the rebound phenomenon)
for the thalamic neurons. For the neurons of the sensorimotor
cortex, Andersen et al. (1964a, 1964c) attribute this facilitation to
thalamic activation.

Role of Inhibition in Generation of Phased Responses of Cortical Neurons, Slow Waves of Evoked Potential, and Depression of Responses after First Stimulus

After application of brief stimuli, immediately after the initial spike discharge, the complex of the primary evoked response,
and the period of early increase of excitability, a long pause in
impulse activity takes place, with a large short wave of the evoked
response, and a phase of depression in the responses of different
parts of the central nervous system.

The close correlation between these processes suggests that
the same mechanisms are concerned in their genesis. Most probably, these mechanisms are inhibitory in character.

The phase of lowered excitability arises during hyperpolarization of the cell membrane, as has been demonstrated for neurons of the ventro-basal nuceli of the thalamus (Andersen et al.,
1964a, 1964c), for the sensorimotor cortex (Stefanis and Jasper,
1964b), for the LGB (Sefton and Burke, 1965), and for the visual
cortex (Li et al., 1960). It is possible that the positive wave recorded extracellularly deep in the cortex (Kondrat'eva, 1965; Polyanskii, 1965a) or in the LGB (Vastola, 1959; Bishop and Davis,
1960) and the corresponding negative wave on the surface of the
cortex or in the optic radiation reflect this hyperpolarization of
the membrane. This fits in with results obtained by Andersen
and Eccles (1962) and Purpura (1966), who showed that the extra-

cellular positive wave reflects hyperpolarization of the cell membrane if this develops synchronously in many neurons.

Evidently this phase in neuronal activity is not entirely due
to postexcitatory depression — to after-hyperpolarization, as some
consider (Marshall, 1941; Vastola, 1959; Bishop and Davis, 1960;
Pearlman, 1963), because it may be observed even without preceding activation (Suzuki and Tukahara, 1963; Stefanis and Jasper, 1964b; Andersen et al., 1964a, b, and c; Kondrat'eva, 1964).
Whether or not after-hyperpolarization plays any part in this process is at present uncertain, but even if it does, its role is not
decisive.

Many investigators consider that the hyperpolarization arising after application of a stimulus reflects the inhibitory postsynaptic potential (IPSP), because first, while it is present the response of the neuron during heterosynaptic or homosynaptic testing or during intracellular stimulation disappears or becomes
weaker (Li, 1956; Suzuki and Tukahara, 1963; Stefanis and Jasper, 1964b; Andersen et al., 1964a, 1964c; Sefton and Burke,
1965; Kondrat'eva, 1965; Polyanskii, 1965a, 1965b; Tolkunov,
1965, and others). Second, it may be increased by the passage of
a depolarizing current applied through a microelectrode, and reversed in sign by the passage of a hyperpolarizing current or
during migration of Cl^- from the electrodes (Andersen et al.,
1964c). Third, while it is present, spike activity is absent.
Fourth, this inhibition is possibly not caused by the same factor
that triggers the activation phase, but by another source, because
no relationship is found between the latent periods of the phases
of activation and inhibition of the same neuron during subsequent
application of the stimulus (Kondrat'eva, 1964). Moreover, according to Li et al. (1960), the initial depolarization recorded in
neurons after stimulation is often cut short by hyperpolarization,
so that the cell cannot discharge with a spike potential. Fifth,
this inhibitory pause may be summated, as is frequently observed
when two stimuli are applied with a short interval between them
(Kondrat'eva and Volodin, 1966).

In the view of Stefanis and Jasper (1964b) and of Andersen
et al. (1964a, 1964c), this inhibition is effected by multisynaptic
pathways, because one powerful stimulus is less effective than

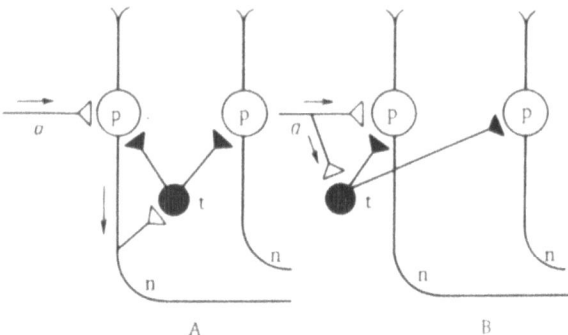

Fig. 4. Scheme of hypothetical pathways for genesis of inhibitory pause in cortex. A) Reciprocal collateral inhibition;
B) afferent collateral inhibition; a) afferent fibers; n) axons;
p) pyramidal neurons; t) hypothetical internuncial inhibitory
neurons.

several subthreshold stimuli; the rate of growth of the IPSP in
unanesthetized preparations is higher than in those anesthetized
with barbiturates, probably because of the earlier action of the
anesthetic on internuncial neurons.

The inhibitory pause after the stimulus is represented at
various levels of the central nervous system. For example, in
the visual system it can be found in the retina, the LGB, and the
cortex. It is also found in isolated areas of nerve tissue: it was
found in the retina by Crapper and Noell (1960), in the ventral
thalamus by Andersen et al. (1964a, 1964c), and in the isolated
cortex by Burns et al. (1957), Creutzfeldt and Struck (1962), and
myself. Consequently, in these parts of the nervous system there
are cells evoking inhibition.

Some workers have found neurons giving discharges of high
frequency and a long latent period after stimulation. The frequency of their spike activity reaches a maximum at the time of
commencing inhibition in other neurons, and they are excited by
antidromic stimulation only synaptically. They have been classed
as inhibitory internuncial neurons, evidently responsible for reciprocal inhibition (Li et al., 1960; Kondrat'eva, 1964; Sefton
and Burke, 1965, and others). However, afferent inhibition likewise cannot be excluded in this case (Eccles, 1965b), for if two

flashes are applied with a short interval between them, the activation phase is absent in the response to the second stimulus while the inhibitory pause is well defined (Kondrat'eva and Volodin, 1966). Schemes of this inhibition according to interpretations given by these authors are shown in Fig. 4.

It has been shown for the neurons of the ventral thalamus (Andersen et al., 1964a, 1964c) and the LGB (Angel et al., 1965) that, besides postsynaptic inhibition, presynaptic inhibition also plays a role. This question remains open in the case of the cortical neurons.

It can thus be postulated that the inhibitory postsynaptic potential is responsible for the pause in spike activity, depression of the response to the testing stimulus applied a certain time after the first stimulus, and the slow wave of the evoked potential. However, there are no simple relationships between these indices, because depression ends sooner than the slow wave and the pause in impulse activity.

Evidently this inhibition, arising in most neurons of the central nervous system after initial activation, performs several functions. It separates the initial volley of impulses from subsequent volleys, preventing fusion of stimuli (Rosenzweig and Rosenblith, 1953; Gershuni, 1962; Donchin et al., 1963), limits the avalanche-like spread of excitation over the cortex (Jung and Baumgartner, 1955; Livanov, 1965a, 1965b), creates a contrast, and strengthens the localization of excitation for precise coordinated activity (Stefanis and Jasper, 1964b). It is also possible that the duration of this inhibition determines the lability of the system. Some workers consider that this inhibition plays a definite role in fixing the rhythm of biopotentials and the genesis of alpha-like waves and driven rhythms (Cohen et al., 1957; Krnjević et al., 1964; Andersen et al., 1964c; Kondrat'eva and Bakaeva, 1966).

Further experimental investigations are needed to elucidate the nature of this inhibition in systems of cortical neurons.

Relationship between Distant Synchronization and Steady Potential Shifts in the Cerebral Cortex[*]

T. A. Korol'kova and T. B. Shvets

Institute of Higher Nervous Activity and Neurophysiology
Academy of Sciences of the USSR
Moscow, USSR

The phenomenon expressed by synchronized changes in potentials at different points of the brain is well known in electrophysiology. In M. N. Livanov's laboratory, electrical activity of this type has been called distant synchronization of potentials. The investigation of this phenomenon on a particularly wide scale began with the introduction of correlation analysis into electroencephalography (Brazier, 1960; Adey, 1961b; Livanov, 1962b; Glivenko et al., 1962a; Sologub, 1964; Tunturi, 1963; Livanov et al., 1965; Grindel', 1965; Boldyreva, 1965, and others). However, no clear idea has yet been formed of the mechanisms lying at the basis of synchronization of brain potentials.

Most probably the phenomenon of distant synchronization (DS) is based on both impulse and electrotonic influences. To investigate the latter mechanism, associated changes in (DS) and the steady potential level (SP) must be investigated.

No reference could be found in the literature to the simultaneous investigation of these phenomena in the same areas of the brain. To examine such changes, it seemed appropriate to study their correlation during formation of a conditioned reflex. From several investigations it is known that both these phenomena undergo precise changes during conditioning. Dumenko (1953a), Livanov (1962d), Glivenko et al. (1962b), and others have shown that in the course of combinations the distant synchronization in

---

[*] Pages 160-167 in the Russian edition.

the cortex was increased, and during stabilization of the reflex it returned to its background level. Changes in the steady potential during conditioning have been investigated by Shvets (1963). She showed that in the course of combinations, slow changes in SP developed in response to stimulation. Initially they were generalized, but later became concentrated at definite cortical points.

The object of the present investigation was to study simultaneously the changes in DS and slow shifts of SP at the same points of the cortex during conditioning.

METHOD

In chronic experiments on nine rabbits, a defensive conditioned reflex was formed to an electric light (100 W in two cases and 60 W in seven) placed above the animal's head at a distance of 50 cm. A reflex was produced with a time lag of 9 sec. The unconditional stimulus was electrical stimulation (square pulses, duration 1 msec, frequency 50/sec, time of application 1 sec) of the skin of the left paw by means of an "Alvar" stimulator. In each experiment the rabbit received six combinations at intervals of 3 to 7 minutes. The experiments were performed on alternate days.

Movements were recorded pneumographically. The cortical potentials were amplified by means of a two-channel VEKS-- Reegoscope dc amplifier. Motion pictures were taken from the Reegoscope screen at a speed of 1 cm/sec.

Potentials were recorded from four rabbits by nonpolarizing calomel electrodes fixed to the skull pared away to make it thinner over the frontal, parietal, and occipital regions of the right hemisphere (Fig. 1A). To obtain a more local recording of the potentials, in five animals platinum wire electrodes were used, 0.3 mm in diameter and insulated with varnish except at the tip, implanted into the bone as far as the inner table. From 9 to 18 electrodes were implanted in each rabbit. Their location is shown in Fig. 1B.

A monopolar technique was used to record the potentials of all animals, the indifferent electrode being placed on the nasal bones. This method gives the fullest information on DS.

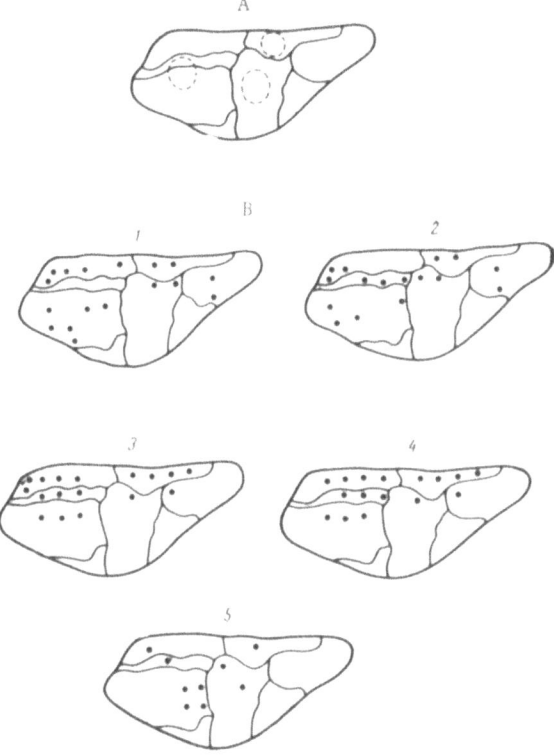

Fig. 1. Arrangements of electrodes. A) Nonpolarizing electrodes; B) polarizing electrodes; 1, 2, 3, 4, 5) different
rabbits. Borders of cytoarchitectonic areas after Rose (1931)
are shown by thin lines. Position of electrodes is shown by
circles and dots.

The cortical electrical activity was recorded successively
from selected pairs of electrodes.

Since the intent was only to establish a parallel between the
increase in DS and the appearance of slow shifts of SP, the evaluation "increased" or "not increased" in response to presentation
of the stimulus by comparison with its initial state was regarded
as adequate when assessing the DS. Visual assessment of DS was
therefore sufficient, without precise quantitative measurement.

To determine the extent to which the two types of response
appeared in association with each other, all types of possible re-

sponses to stimulation were calculated. The results of these calculations are given in Table 1. The data obtained were analyzed
by statistical methods. Using the χ^2 criterion with a 1% level of
significance,* the presence or absence of a significant difference
was determined between cases of appearance of SP shifts with an
increase in DS and in the absence of such an increase. The following formula was used for comparing two probabilities:

$$\frac{(x_1 \cdot n_2 - x_2 \cdot n_1)^2 (n_1 + n_2 - 1)}{n_1 \cdot n_2 (x_1 + x_2)(y_1 + y_2)},$$

where n_1 represents the number of cases of increase in DS, n_2 the
number of cases of absence of increase, x_1 the number of cases of
appearance of slow fluctuations of SP with an increase in DS, x_2
the number of cases of appearance of slow fluctuations of SP in
the absence of an increase in DS, y_1 the number of cases of absence of slow fluctuation of SP with an increase in DS, y_2 the number of cases of absence of slow fluctuation of SP in the absence of
an increase in DS, and one represents the number of degrees of
freedom.

RESULTS AND DISCUSSION

In response to indifferent photic stimulation an increase in
DS with the appearance of slow fluctuations of SP took place only
for a few applications. As Table 1A shows, on the one hand slow
changes of SP took place in response to stimulation in both the
presence and absence of an increase in DS. On the other hand, an
increase in DS was not always accompanied by the appearance of
slow fluctuations of SP. Statistical analysis showed that the difference between the number of slow shifts of SP appearing with
and without an increase in DS is not statistically significant.

A response to the isolated action of electrical stimulation of
the skin arose more often than to indifferent photic stimulation.
The results of calculations of various types of response to this
stimulus are given in Table 1B. Analysis showed that no statistically significant difference likewise was found between the number
of slow changes of SP during the action of this stimulus in the
presence or absence of an increase in DS.

---

*B. L. Van der Waerden, Mathematical Statistics, p. 272.

TABLE 1. Quantitative Relationship between magnitude of DS and Slow Changes in SP on the Surface of the Rabbit's Cortex in Response to Isolated Stimulation

	A Total number of applications of indifferent photic stimulus: 168				B Total number of applications of isolated electrodermal stimulation: 8			
Nonpolarizing electrodes	increase in DS in response to stimulation: 6		no increase in DS in response to stimulation: 162		increase in DS in response to stimulation: 5		no increase in DS in response to stimulation: 3	
	change in SP present: 3	change in SP absent: 3	change in SP present: 29	change in SP absent: 133	change in SP present: 3	change in SP absent: 2	change in SP present: 3	change in SP absent: none
	50%	50%	18%	82%	60%	40%	100%	0%
	Total number of applications of indifferent photic stimulus: 190				Total number of applications of isolated electrodermal stimulation: 60			
Polarizing electrodes	increase in DS in response to stimulation: 16		no increase in DS in response to stimulation: 174		increase in DS in response to stimulation: 34		no increase in DS in response to stimulation: 26	
	change in SP present: 3	change in SP absent: 13	change in SP present: 10	change in SP absent: 64	change in SP present: 19	change in SP absent: 15	change in SP present: 17	change in SP absent: 9
	19%	81%	6%	94%	56%	44%	65%	35%

In response to the conditional stimulus, changes in SP appeared most frequently at the time of appearance of the conditioned movements. At these same times the conditional stimulus began to increase the DS. This time, however, the two types of response did not always arise simultaneously. Examples of the possible variations are given in Fig. 2A, C, D, and F. As a rule the amplitude of the slow changes in SP was small, and the leading edge of their increase was sloping. As a result, the changes in DS and the slow changes in SP could be assessed at the same time. The quantitative relationships between the various types of response are given in Table 2A. As a result of statistical analysis of the calculated figures, no statistically significant difference between the number of slow fluctuations of SP in the presence and in the absence of an increase in DS was found.

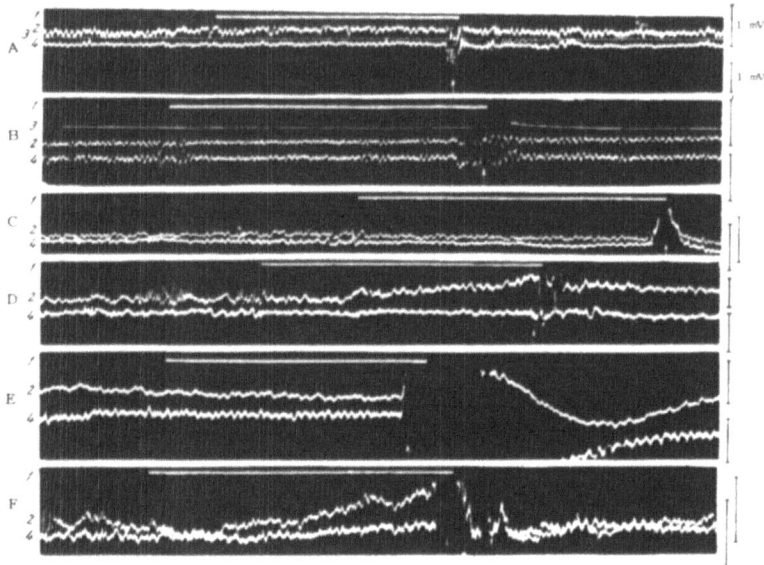

Fig. 2. Changes in DS and SP in response to conditional and unconditional stimuli. 1) Conditional stimulus; 3) movement. Arrow denotes moment of reinforcement. A) Absence of increase in DS and slow changes in SP in response both to conditional stimulus and to reinforcement: 2) EEG of frontal region; 4) EEG of occipital region; B) increase in DS in response to reinforcement with absence of slow changes in SP: 2) occipital EEG; 4) frontal EEG; C) increase in DS in response to conditioned stimulus in absence of slow changes in SP: 2) parietal EEG; 4) frontal EEG; D) absence of increase in DS in response to conditional stimulus with appearance of a slow change in SP: 2) frontal EEG; 4) occipital EEG; E) absence of increase in DS in response to reinforcement with appearance of slow changes in SP: 2) occipital EEG; 4) parietal EEG; F) presence of increase in DS and slow changes in SP in response both to conditional stimulus and to reinforcement: 2) frontal EEG; 4) parietal EEG.

In response to electrodermal reinforcement, after the first combinations the slow changes of SP appeared more regularly than before formation of the conditioned reflex. The DS also increased in the same period in the aftereffect of the combinations. This time, however, an increase in DS and slow fluctuations in SP did not always occur in the same traces. Examples of possible variations are given in Fig. 2A, B, E, and F. As this figure shows, slow changes in SP in this case had a steep leading edge of increase. In most cases, they also were of large amplitude, so that the beam aften went off the edge of the film. The result of all this was that the DS was assessed during the aftereffect of the combinations, coinciding with a decline in the SP.

The quantitative relationships between DS and slow changes in SP in response to reinforcement are shown in Table 2B. Statistical analysis of the results showed no significant difference between the number of slow fluctuations of SP in the presence or absence of an increase in DS.

Besides the quantitative assessment of the relationship between changes in DS and slow changes in SP in response to stimulation obtained from all leads taken together, they also were assessed for several regions of the cortex most concerned with formation of the conditioned reflex: frontal with occipital and parietal with occipital. The results of the calculations for these leads are given in Table 3.

Table 3 shows that the following observations were made during application of the conditional stimulus. For rabbits with nonpolarizing electrodes, with recordings from the frontal and occipital regions, a statistically significant difference was found between the number of slow changes in SP during an increase in DS and their number in the absence of such an increase. With recordings from the parietal and occipital regions, no such significant difference was present. For rabbits with polarizing electrodes, the difference was very significant when the recordings from the parietal and occipital regions were compared. No significant difference was found by assessment of recordings from the frontal and occipital regions. The difference between the results obtained by the use of different electrodes may possibly be connected with some variation in their localization (Fig. 1).

To the unconditional stimulus, the same patterns were observed as when the results of all leads taken together were assessed.

Hence, the following facts emerged from this investigation: 1) no statistically significant difference was found between the number of slow changes of SP appearing both in the presence and in the absence of an increase in DS during application of indifferent photic stimulation, isolated electrodermal stimulation, or electrodermal reinforcement; 2) a statistically significant difference was

TABLE 2. Quantitative Relationships between Magnitude of DS and Slow Changes in SP on the Surface of the Rabbit's Cortex in Response to Conditioned and Unconditioned Stimuli

	A — Total number of applications of conditional stimulus: 204				B — Total number of applications of reinforcement: 204			
Nonpolarizing electrodes	increase in DS in response to stimulation: 34		no increase in DS in response to stimulation: 170		increase in DS in response to stimulation: 124		no increase in DS in response to stimulation: 80	
	change in SP present: 8	change in SP absent: 26	change in SP present: 21	change in SP absent: 149	change in SP present: 82	change in SP absent: 42	change in SP present: 59	change in SP absent: 21
	24%	76%	12%	88%	66%	34%	74%	26%

	Total number of applications of conditional stimulus: 313				Total number of applications of reinforcement: 291			
Polarizing electrodes	increase in DS in response to stimulation: 70		no increase in DS in response to stimulation: 243		increase in DS in response to stimulation: 108		no increase in DS in response to stimulation: 183	
	change in SP present: 45	change in SP absent: 25	change in SP present: 123	change in SP absent: 120	change in SP present: 73	change in SP absent: 35	change in SP present: 137	change in SP absent: 46
	64%	36%	51%	49%	68%	32%	75%	25%

present, however, during the action of the conditional stimulus when certain leads were analyzed.

The absence of a statistically significant difference may be interpreted as indicating absence of a connection between the two observed phenomena, and its presence may be regarded as indicating such a connection. Consequently, it may be concluded from the facts obtained that slow changes in SP and the increase in DS taking place during the action of indifferent and isolated electrodermal stimulation are not interconnected. The slow fluctuations of SP arising in response to reinforcement and the increase in DS occurring during the aftereffect of the combinations likewise were not interconnected. However, the slow changes in SP and the in-

TABLE 3. Quantitative Relationships between Magnitude of DS and Slow Changes in SP in Regions of the Rabbit's Cortex between Which a Connection is Formed in Response to Conditional and Unconditional Stimulation

		A Total number of applications of conditional stimulus: 80				B Total number of applications of reinforcement: 80			
		increase in DS in response to stimulation: 14		no increase in DS in response to stimulation: 66		increase in DS in response to stimulation: 44		no increase in DS in response to stimulation: 36	
Nonpolarizing electrodes	frontal – occipital	change in SP present: 6	change in SP absent: 8	change in SP present: 8	change in SP absent: 58	change in SP present: 31	change in SP absent: 13	change in SP present: 27	change in SP absent: 9
		43%	57%	12%	88%	70%	30%	75%	25%
		Total number of applications of conditional stimulus: 72				Total number of applications of reinforcement: 71			
	parietal – occipital	increase in DS in response to stimulation: 11		no increase in DS in response to stimulation: 61		increase in DS in response to stimulation: 42		no increase in DS in response to stimulation : 29	
		change in SP present: 2	change in SP absent: 9	change in SP present: 7	change in SP absent: 54	change in SP present 27	change in SP absent: 15	change in SP present: 20	change in SP absent: 9
		18%	82%	11%	89%	64%	36%	69%	31%
		Total number of applications of conditional stimulus: 48				Total number of applications of reinforcement: 49			
Polarizing electrodes	frontal – occipital	increase in DS in response to stimulation: 14		no increase in DS in response to stimulation: 34		increase in DS in response to stimulation: 17		no increase in DS in response to stimulation: 32	
		change in SP present: 9	change in SP absent: 5	change in SP present: 16	change in SP absent: 18	change in SP present: 10	change in SP absent: 7	change in SP present: 23	change in SP absent: 9
		64%	36%	47%	53%	60%	40%	71%	29%
		Total number of applications of conditional stimulus: 61				Total number of application of reinforcement: 53			
	parietal – occipital	increase in DS in response to stimulation: 11		no increase in DS in response to stimulation: 50		increase in DS in response to stimulation: 13		no increase in DS in response to stimulation: 40	
		change in SP present: 10	change in SP absent: 1	change in SP present: 19	change in SP absent: 31	change in SP present: 9	change in SP absent: 4	change in SP present: 35	change in SP absent: 5
		90%	10%	38%	62%	68%	32%	87%	13%

crease in DS in response to the conditional stimulus in regions of
the brain concerned in conditioning were interrelated phenomena.
Yet even though these phenomena were interconnected, they did not
always arise at the same time. On the one hand, this could be because each of these phenomena are part of a complex, integral
process which, in some cases, masks the interconnection under investigation. Another possible explanation is that these two phenomena are not mutually dependent upon each other, but are possibly
the result of a third process, associated with formation of the temporary connection. This is the more probable explanation because
the interconnection was apparent in those parts of the cortex most
concerned with the process of formation of the conditioned reflex,
and it came to light at times connected with the appearance of the
temporary connection (in response to the conditional stimulus).

Hence, this investigation did not answer the question of the
role of electrotonic influences in formation of DS. It simply demonstrated the presence of a connection between the increase in DS
and the appearance of slow fluctuations in SP in response to a conditional stimulus. The question whether these phenomena are dependent on each other or whether they reflect some common third
process remains unanswered. However, the results do serve as a
basis for further research in this direction.

SUMMARY

1. No connection was found between the increase in distant
synchronization of potentials (DS) and slow fluctuations in the
steady potential level (SP) arising in response to indifferent photic
and isolated electrodermal stimulation.

2. No connection was found between slow fluctuations in SP
in response to reinforcement and the increase in DS during the aftereffect of combinations, although both these phenomena appeared
at the same period of conditioned-reflex formation.

3. A connection was found between the slow changes in SP
and the increase in DS arising in regions between which a temporary connection is formed in response to the action of the conditional stimulus.

In conclusion the authors record their gratitude to V. N. Bog-danovich for his help with the statistical analysis of the results.

Electrophysiological Investigation of Specific and Nonspecific Interoceptive Synaptic Influences on Spinal Neurons [*]

P. G. Kostyuk, N. N. Preobrazhenskii, and Z. A. Tamarova

Bogomolets Institute of Physiology
Academy of Sciences of the Ukrainian SSR
Kiev, USSR

Recent advances in microelectrode techniques of investigation of individual nerve cells have made it possible not only to obtain accurate information regarding the course of the fundamental physiological processes in different types of central neurons, but also to study the mechanism by which processes in individual cells are integrated into the activity of a neural center and to study coordination between the activities of different centers.

Before such a problem can be solved, knowledge of processes in single neurons is of course essential. However, it is no less important to discover whether the principles governing the course of neural processes which have been discovered are characteristic of the whole population of neurons forming a given system. This is a perfectly real problem, even when a technique such as microelectrode searching, which is to a certain extent random, is used and which by no means requires an unlimited increase in the number of recorded units. Experience shows that during a search in a particular direction, in every case sufficient neurons are recorded to enable a definite conclusion to be drawn regarding the homogeneity of that particular population with respect to the course of their fundamental neural processes or it can be divided into subgroups, each of which can be investigated independently. It is very helpful in such cases to compare the activity of individ-

---

[*] Pages 168-180 in the Russian edition.

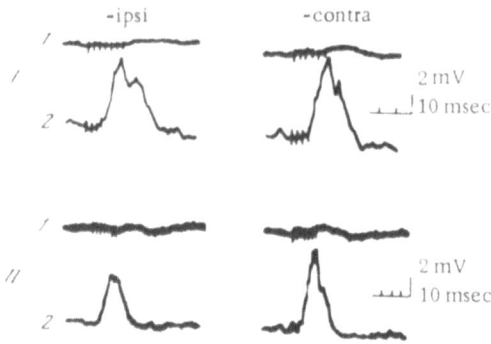

Fig. 1. PSP of flexor (I) and extensor (II) motoneuron during stimulation of ipsilateral and contralateral splanchnic nerves. 1) Potential on dorsal
surface of investigated segment; 2) PSP of motoneuron.

ual elements with certain characteristics of the combined activity
of the whole population.

It is no less important to determine the characteristic properties of the connections between neurons in the system under investigation; modern morphological and electrophysiological methods allow this to be done fairly successfully.

The combination of all these approaches lays the foundations
for understanding the principles governing integration of neurons
in a center.

The organization of various functional systems of the brain
is being investigated in accordance with this principle accepted at
the present time in many neurophysiological laboratories. However, so far they have left practically untouched the question of
neuronal organization of the system of communication between the
visceral and somatic structures of the central nervous system.
The theoretical and practical importance of precise information
regarding the neuronal organization of visceromotor connections
needs no special explanation, and for that reason a detailed investigation of the synaptic organization of interoceptive synaptic influences on spinal neurons has been made.

The analysis begins with simpler conditions: with the spinal cord isolated from suprasegmental influences. A synchronous

afferent volley reaching the spinal cord under these conditions along one of the visceral nerves can be transmitted to segmental interneurons and motoneurons not only of those segments which are close to the point of entry, but also of very distant segments (lumbar, and even sacral). This fact was demonstrated some years ago both myographically (Merkulova, 1952, and others) and electrophysiologically (Downman, 1953; Evans and McPherson, 1958; Duda, 1960). However, to obtain absolutely accurate information concerning the temporal characteristics of this spread and the nature of the processes evoked by it in the spinal neurons, their potentials must be recorded intracellularly. The results of such an investigation are illustrated in Fig. 1. Recordings were made from motoneurons of the 5th–7th lumbar segment of spinal cats, the neurons being identified by antidromic excitation of their corresponding muscle nerves.

The mean latent period of the postsynaptic potentials (PSP) evoked in the investigated group of motoneurons by impulses produced by electrical stimulation of the central end of the divided splanchnic nerve was 17.4 ± 0.6 msec for ipsilateral and 16.2 ± 0.5 msec for contralateral stimulation. In both cases the duration of the PSP was about 50 msec. Characteristically the postsynaptic processes evoked by the interoceptive wave in such animals were uniform in type. In both flexor and extensor motoneurons the afferent wave from the splanchnic nerve generated depolarizing excitatory PSPs. Only in two cases were weak hyperpolarizing potentials found. However, because of the large extracellular fields in the region of the motor nuclei, we could not confidently differentiate them from the effects of the extracellular electric field.

Comparison of the threshold strength of stimulation of the splanchnic nerves required to generate a synaptic response with the action potentials of the isolated nerve and the sympathetic chain showed that this response is associated with excitation of fibers of the $A\gamma\delta$ group. Fast conducting afferents of $A\beta$ type, which play an important role in the transmission of interoceptive information to the brain (Amassian, 1961) do not evoke it.

Hence, in the spinal cord there is a system of relatively fast interoceptive activation of motoneurons of distant segments. Synaptically speaking, this system is non-specific: it has no precise

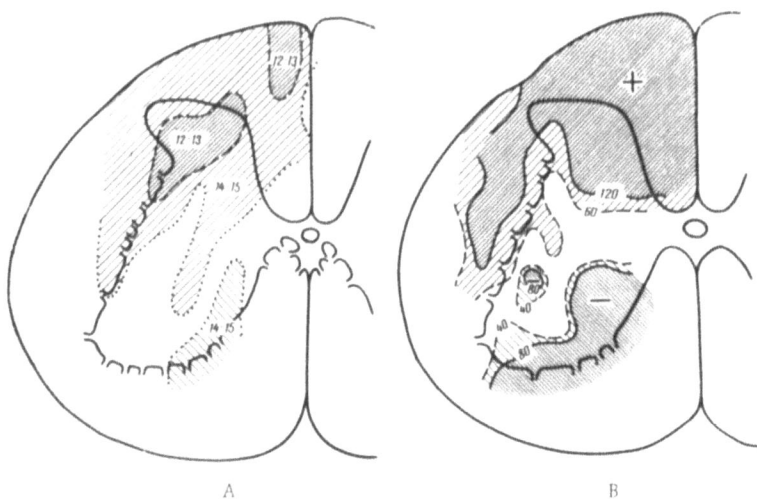

Fig. 2. Order of activation of various neuronal structures of the fifth lumbar segment of the spinal cord (A) during visceromotor reflexes in spinal cats (mean values of latent periods in msec from results of several experiments) and schemes of distribution of amplitude values of electrical responses (in µV) over transverse section of fifth lumbar segment of spinal cord 22 msec after stimulation of ipsilateral splanchnic nerve in a spinal cat (B).

organization on the reciprocal principle and it may lead to excitation of both extensor and flexor lower limb motoneurons.

To determine the neuronal organization of this system, it is first necessary to discover the pathways of intersegmental transmission of interoceptive activity. On the basis of existing data for pathways concerned with its upward elaboration (Amassian, 1951; Downman and Evans, 1957; Durinyan, 1961), it can be postulated that these intersegmental influences are also transmitted by collaterals of fibers belonging to both the dorsal and the ventrolateral tracts, forming synaptic endings on the interneurons and motoneurons of the corresponding segments. During focal recording of potentials of a very large number of points of the spinal cord with low-resistance microelectrodes, it is accurate enough to determine the localization of the activated structures in the spinal cord from the intensity of the electric field they create.

Analysis of a large number of records shows that the latent periods of electrical responses recorded in different parts of the spinal cord vary; the distribution of latencies over the transverse section of the cord is illustrated in Fig. 2A. The earliest electrical responses arise in the medial part of the dorsal column (12-13 msec). After a very slight delay, electrical activity also develops in the deep part of the lateral column, adjacent to the dorsal horn, and also in the dorsal horn itself. After 14-15 msec, it spreads to the region of the intermediate column of gray matter of the spinal cord. Meanwhile electrical responses appear in the medial part of the ventral horn, travelling toward the motor nuclei. After 15-17 msec a local potential is recorded in the region of the motoneuron, in full agreement with data for latent periods of development of their postsynaptic potentials obtained by intracellular microelectrodes (see above).

On the basis of these records charts were made of distribution of the amplitudes of focal potentials (isopotential maps) for the cross-section of the spinal cord after different time intervals. An example of such a chart for an interval of 22 msec is shown in Fig. 2B. At this period the region of the motoneurons is an area of electronegativity (i.e., an area of excitatory postsynaptic activity).

These results suggest that the intersegmental conduction of interoceptive activity takes place both along the fibers of the dorsal column (i.e., along direct collaterals of the afferent fibers), and along the propriospinal fibers of the lateral and ventral columns (i.e., via intersegmental interneurons) with an additional delay. However, even in the case of direct transmission, activity spreads to the motoneurons only via the segmental interneurons (and also, perhaps, via chains of several interneurons, because the difference between the latent periods of the responses in the dorsal column and in the motor nuclei is 3-5 msec).

Microelectrode recordings of activity of single interneurons in different parts of the gray matter confirm this conclusion very well. The latent periods of responses of these cells indicate that the interoceptive wave can activate an interneuron excited by somatic afferents with a shorter latent period than that of activation of motoneurons; at the same time, a much longer latency of interoceptive activation of interneurons is also possible.

Hence, the system of activation of somatic elements by interoceptive impulses is fairly complex. Its complexity is increased by the fact that the action of the interoceptive volley is not limited to generation of postsynaptic processes in the corresponding interneurons and motoneurons. This volley causes considerable and prolonged depolarization of the central endings of somatic afferent fibers, which is associated with presynaptic inhibition of the afferent volley. The mean latent period of depolarization is 20 msec; it reaches a maximum 13.6 msec after its beginning and the mean time constant of its decline is 53 msec. Direct verification shows that in the period of development of this depolarization an essential weakening of the intensity of the transsynaptic action of the somatic afferent endings in fact takes place.

Special investigations of the functional importance of presynaptic inhibition show that it may be a mechanism of effective suppression of the central action of weak ("background") afferent influences without, at the same time, preventing the entry of intense (synchronous) afferent volleys into the spinal cord. Evidently this partial "liberation" of the central nervous system from all manner of unimportant information, along with the diffuse increase in excitability of the somatic neurons, is the essence of the functional role of this system of spinal viscero-somatic connections.

The same mechanism of presynaptic inhibition can evidently cause weakening, and even complete blocking, of interoceptive influences themselves in cases when repeated intense (synchronous) interoceptive volleys arrive. The synaptic effects produced in such cases in motoneurons rapidly diminish with each successive volley. This weakening is due, not to the development of inhibitory postsynaptic processes in the segmental interneurons or motoneurons, but to the cessation of arrival of intersegmental influences.

There are strong grounds for considering that presynaptic depolarization is the result of the action of special "depolarizing" interneurons (Eccles et al., 1962), most probably neurons of the substantia gelatinosa of the dorsal horn. In fact, as mentioned above, on the arrival of the interoceptive volley, very intense activation of the dorsal horn neurons takes place, including those in the region of the substantia gelatinosa Rolando.

This is how events develop when the spinal mechanisms are isolated from suprasegmental influences. It may be assumed that the results obtained are not significantly altered by such side influences as the shock after transection of the spinal cord. For

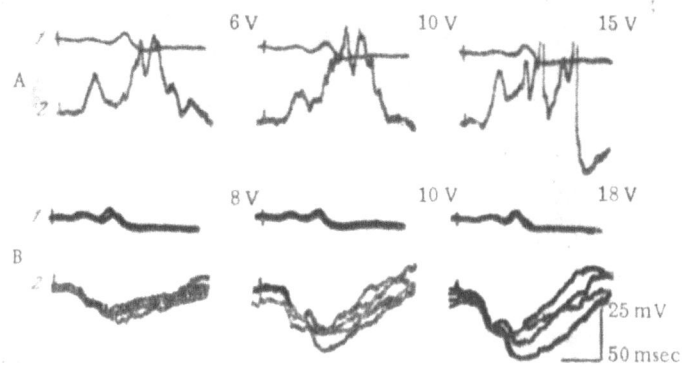

Fig. 3. Postsynaptic potentials of flexor (A) and extensor (B) motoneurons during stimulation of splanchnic nerve in cats with intact
central nervous system. 1) Potential of dorsal surface of spinal cord;
2) synaptic potential of motoneuron. B records obtained by superposition of several sweeps, synchronized with stimulation.

example, injection of strychnine, significantly increasing reflex
excitability, into such animals caused no appreciable changes in
PSPs generated in the lumbar neurons by an interoceptive volley.
The mean latent period of the PSPs was practically identical with
that in animals not receiving strychnine. Strychninization merely
increased the amplitude of the PSPs and facilitated the appearance
of intense PSPs or discharges of motoneurons in response to fewer stimuli.

If a similar analysis is carried out for animals with intact
connections between the spinal cord and brain, the picture is very
different.

The first feature to draw attention is that the postsynaptic
potentials in the spinal motoneurons in response to an afferent
volley from the splanchnic nerve in this case appear after a much
longer latent period than in spinal animals. In the flexor motoneurons the earliest EPSP appeared after a latent period of 25-
30 msec, but this postsynaptic activity was of very low amplitude
and rarely generated a propagated spike. Not until after 50-60
msec (when under spinal conditions synaptic activity had ceased
altogether), did an intense wave of postsynaptic activity appear,
of considerable duration (up to 100 msec or more). On reaching
the threshold, the synaptic depolarization at this period frequently

generated one, or sometimes two or three spikes (Fig. 3A). The appearance of an action potential was followed by weakening of synaptic depolarization, but not usually by its complete disappearance, and soon after it began to increase once more. It is important to note that the changes produced in the same motoneuron by stimulation of the ipsilateral and contralateral splanchnic nerve were identical in sign. In some cases the latent period of the PSP evoked by impulses from the contralateral nerve was a few milliseconds shorter than the latent period of the ipsilateral changes, while their amplitude was somewhat larger; however, no statistically significant difference between these changes could be found.

The second important difference is seen if changes in the extensor motoneurons are investigated. The PSP in these cells consisted of the same phases as in the flexor motoneurons, but the type of polarization change in this case differed significantly from that seen before. The weak first component of the PSP, with a somewhat longer latent period than the analogous component in the flexor motoneurons, could be either of depolarizing or of hyperpolarizing type. However, the principal, second component of the PSP, unlike the first, was always of the hyperpolarizing, inhibitory type. The IPSPs in this case were of very considerable amplitude and duration. Corresponding examples are given in Fig. 3B.

The results show conclusively that the intraspinal system of interneurons responsible for the rapid nonspecific transmission of interoceptive influences to motoneurons is under the tonic inhibitory control of certain brain structures, and when the nervous system is intact, its activity is considerably inhibited or even completely blocked. Data indicating the possibility of this tonic inhibitory suprasegmental control of the transmission of influences from the splanchnic nerve to intercostal motoneurons has been obtained (Downman and Hussain, 1958). This is evidently a case of closely similar, or even identical inhibitory systems.

Answers must now be found to two questions: 1) is inhibition associated with difficulty in transmission from interoceptive fibers to subsequent neurons at the point of entry of the former into the spinal cord (i.e., in the thoracic segments), or does the spread of afferent impulses along intraspinal pathways take place without hinderance and only its transmission to the segmental interneurons or motoneurons is inhibited? 2) What suprasegmen-

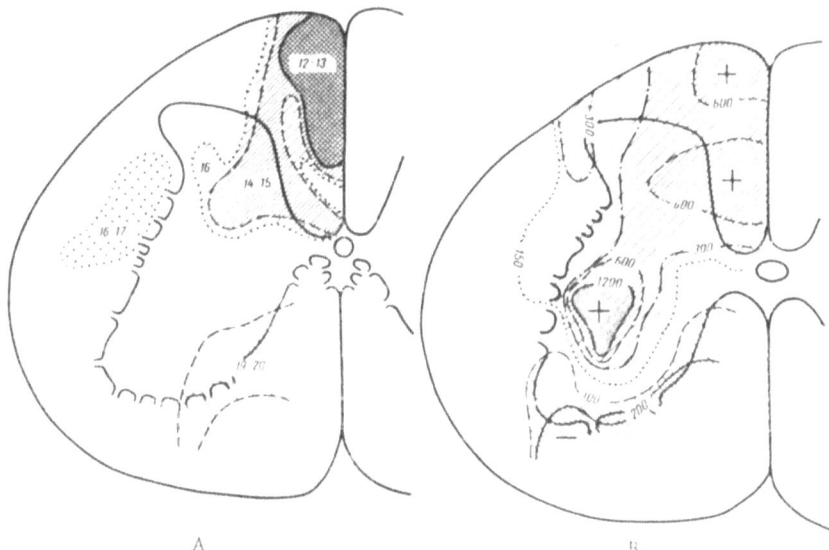

Fig. 4. Latent periods of electrical responses arising in different parts of the spinal cord during stimulation of the ipsilateral splanchnic nerve in cats with an intact nervous system (A) and distribution of amplitudes of focal potentials 35 msec after stimulation of the ipsilateral splanchnic nerve in a cat with an intact central nervous system (B). Legend as in Fig. 2.

tal structures are responsible for this inhibtion and how is their tonic acitvity effected?

To answer the first question, in experiments on animals with an intact nervous system we made a detailed investigation of the distribution of electric fields throughout the transverse section of one segment of the spinal cord just as in the investigation on spinal animals described above. This investigation clearly showed that when connections between the brain and spinal cord are intact, the pattern of the temporal sequence and character of the focal potentials evoked by the interoceptive volley changes sharply.

The earliest focal potentials in both intact and spinal animals are found in the region of the dorsal columns (12-13 msec). However, signs of activity in the region of the lateral columns do not appear until after 16-17 msec, and in the medial part of the ventral horns and the adjacent white matter, until after 19-20 msec

(Fig. 4A). Obvious delay in arrival of impulses along the intersegmental pathways is thus found. Only in the dorsal horn, containing direct descending collaterals of the afferent fibers, which
cannot be inhibited, is the spread of interoceptive impulse activity
unchanged. Consequently, inhibition of transmission of these impulses to descending internuncial neurons is firmly established.
It is also clear that interneuronal transmission is also inhibited
in the spinal segments investigated. Although the spread of activity remains unchanged in the dorsal columns and considerable
activity is observed after 14–15 msec in the region of the interneurons also, no activity arises under these circumstances in the
motoneurons.

Intense activity in the motor nuclei is clearly demonstrable
only after 30 msec (Fig. 4B). At this period two characteristic
foci develop, the negative focus coinciding reasonably accurately
with the histological localization of the flexor motoneurons and the
positive with the localization of the extensor motoneurons. These
powerful foci take a long time to develop, and after 50 msec they
are still not very intense, in complete agreement with the prolonged character of reciprocal postsynaptic processes in motoneurons observed in response to visceral impulses when intracellular
recording is used. In the region of the dorsal horn and dorsal columns, however, activity is now much weaker.

Hence, a definite answer can be given to the first question:
the descending tonic control depresses the function of the internuncial neurons at all levels, thereby preventing the onset of postsynaptic changes in the motoneurons. This inhibition is nonspecific, extending to the transmission of influences to both flexor and
extensor neurons.

The answer to the second question is much more complicated and laborious.

Comparison of the course of "interoceptive" postsynaptic
processes in motoneurons in animals with an intact central nervous system and in decerebrate animals shows that in the latter
the tonic inhibitory control is completely preserved. Accordingly,
the most probable hypothesis for subsequent experimental verification is that this control is effected by the reticular formation of
the medulla. To verify this hypothesis it must be established what

changes are produced in visceromotor transmission by activation of reticular structures.

Magoun and Rhines (1946) found that stimulation of the medial reticular formation of the medulla causes diffuse inhibition of various spinal reflexes, both extensor and flexor. However, some have cast doubt on this conclusion. According to these workers, during stimulation of the medullary reticular formation, very varied descending effects, often unpredictable, may be obtained: both diffuse and reciprocal, or mixed (Gernandt and Thulin, 1955; Sprague and Chambers, 1954; Brooks et al., 1956). None of these investigators made a precise analysis of the latent periods and temporal characteristics of the various descending effects. We therefore carried out a special series of measurements of processes in spinal neurons during direct stimulation of certain points of the cross section of the medulla.

If threshold stimuli of short duration are applied, diffuse inhibitory influences on spinal neurons can be separated from reciprocal influences, and it is possible, moreover, to determine those parts of the brain stem whose stimulation produces either effect. With an increase in the strength of the stimuli, naturally both effects will be produced by electrodes in either situation (because of branching of the current in the brain tissue), but even in this case their relative intensity will obviously depend on the region of stimulation. All measurements were made during stimulation by a bipolar electrode with a very short interelectrode distance and insulated as far as the tips. According to our observations it is this type of stimulation which gives most precisely localized results.

An example of an experiment illustrating these remarks is given in Fig. 5. Some characteristic points of stimulation are shown on the diagram. For each point curves of measurements of synaptic excitability were obtained for spinal flexor and extensor motoneurons developing after stimulation of that point by a series of 15 pulses with a frequency of 500/sec (the curves were plotted on the basis of assessment of changes in the corresponding monosynaptic reflexes). In cases when the region of the gigantocellular nucleus in the medial reticular formation was stimulated selectively, motoneurons of both flexor and extensor muscles were subjected to identical and purely inhibitory influences.

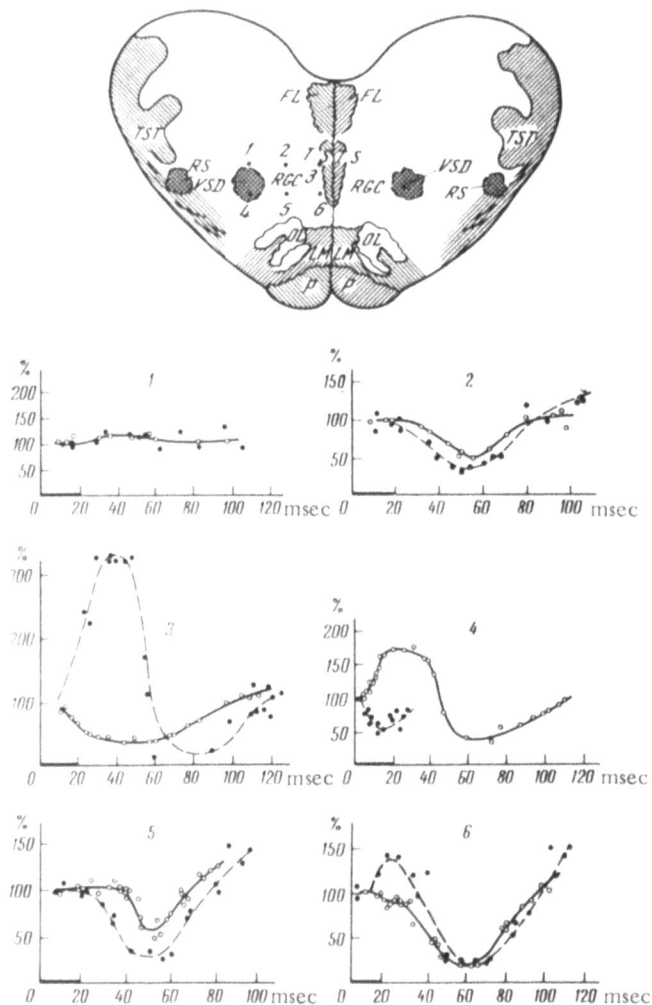

Fig. 5. Changes in monosynaptic responses of extensor (continuous line) and flexor (broken line) motoneurons of the spinal cord in percentages of their initial value (ordinate) at various time intervals (abscissa) after stimulation of six points in the medulla. Location of the points is shown on a diagram of a section through the brain stem. Stimulation of brainstem with 15 pulses at a frequency of 500/sec. Decerebrate and decerebellate cat.

The course of this inhibition was constant in all the experiments. It appeared after a long latent period (about 20 msec), reached a maximum after approximately 30 msec, and its total duration was 70-80 msec. The effectiveness of inhibition largely depended on the character of stimulation of the gigantocellular nucleus; it was deepened considerably by an increase in the number of stimuli in the series, although the total duration remained roughly the same.

Simultaneous intracellular recordings of the potentials of the flexor and extensor motoneurons showed that during this inhibition a postsynaptic hyperpolarization developed in them, its course corresponding precisely to that of inhibition of the monosynaptic discharge. This hyperpolarization is converted into depolarization during diffusion of chloride anions inside the cell, and in its ionic mechanisms it is therefore indistinguishable from the ordinary IPSP of motoneurons which have now been adequately studied. Our results in this respect agree with those recently published by Llinas and Terzuolo (1964, 1965). Hence, the mechanism of the synaptic action of the principal structure in the medullary reticular formation (the gigantocellular nucleus) is unquestionably a nonspecific, diffuse, postsynaptic inhibition of the motoneurons.

However, the action of the reticular formation on spinal neurons is not limited to these postsynaptic effects on the motoneurons. During stimulation of its ventromedial portions in the region of the medulla and comparison of changes in monosynaptic and polysynaptic responses of the spinal motoneurons arising after stimulation, the characteristic features of these changes are revealed (Fig. 6). Synaptic potentials and reflex discharges evoked in the same motoneuron by impulses arriving along both monosynaptic and polysynaptic pathways are inhibited. However, inhibition of the polysynaptic responses is always much longer in duration, although sometimes less deep. This fact undoubtedly suggests that, besides its influences on motoneurons, the bulbar reticular formation has a special inhibitory effect directly on spinal interneurons. The fact that the temporal course of this inhibition differs from that of the "reticular" IPSP in the motoneurons indicates its association with a specific neuronal mechanism. This is also suggested by the fact that this inhibition is not diminished at all by strychnine; on the contrary, after injection of strychnine it is considerably strengthened. Evidently the inhibition of

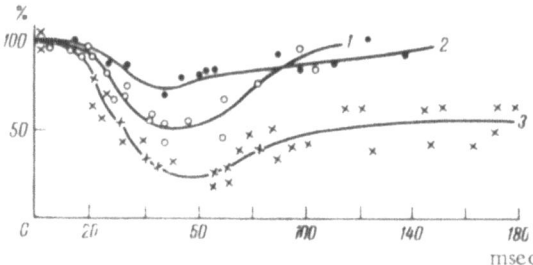

Fig. 6. Changes in monosynaptic responses of motoneurons evoked by stimulation of the gastrocnemius nerve
(1) and in polysynaptic responses evoked by stimulation
of the sural nerve (2) and greater splanchnic nerve (3)
at various intervals after stimulation of reticular formation of medulla at a point at a depth of 3.5 mm from
floor of fourth ventricle. Reticular formation stimulated
with five pulses at a frequency of 500/sec. Decerebrate
and decerebellate cat, injection of 1.2 mg/kg strychnine. Legend as in Fig. 5

the spinal internuncial neurons just described corresponds to the
tonic inhibition of polysynaptic reflex arcs whose existence was
postulated by Lundberg and co-workers (Eccles and Lundberg,
1959; Holmqvist and Lundberg, 1959; Holmqvist et al., 1960) on
the basis of experiments studying changes in polysynaptic reflexes
after transection of the brain stem at various levels.

It is especially interesting that it is the viscero —somatic
transmission which undergoes prolonged and deep inhibition. In
some cases during stimulation producing barely perceptible inhibition of polysynaptic transmission from somatic afferents (for
example, from the afferents of a flexor reflex), almost complete
suppression of viscero-somatic inhibition develops and lasts for
200–400 msec or longer after the end of stimulation (a corresponding example is shown in Fig. 6). Injection of strychnine considerably strengthens the inhibitory action of the reticular formation on viscero-somatic transmission. The inhibitory action is
manifested regardless of the motoneurons to which the interoceptive volley spreads — flexor or extensor. Its synaptic organization
is thus nonspecific just like the synaptic organization of the reticular inhibitory action on motoneurons.

Consequently, the reticular formation of the medulla can be considered as a highly effective antagonist of the spinal system of visceromotor connections, under normal conditions preventing substantial activity of the latter mainly through inhibitory influences on certain special interneurons of these connections and also through nonspecific influences on segmental interneurons and motoneurons. It is difficult as yet to make any definite pronouncement regarding the mechanism maintaining the constant tonic inhibitory activity of the reticular formation. However, it must be mentioned that according to our findings, removal of the cerebellum appreciably shortens the latent period of synaptic processes evoked in motoneurons by an interoceptive wave in the experimental animals, and the earlier components of these processes, which are usually suppressed, now appear. It can be assumed that the tonic inhibitory action of the bulbar reticular formation on viscero – somatic transmission is largely maintained by cerebello –reticular influences. In this connection it should be recalled that electrophysiological investigations in several laboratories (Dell and Olson, 1951; Kullanda, 1959; Bratus', 1960) demonstrated the ability of the interoceptive volley to activate cerebellar structures extremely effectively. Such activation may possibly lead to regulation of the tonic inhibitory control of the reticular formations, abolishing it, for example, should a particularly intense flow of afferent impulses arrive from the internal organs.

As was pointed out above, inhibition of spinal mechanisms of transmission of viscero –somatic influences is only one manifestation of the action of suprasegmental structures in the system of communication between the visceral and somatic elements of the central nervous system. The interoceptive volley, if connections are intact between the spinal cord and hind brain, evokes very intense processes with a complex synaptic organization in the motoneurons, taking place with the participation of brain structures. A result of the activity of these structures is intense excitation of flexor motoneurons, accompanied by inhibition of extensors.

In order to decide which structures of the medulla may have this type of action on motoneurons, the experiments described above may be considered in which the effects produced in motoneurons from different points of the brain stem were investigated systematically. Stimulation of dorsal portions of the reticular formation leads to slight nonspecific facilitation of synaptic processes in motoneurons. This facilitation is strengthened by stimulation of more rostral parts of the brain, and it undoubtedly corresponds to the effect of excitation by the nonspecific descending facilitatory system discovered by Rhines and Magoun (1949). The nonspecific character of the effect of these structures on motoneurons possessing different functions rules out the possibility of their participation in the influences being examined here.

Intense reciprocal influences on spinal motoneurons, including excitation of flexor and inhibition of extensor cells, arise constantly during stimulation of part of the brain stem close to the midline at a depth of about 3 mm from the floor of the fourth ventricle (Fig. 5). The latent period of the PSPs in the motoneurons during stimulation of these areas averages 10 msec, and reciprocal PSPs appear in the motoneurons after summation of a very small number of stimuli.

Hence, from the region of median structures of the medial reticular formation of the medulla, a specific descending pathway with a fairly rapid rate of conduction passes to the spinal motoneurons, its function differing sharply from that of the gigantocellular nucleus. Excitation of this pathway causes a significant increase in excitability of flexor motoneurons and, possibly, to direct "trigger" effects on these motoneurons.

No distinct reciprocal synaptic changes in spinal motoneurons, including excitation of flexor cells, can be obtained by stimulation of any other parts of the cross section of the medulla. During stimulation of lateral areas next to the region of the direct vestibulospinal tract, intense reciprocal synaptic processes, although opposite in direction (excitation of extensor motoneurons and associated inhibition of the flexor) develop in the motoneurons. These responses are clearly differentiated by their very short latent period (6 msec) and are consequently associated with excitation of a descending system with a very high speed of conduction. This system can only be the vestibulospinal tract.

It may accordingly be concluded that the most probable structure carrying specifically organized descending influences evoked by an interoceptive volley to spinal neurons is the system localized in the median region of the medial reticular formation of the medulla. The latent period of reciprocal synaptic processes evoked by an interoceptive volley through the participation of the brain stem is in full agreement with this hypothesis.

The neuronal organization of interaction between the visceral and somatic structures of the brain is thus quite complex even in the spinal cord and brain stem. This interaction takes place through neuronal systems with a relatively nonspecific, diffuse synaptic organization and also through more precisely directed systems with a high level of synaptic organization. Evidently the participation of these different systems, which under our experimental conditions are far removed from those of natural activity, are often activated simultaneously, may presumably be differentiated when visceromotor interaction takes place under conditions of natural stimulation. However, special and complex analysis of these matters is required. It is difficult at present to say to what extent these neuronal mechanisms correspond to the division of the final visceromotor effects into "trigger" and "correcting," as established by the work of Merkulova (1952), Bulygina (1959), and others. Whatever the true state of affairs, however, the differences in effectiveness of synaptic activation brought about through the propriospinal and bulbar systems are clear enough, and these differences are undoubtedly reflected in the final motor effect produced by activity of the corresponding systems.

The problem of viscero−somatic integration in the intact organism of course includes much more than the mechanisms analyzed here. The higher levels of the brain, especially the thalamic and hypothalamic regions and the cerebral cortex, play an even more significant role in this integration. Interoceptive information arriving along fast-conducting ascending systems composed of fibers of the $A\beta$ group may be of considerable importance in this respect. However, the problem of viscero−somatic integration at higher levels of the central nervous system is exceptionally complex, and its neuronal mechanisms still await elucidation.

Direct and Indirect Cortical Influences on Thalamic Nuclei *

S. P. Narikashvili

Institute of Physiology
Academy of Sciences of the Georgian SSR
Tbilisi, USSR

M. N. Livanov conducted many investigations studying the principles governing cortical electrical activity. In the course of his work he probably became more and more convinced that cortical electrical activity is closely connected with the state of various subcortical structures, whose activity cannot take place without intervention and correction by cortical elements. Many facts have now been accumulated which demonstrate the role of subcortical structures in the genesis of cortical electrical rhythms, but comparatively little is known of the mechanisms by which the cortex exerts its influence on subcortical structures. The main reason for this is the great complexity of the phenomena, so that its study requires more precise methods of investigation and great ingenuity on the part of the experimenter.

Beginning with the work of Cajal (1909) and Mettler (1935), many investigators have demonstrated anatomical connections between the cortex and subcortex (see reviews by: Rossi and Brodal, 1956; Brodal, 1957; French, 1958; Shkol'nik-Yarros, 1958; Narikashvili, 1961, 1966). Corresponding to these anatomical findings, physiological experiments revealed the influence of certain parts of the cortex on phenomena taking place in the subcortex (McCulloch et al., 1946; Niemer and Jimenez-Castellanos, 1950; Bremer and Terzuolo, 1952, 1953a, 1954; Hugelin et al., 1953; French et al., 1955; Segundo et al., 1955; Adey et al., 1957; Hugelin and Bonvallet, 1957; Jouvet and Michel, 1958; Serkov et al., 1960; Ogden, 1960; Wieden and Ajmone-Marsan, 1960; Iwama

---

* Pages 181-187 in the Russian edition.

and Yamamoto, 1961; Gustson, 1953; Meshcherskii and Okujawa, 1963; Meshcherskii and Gustson, 1964; Meshcherskii, 1965; Rabin, 1964a, 1964b, 1965, and others).

Since the cerebral cortex is connected with many subcortical structures which, in turn, are in two-way communication with each other, the possibility of a corticofugal influence on the thalamic nuclei, both directly and indirectly through the brain stem structures, for example, through the reticular formation, cannot be ruled out.

It is interesting to examine the interaction between these two types of influences on the thalamic nuclei and to establish, in particular, the degree to which each pathway is used for cortical regulation of the activity of the thalamic nonspecific nuclei. I give below certain facts and conclusions obtained from experiments carried out jointly with S. M. Butkhuzi, D. V. Kadzhaya, and É. S. Moniava, which demonstrate the character of this cortical influence on the thalamic nuclei and the ways in which this influence is brought about.

Experiments were performed on unanesthetized cats, immobilized by intravenous injection of tubocurarine, and on encéphale isolé preparations. The operation (exposure of both hemispheres and in some experiments extirpation of the cerebellum, etc.) was performed under ether anesthesia. The thalamic responses chosen to reflect by their changes the character of the influence of corticofugal impulses were the recruiting response (for the nonspecific nuclei), the augmenting response, and responses to peripheral stimulation (for the specific nuclei). The recruiting and augmenting responses were evoked by slow (5-10/sec) repeated square pulses (0.5-1 msec) acting through bipolar electrodes introduced stereotaxically into the thalamic central medial and ventral posterolateral nuclei. Bipolar and monopolar leads were used to record potentials from the surface of the cortex and from the subcortical structures. Of the various parts of the cortex, the sensorimotor area was stimulated most commonly (frequency from 2 to 100/sec or higher, 10-20 V, for 10-20 sec). In some cases the cortex was stimulated against a background of existing thalamic responses, while in others thalamic responses were evoked before and immediately after the end of tetanic stimulation of the cortex.

Effect of Cortex on the Thlamic

Nonspecific Response

In every case with stimulation of the sensorimotor cortex whether against a backgroud of a continuous recruiting response or when the response was evoked separately before and immediately after cortical stimulation) complete inhibition of the recruiting response is observed in the parts of the cortex and subcortical structures from which recordings are being made (Narikashvili et al., 1960a). This state persists for a long time (several tens of seconds) after stimulation of the cortex ends, after which the recruiting response is gradually restored to its initial form. The important feature is that under the influence of cortical stimulation the recruiting response changes its character considerably, taking place without a periodic increase of potentials, i.e., apparently persisting all the time at the level of the "waning" phase of the potentials.

Under optimal experimental conditions, this effect of stimulation of the sensorimotor cortex is expressed by total inhibition of the recruiting response. Under less favorable conditions (deterioration of the state of the cortex or the preparation as a whole), however, and also if other areas of the cortex having a weak action are stimulated, the effect may be manifested as inhibition in other ways (the later onset of the first phase of "waxing" or a marked decrease in amplitude of the potentials in this phase).

In other cases, during cortical stimulation only some decrease in the duration of the "waxing" phase of the potentials and an increase in the interval between them are observed in response to cortical stimulation. This change in the recruiting response expresses weaker cortical action. This is clear from the fact that such a response is usually observed during stimulation of the middle and posterior parts of the suprasylvian and marginal gyri, and also of the ectosylvian gyrus, i.e., cortical regions with a weaker influence.

Hence, whatever part of the cortex is stimulated, and whatever the manifestation of this effect (belonging to the changes described above), in every case inhibition of the recruiting response is obtained, although expressed to a varied degree.

The inhibition is generalized, i.e., it is observed simultaneously in different parts of the cortex and subcortex, and also in the opposite hemisphere. It evidently is not due to changes at the cortical level, but to the action of corticofugal impulses on activity of subcortical structures.

It may be assumed that inhibition of the recruiting response arises as a result of direct cortical action on those thalamic nonspecific nuclei whose stimulation evokes the recruiting response. However, the possibility also is not ruled out that changes in the recruiting response may take place through activation of the reticular formation, because the possibility of corticofugal excitation of the reticular formation, on the one hand, and the inhibitory action of direct stimulation of the reticular formation on the recruiting response are both well known.

To verify these hypotheses, experiments were carried out on some of the same animals to test the effect of stimulation of the reticular formation on the recruiting response and the effect of division of the brain stem on the influence of cortical stimulation. By stimulation of the reticular formation, approximately the same picture of inhibition of the recruiting response was observed as during cortical stimulation. After division of the brain stem, stimulation of the sensorimotor cortex no longer had this effect. Hence it may be concluded that the cortical influence is mediated entirely through the reticular formation. However, other experiments, and also a more detailed study of the problem, showed conclusively that the cortical influence on the recruiting response cannot be mediated entirely through activation of the reticular formation. It is clear from many experiments that after stimulation of the cortex, although changes of such magnitude in the recruiting response do not take place as when the brain is intact, nevertheless certain changes are found, and in particular, the interval between the phases of waxing and waning is increased, the duration of these phases themselves is shortened, and the amplitude of their potentials is reduced, indicating some residual, weak inhibitory action of stimulation of the sensorimotor cortex. This effect may undoubtedly be mediated through direct corticothalamic connections.

Cortical Influences on the

Thalamic Specific Response

Under the same experimental conditions the influence of the cortex was studied on the augmenting response evoked by slow, repeated stimulation of the ventral posterolateral nucleus (Narikashvili, Moniava, and Butkhuzi, 1961). As in the case of the recruiting response, the potentials of the augmenting response are diminished by stimulation of the sensorimotor cortex. At the same time, compared with the recruiting response, it is much more difficult to produce changes in the augmenting response. This is shown by the fact that inhibition is observed much less frequently, it is weaker in degree, it lasts for a shorter time, and it requires the use of more intensive stimulation of the cortex. If the cortex is stimulated in the same preparation after division of the brain stem at the collicular level, i.e., after exclusion of the reticular formation, the same effect of cortical stimulation is obtained as before division. In other words, the reticular formation of the brain stem apparently has no significant role in this inhibition of the augmenting response. Since an augmenting response was recorded in those parts of the cerebral cortex that were subjected to direct electrical stimulation, it is clearly difficult in this case to exclude completely the significance of changes in the sensorimotor cortical neurons themselves (under the influence of stimulation). On the other hand, significant changes in activity of the somatosensory relay nucleus could not be expected in response to stimulation of other sensory areas of the cortex.

To make a more detailed investigation of the problem, experiments were also carried out to study changes in the response of the lateral geniculate body (Kadzhaya and Narikashvili, 1962, 1965; Narikashvili and Kadzhaya, 1963, 1965). If the visual cortex or other areas (particularly the sensorimotor) are stimulated against the background of slow repetitive photic stimulation, and the corresponding responses of the lateral geniculate body and visual cortex are recorded, different effects are observed after the end of stimulation depending on its strength. With weak stimulation, a generalized inhibition of cortical and thalamic responses is observed very similar to that obtained by stimulation of the re-

ticular formation (Narikashvili et al., 1960b). This effect is due
to corticofugal excitation of the reticular formation. This is
clear from the fact that after division of the brain stem at the collicular level, this effect is no longer observed. Hence, just as in
the experiments described above to study the effect of cortical
stimulation on the recruiting and augmenting responses, during
weak stimulation the direct action of the cortex on the thalamic
nuclei is seen first. Corticofugal pathways leading to the reticular formation are evidently the most easily excited, and any
changes in the thalamic nuclei, (and also in other subcortical
structures) following stimulation of the cortex develop most easily
by activation of the reticular formation.

With more intense stimulation (both in strength and in duration), bringing the cortex into a preconvulsive state, a general increase in amplitude of the thalamic and cortical responses may
be observed. It is too early at this stage to be sure what changes
bring about this general augmentation of responses, or the way in
which they do so, but it is quite clear that in this case there is a
general increase in excitability of subcortical and cortical elements.

Finally, in response to cortical stimulation of even greater
intensity, paroxysmal activity arises, followed by prolonged depression. In the period of cortical depression the responses of
the lateral geniculate body increase considerably in amplitude. As
the depression gradually passes off, the amplitude of the thalamic
responses diminishes and returns to its initial level.

A general increase in amplitude of the cortical and thalamic
responses and an increase in the responses of the lateral geniculate body (during cortical depression) are also observed after division of the brain stem at the collicular level. In other words,
these phenomena do not involve the participation of the reticular
formation and they fall completely into the category of corticothalamic relationships.

I regard this increase in amplitude of the responses of the
lateral geniculate body during postconvulsive depression of cortical activity as the result of removal (during depression) of the
tonic inhibitory action of the cortex on the thalamic neurons, just
as is observed in other cases (during depression of cortical activity through cooling, after thermocauterization of the pial vessels,

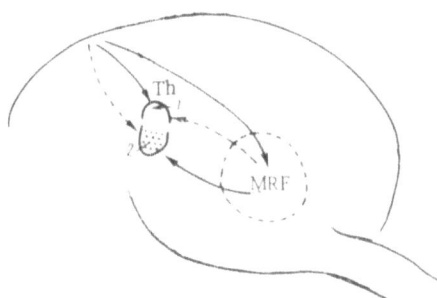

Fig. 1. Scheme representing direct and indirect cortical influences on thalamic specific
and nonspecific nuclei. Th—thalamus;
MRF—mesencephalic reticular formation.
1) Specific thalamic nuclei; 2) nonspecific.

or severe disturbance of the cortical circulation after rapid intravenous injection of a large dose of succinylcholine chloride).

On the basis of the facts described above it may be concluded that in every case of cortical stimulation the main flow of corticofugal impulses is directed primarily toward the reticular formation. This flow is evidently intended not only, or perhaps not
so much for regulating the activity of the neurons of the reticular
formation itself, as for regulating the activity through the reticular formation of other subcortical structures of the brain, and
also spinal activity. In this sense, the reticular formation may be
considered as a powerful intermediate mechanism of cortical regulation widely and rapidly modifying the activity of many subcortical structures in accordance with the character of the corticofugal volley of impulses.

As our experiments showed, it is mainly the thalamic nonspecific nuclei which are regulated by this mechanism, whereas
the activity of the specific relay nuclei is regulated mainly by direct thalamocortical pathways. The cortical regulation of the thalamic nuclei, on the basis of our experimental findings, may be
represented schematically as in Fig. 1. The continuous line shows
the path of the more powerful and predominant effect of cortical
impulses of the thalamus, while the broken line represents the
weaker influence. Judging by this scheme, the thalamic nonspecific nuclei are controlled principally by the cortex through activa-

tion of the reticular formation, while the specific relay nuclei are controlled mainly through direct corticothalamic pathways.

Some impression of thalamo—cortical relationships as a whole may be obtained. The nonspecific nuclei, influencing the cortex by a roundabout way (having no direct connections with it), are controlled by the cortex mainly by a roundabout way also — through activation of the reticular formation. So far as the thalamic specific nuclei, connected by direct pathways with the cortex, are concerned, they are regulated mainly by direct corticofugal fibers. Bearing all this in mind, it seems that there may be grounds for speaking of a corticofugal influence of dual character: indirect, diffuse, or nonspecific and direct, or specific. The discrepancy between results obtained by different workers, who have observed either a diffuse or a purely tonic effect of corticofugal impulses, is evidently attributable to the fact that they did not take account of the corticofugal influence of both aspects of this dual character, which may be exhibited to a greater or lesser degree depending on the experimental conditions and the actual subcortical structures studied.

SUMMARY

The effect of brief tetanic stimulation of different areas of the cortex (principally sensorimotor) on the recruiting response and augmenting response evoked by slow repetitive stimulation of the thalamic nonspecific and specific relay nuclei was studied in unanesthetized, curarized cats.

With a preparation in good condition and with choice of optimal conditions of stimulation (frequency, voltage, duration) of the sensorimotor cortex, the phase of "waxing" of the recruiting response is inhibited in all parts of the cortex and subcortical structures. With stimulation of the sensorimotor cortex of lower intensity or with stimulation of other parts of the cortex with comparatively weak action, other and less marked signs of inhibition of activity of the thalamic nonspecific nuclei may be observed, namely: 1) delay in the onset of the first waxing phase; 2) a decrease in amplitude of the potentials of the waxing phase; 3) lengthening of the interval between the waxing phases; 4) shortening of the waxing phase.

This influence of the sensorimotor cortex on the recruiting response is mediated largely through activation of the reticular formation, as the following facts show: 1) direct electrical stimulation of the mesencephalic reticular formation causes changes of the same character in the recruiting response; 2) after division of the brain stem at the level of the superior colliculi cortical influences on the recruiting response are considerably reduced (almost abolished). Whatever slight influence of the sensorimotor cortex which can still be observed after division of the brain stem is evidently due to the action of direct corticofugal fibers running toward the thalamic nonspecific nuclei.

Under the influence of stimulation of the sensorimotor cortex the augmenting response is also inhibited, but much less strongly and for a shorter time than the recruiting response. After division of the brain stem at the collicular level the effect of stimulation of the sensorimotor cortex on the augmenting response is changed only little if at all. This demonstrates that corticofugal impulses in this case act mainly by their direct influence on the thalamic relay nuclei and not through activation of the reticular formation.

The sensorimotor cortex thus influences the thalamic nuclei in two ways: directly (the direct corticothalamic pathway) and through activation of the reticular formation (indirect corticothalamic pathway). The direct corticothalamic pathway, as just described, is intended primarily and principally for the selective regulation of activity of the specific thalamic relay nuclei, whereas the indirect pathway exerts its generalized influence mainly on the thalamic nonspecific nuclei.

The direct inhibitory corticofugal effect is best demonstrated after temporary exclusion of cortical activity (postconvulsive depression, thermocauterization of the pial vessels, cooling of a large area of the cortical surface, disturbance of the cortical blood supply by rapid intravenous injection of a large dose of curarizing drugs). In this case the responses of the thalamic relay nucleus arising during peripheral stimulation increase considerably in amplitude.

Electrographic Correlates of Intersignal Responses during Defensive Conditioning in Dogs[*]

T. S. Naumova

Brain Institute
Academy of Medical Sciences of the USSR
Moscow, USSR

The preservation of traces of conditioned-reflex excitation in the central nervous system is a central problem in the physiology of higher nervous activity. Prospects for the solution of this highly complex problem have improved particularly as a result of the application of electrophysiological techniques.

With pride it can be said that those laying the foundations for the study of the dynamic organization of traces of conditioned-reflex excitation in the central nervous system belong to the school of the eminent Soviet electrophysiologist Mikhail Nikolaevich Livanov (Livanov and Polyakov, 1945; Livanov and Rabinovskaya, 1947; Livanov and Korol'kova, 1948; Livanov and Frenkel', 1951; Livanov et al., 1951; Livanov, 1962). These investigations showed that in the cerebral cortex of rabbits conditioned-reflex formation is accompanied by the preservation of traces of excitation which change their location in accordance with the stages of conditioning, as follows. Initially they are generalized in character, later they are preserved in the cortical zones of interacting analyzers, gradually concentrating at the cortical focus of the unconditioned reflex and finally disappearing altogether. These observations have been fully confirmed by Soviet and Western workers.

In this article an attempt is made to analyze one form of preservation of traces of conditioned-reflex excitation known as intersignal responses. Intersignal responses are known to accom-

---

[*] Pages 188-199 in the Russian edition.

pany the formation of any conditioned reflex. It is claimed that they are due to after-processes in the structures of the central nervous system which persist after each successive combination of conditional and unconditional stimulation (Pavlov, 1949; Kupalov, 1933; Abuladze, 1951, and others).

The results so far obtained do not, however, enable one to judge whether all responses observed in intersignal periods and at various stages of development of the reflex are of the same nature, to state what are the "trigger" processes giving rise to intersignal responses at different stages of formation of the reflex, and finally, to decide the significance of intersignal responses for formation of the conditioned connection.

To study all these problems the electrical activity of specific (skin-motor and auditory) and nonspecific systems at the levels of the rhomb-, mes-, di-, and prosencephalon was investigated in seven dogs during intersignal responses appearing during defensive conditioning, mainly to acoustic stimulation.

METHOD

Chronic experiments were carried out on dogs. Electrodes (25 for each animal) were implanted as follows: in the cortex by the method of Luria and Trofimov (1956) and by a technique specially developed by myself (implanted electrodes); into the mesencephalon by the method of Lyubimov and Trofimov (1958), and into the medulla, also by my own special technique. The EEG was recorded by the "bipolar" method. The interelectrode distance of the nichrome electrodes inserted into the subcortex and cortex did not exceed 2 mm, and their diameter was 80 μ. The EEG of the cortex was recorded by means of surface silver electrodes, 4-5 mm apart and 0.3 mm in diameter. Defensive conditioning was carried out by Protopopov's method (1909). The conditional stimulus was a tone of 500 cps and intensity 48 dB above human threshold and clicks with an intensity of 56 dB. For reinforcement, threshold electrodermal stimuli were applied to the right forelimb. The conditional stimulus acted in isolation for 2-3 sec. Combinations succeeded each other at intervals of not less than 2 min. Intersignal responses were recorded in the intervals between combinations. The EEG was recorded by means of 8- and 16-channel ink-writing electroencephalographs.

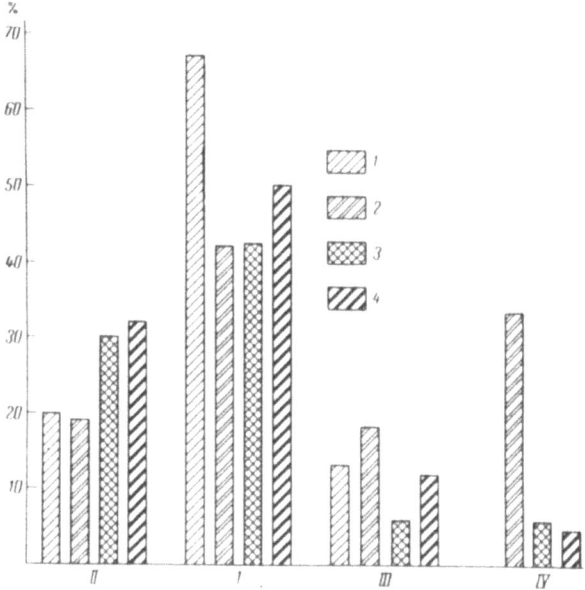

Fig. 1. Relative proportion of types of intersignal responses in four dogs. Abscissa: intersignal responses developing against a background of changes in electrical activity characteristic of signal responses: I) in a given stage of reflex formation; II) in the next stage of reflex formation; III) in the preceding stage of reflex formation; IV) against the background of absence of visible EEG changes. Ordinate: number of intersignal responses of a particular type as a percentage of their total number. Dogs: 1) Belyi, 2) Dzhil'da, 3) Rika, 4) Tera.

The EEG changes during 800 intersignal responses in four dogs (from 100 to 300 responses in each animal) were analyzed.

RESULTS

The study of electrical phenomena in various structures of the auditory system (cochlear nuclei, superior olive, medial geniculate bodies, inferior colliculi, cortex of the sylvian and ectosylvian gyri), the skin—motor analyzer (nucleus gracilis and nucleus cuneatus, ventral nucleus of the thalamus, cortex of the sig—

moid gyrus), and also in the nonspecific systems of the brain stem (reticular formation of the medulla, mesencephalon, and diencephalon), and in other structures during intersignal responses of the limb showed that changes in electrical activity accompanying such responses are divisible into two groups.

In some cases intersignal responses are unaccompanied by visible changes in electrical activity. This type accounts for between 1 and 35% of all intersignal responses in the different dogs (Fig. 1-IV), whereas the conditional stimuli are unaccompanied by visible EEG changes in 10-17% of all responses.

The second group of intersignal responses is accompanied by changes in electrical activity which repeat the EEG changes characteristic of a particular electrographic stage of defensive conditioning.

In previous publications (Naumova, 1962, 1965a, 1965b) it was stated that defensive conditioning in dogs goes through five stages. The first two stages are characterized by generalized changes in electrical activity in the brain: an initial generalized depression (stage I), followed by generalized exaltation (stage II) of electrical activity. In the first period of concentration (stage III), the effect of local activation is recorded in all structures of the auditory analyzer and the cortical components of the skin − motor analyzer. In the second period of concentration (stage IV) the conditioned response takes place against a background of selective strengthening and quickening of activity, mainly in the system of structures responsible for the conduction and perception of nociceptive afferent impulses (reticular formation, central gray matter of the brain stem, sigmoid cortex) and, finally, in stage V essential changes in electrical activity are no longer recorded either in the system of interacting analyzers or in intermediate formations.

Fig. 2. EEG changes in various structures of the cerebral cortex and medulla of the dog Dzhil'da taking place in the first (A), second (B), and third (C) stages of defensive conditioning. A) 91st combination of tone 500 cps, 48 dB with electrodermal stimulation of right forelimb; B) 105th and C) 123rd application of conditional tone. B, C) the same leads as in A. Legend to this and subsequent figures: ......; - cortical implanted electrodes. 1) Electromyogram of right forelimb; 2) electrodermal stimulation; 3) conditional acoustic stimulation; 4) time 1 sec.

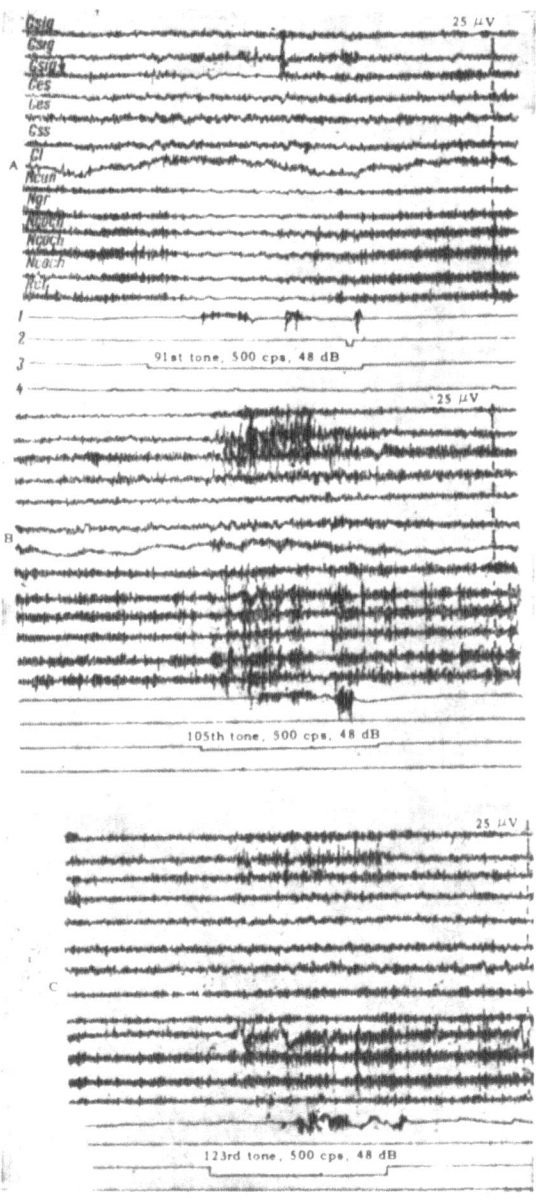

Fig. 2.

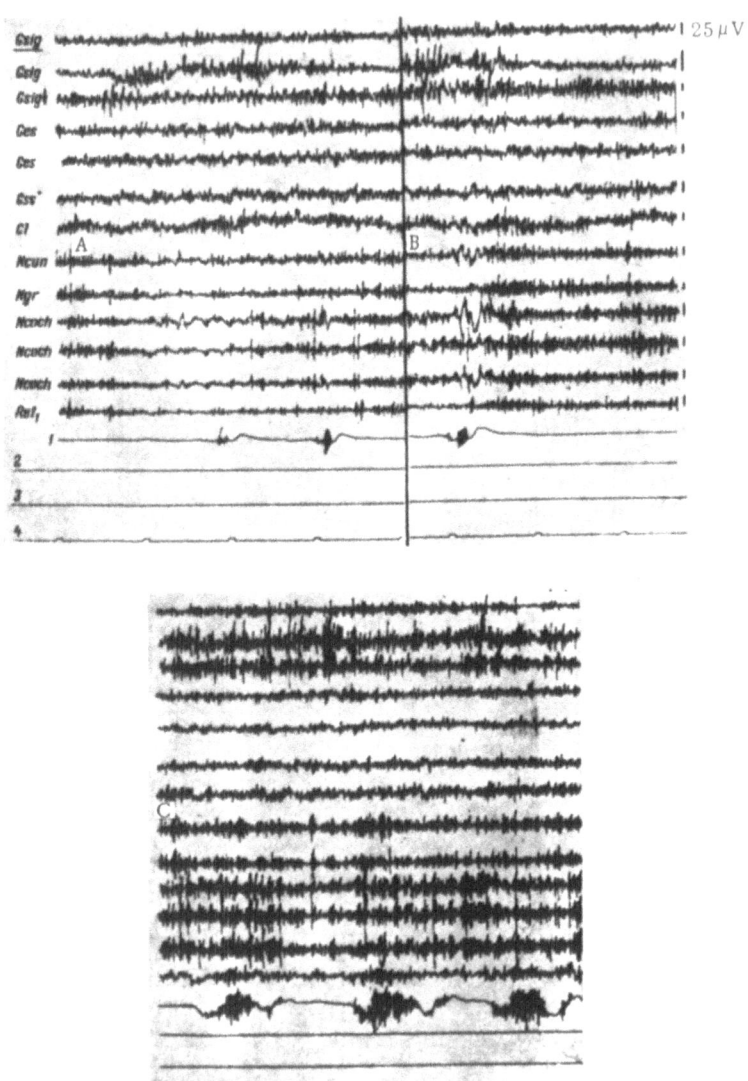

Fig. 3. Electrographic patterns of activity accompanying intersignal responses at various stages of defensive conditioning to a tone of 500 cps, 48 dB in the dog Dzhil'da. Legend as in Fig. 2.

An example of the EEG changes in the dog Dzhil'da in the first three stages of defensive conditioning to a tone of 500 cps, 48 dB is given in Fig. 2.

Similar patterns may accompany intersignal responses also. Electrographic pictures of the activity accompanying intersignal responses of the dog Dzhil'da at various stages of defensive conditioning to a tone of 500 cps, 48 dB are shown in Fig. 3. It is clear from Fig. 3A that the intersignal response develops against a background of changes in electrical activity characteristic of the first stage of reflex formation, i.e., against a background of generalized depression of electrical potentials in the brain stem (nucleus gracilis and nucleus cuneatus, cochlear nuclei, and gigantocellular nucleus of the medullary reticular formation), an exaltation effect in the sensorimotor cortex, and its desynchronization in other cortical areas in recordings made by surface electrodes. The intersignal response shown in Fig. 2B is accompanied by changes in electrical activity characteristic of the second stage of reflex formation, i.e., an effect of generalized exaltation of electrical activity in the same structures. In Fig. 2C the intersignal response takes place against a background of EEG changes characteristic of the third stage of reflex formation, i.e., of activation of potentials primarily in the sensorimotor cortex and cochlear nuclei, i.e., in the auditory and skin —motor analyzer systems.

Intersignal responses developing in the later stages of reflex formation may take place in the same way against a background of changes in electrical activity characteristic of those stages.

Quantitative analysis of the intersignal responses in the four dogs showed the existence of three types of electrographic patterns accompanying intersignal responses of the second group. The first type repeats the patterns of activity accompanying changes in potentials characteristic of conditioned responses at each particular stage of reflex formation. These changes accompany 40-70% of all intersignal responses (Fig. 1-I). The second type consists of changes in electrical activity preceding the EEG changes observed during the reflexes in the next stage of their formation. These changes accompany 20-30% of all intersignal responses (Fig. 1-II). Finally, the third type reflects changes in

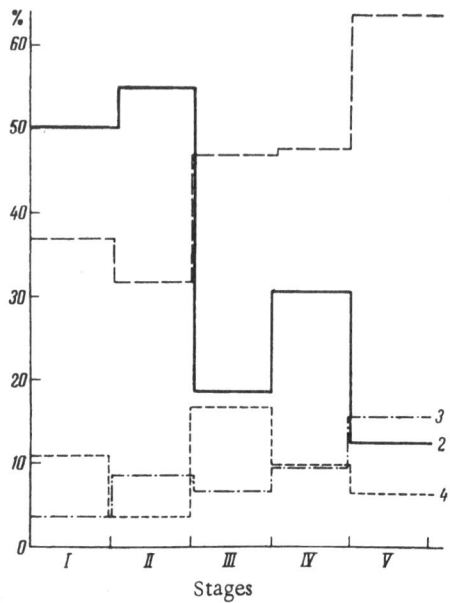

Fig. 4. Relative proportion of types of intersignal responses at various stages of formation of
a defensive conditioned reflex (combined data
for four dogs). 1) I, Fig. 1, 2) II, Fig. 1,
3) III, Fig. 1, 4) IV, Fig. 1.

electrical activity accompanying responses at one of the preceding
stages of reflex formation. Such patterns were observed in 5-20%
of all intersignal responses (Fig. 1-III).

Comparison of these types of changes with changes in activity arising at the moment of action of the conditional stimulus
showed the following differences. There were 70-80% of responses of the first type, 3-7% of the second type, and 4-12% of
the third type. Hence, during intersignal responses changes in
potentials anticipating future events, i.e., EEG changes accompanying conditioned responses in subsequent stages of reflex formation, appear in a much higher percentage of cases (20-30%) than
during responses to conditional stimuli (3-7%).

The relative proportions of electrographic types of activity
accompanying intersignal responses depend on the stage of defensive conditioning to acoustic stimulation. In Fig. 4 the electro-

graphic stages of conditioning are plotted along the abscissa, and
the number of intersignal responses of a particular type in four
dogs is plotted along the ordinate as a percentage of the total number of intersignal responses of these animals in that particular
stage. The highest percentage of intersignal responses in the
first stages of conditioning are of the second type. In other words,
in the first two stages of conditioning (stages of generalization),
the changes in electrical activity accompanying "spontaneous"
limb movements in the highest percentage (55) of cases reflect
patterns of activity accompanying conditioned responses at subsequent stages of reflex formation. In the course of formation and
stabilization of the conditioned reflex, the percentage of intersignal responses of the second type falls to 13%. In contrast to this,
in the course of conditioning there is an increase in the percentage of intersignal responses of the first type, i.e., responses accompanied by changes in electrical activity characteristic of conditioned responses at that particular stage of reflex formation.
The percentage of intersignal responses of the first type show a
particularly marked increase toward electrographic stages IV and
V of reflex formation (from 32-37 to 64%). So far as intersignal
responses of the third type are concerned, i.e., responses accompanied by changes in electrical activity analogous to the changes
accompanying conditioned responses in one of the preceding stages
of reflex formation, the percentage of responses of this type increases somewhat in the course of reflex formation (from 4-9 to
16, Fig. 4). Finally, the percentage of intersignal responses not
accompanied by visible changes in electrical activity shows no
regular changes in the course of conditioning.

Comparing the dynamics of the types of activity during intersignal responses with the dynamics of the same patterns of activity during responses to conditional stimuli in the various stages
of conditioning, the character of distribution of the types of electrical phenomena accompanying conditioned responses in each
stage of defensive conditioning in the same animals remains basically the same, although in absolute values these differences are
less marked. Besides the electrographic patterns of activity accompanying intersignal responses described above, the character
of their traces must be discussed. Sometimes although the intersignal response itself took place against a background of EEG
changes characteristic of that stage of reflex formation, or was

unaccompanied by visible changes in electrical activity, processes
characteristic of the next stage of defensive conditioning developed on its traces. In Fig. 5, for instance, is the EEG of the dog
Ngon'yam, recorded during intersignal responses taking place
against the background of the first stage of defensive conditioning
to rhythmic clicks with an intensity of 56 dB (the stage of generalized depression of electrical processes in the brain structures).
Clearly the first of the intersignal responses demonstrated
(Fig. 5A) developed against a background of generalized depression of cortical, subcortical, and bulbar structures. In the traces
following this response activation appears, localized to the auditory (medial geniculate body, ectosylvian cortex) and skin—motor
(sigmoid cortex) analyzer systems. These changes persisted
longest in the auditory analyzer system. The aftereffect was still
more marked in the system of interacting analyzers after the
next intersignal response, also developing against a background
of generalized depression of electrical processes (Fig. 5B).
Hence, although the intersignal response developed against a background of EEG changes characteristic of stage I of conditioning,
in its traces changes characteristic of stage III of conditioning
appeared.

A similar phenomenon may be observed at the level of the
first relay structures of the auditory system, the cochlear nuclei.
As Fig. 6 shows, the intersignal response in the dog Tuzik (conditioning to a tone of 500 cps, 48 dB) is accompanied by a generalized increase (more marked in the gigantocellular nucleus of the
reticular formation) of potentials, changes characteristic of the
second stage of reflex formation. A powerful effect of exaltation
of activity developed in the cochlear nuclei on the traces of the
first limb movement. It is interesting to note that the next two
intersignal responses led merely to an increase of this activation.
In the auditory and sensorimotor cortex this activation effect was
less marked. Hence, although the intersignal response developed
against a background of EEG changes typical of stage II of reflex
formation, changes characteristic of stage III of conditioning appeared on its traces. In other words, processes characteristic of
future stages of conditioned reflex development are formed on the
traces of intersignal responses of this type.

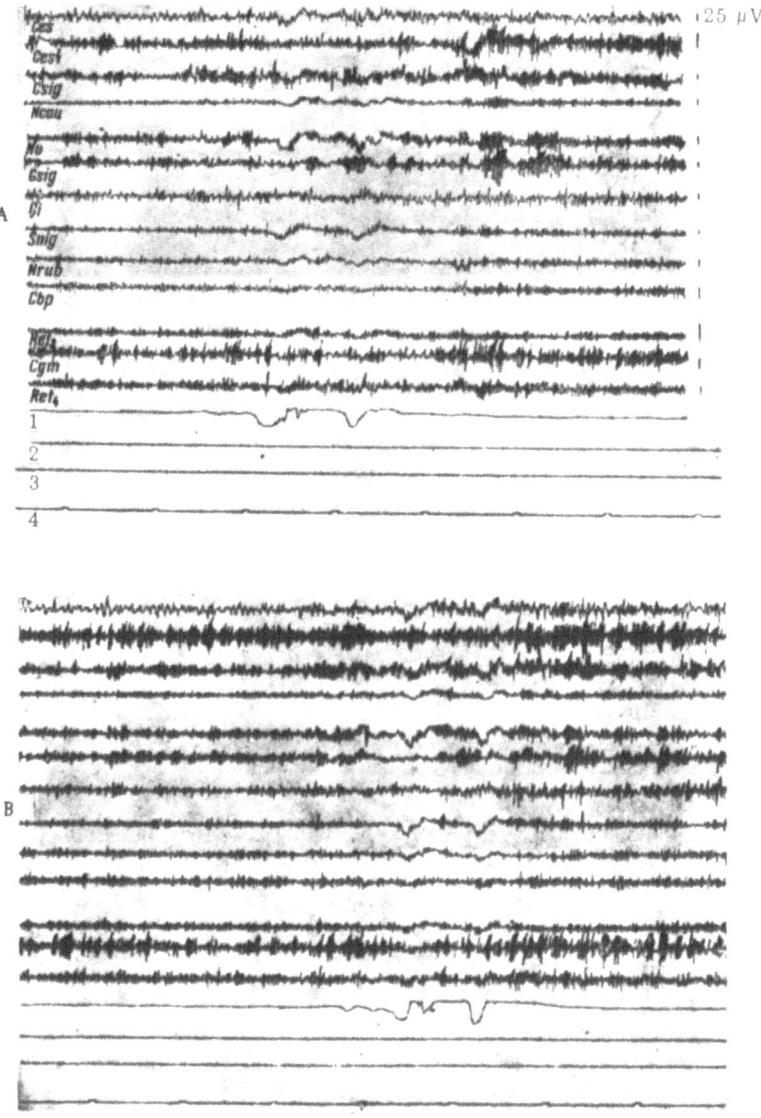

Fig. 5. Electrographic patterns of activity accompanying intersignal responses against a background of the first stage of defensive conditioning in the dog Ngon'yam to regular clicks with an intensity of 56 dB. Legend as in Fig. 2. B: A direct continuation of recording A.

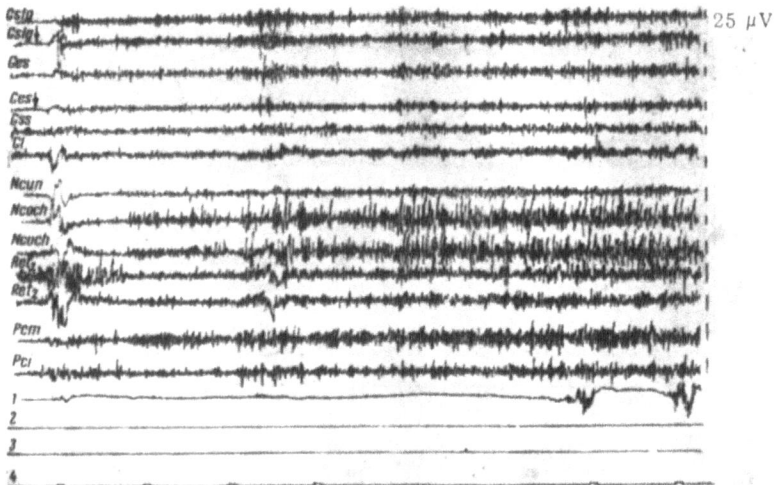

Fig. 6. Intersignal response during defensive conditioning to a tone of 500 cps, 48 dB, in the dog Tuzik accompanied by the same changes in electrical activity as in stage II of reflex formation (effect of generalized increase in activity). Legend as in Fig. 2.

In an earlier paper (Naumova, 1963) I reported that analogous changes in electrical activity, anticipating the EEG changes in a future stage of defensive conditioning, are observed on the traces of conditioned defensive reflexes. These after-processes in the auditory system last for many minutes or even hours.

DISCUSSION

Analysis of the electrical phenomena accompanying intersignal responses during defensive conditioning in dogs indicates that these responses may differ in nature. Intersignal responses unaccompanied by essential changes in electrical activity evidently bear little relationship to the mechanism of formation of the temporary connection. Intersignal responses of this type are more postural in nature, and are expressed by elevations of the limb of smaller amplitude. These responses are grossly individual in character: in some animals they are not observed at all, while in others they constitute more than one-third of all intersignal responses.

Intersignal responses of the second type are closely connected with the mechanism of defensive conditioning. These responses take place in association with the reproduction of electrographic patterns of activity characteristic of a given stage of reflex formation. Three types of these responses are distinguished: the first corresponds to the present stage; the second anticipates a future stage of reflex formation, and finally, the third type reflects a preceding stage of reflex formation. Since each type of intersignal response during reflex formation undergoes definite evolution, similar to the evolution of the electrical phenomena accompanying responses to conditional stimuli, it may be assumed that the mechanisms of these two responses are similar. This is concordant with the opinion of Kupalov (1933), of Konorski and Miller (1936), Rabinovich and Trofimov (1957), Sokolov (1958), Plonskaya (1959), Lat (1960), and others who regard intersignal responses as the result of an increase in excitability (the formation of dominant excitation) in the links of the conditioned-reflex arc arising as a result of preservation of the traces of conditioned-reflex excitation.

So far as the mechanisms of formation of intersignal responses at different stages of reflex formation are concerned, it is evidently difficult to assume that they are completely identical. One must agree with Sokolov (1958) that every intersignal response is due to an increase in excitability mainly of the central links of the analyzer receiving the conditional stimulus. But is this so at all stages of formation of the reflex?

Observations made by Livanov and his pupils show that during formation of a temporary connection the state of the interacting analyzers in the intersignal periods undergoes essential evolution. In the first stages of defensive conditioning potentials synchronous with the conditional stimulation are recorded in the intersignal periods throughout the rabbit's cortex, later they are concentrated in the interacting analyzers, and finally, they persist only in the cortical end of the unconditioned-reflex analyzer. In other words, traces of conditioned-reflex excitation are preserved initially by all cortical structures (starting from the analyzer receiving the conditional stimulus) and later they are found only in the unconditioned-reflex center.

As my investigations have shown, a similar evolution takes

place in the cortical and also subcortical and bulbar structures of
the dog during intersignal responses. At the beginning of reflex
formation, the highest percentage of these responses takes place
during or on the traces of those changes in electrical activity in
which activation is clearly seen in all links of the conditioned analyzer. In the final stage of reflex formation, when recordings of
the evoked and combined electrical activity in most structures of
the auditory system (other than the nuclei of the superior olivary
complex) revealed no visible EEG changes at the time of intersignal limb movements, it is difficult to suggest that the intersignal
responses are due to an increase in excitability of the conditioned
analyzer. Investigations conducted by Livanov's school and my
own findings suggest that conditioned-reflex excitation is preserved mainly in the unconditioned-reflex analyzer, from which
it may readily go out to the effector under the influence of any
form of stimulation.

It must be concluded from these findings that the mechanisms of formation of intersignal responses at different stages of
reflex formation are not identical and that they embody a dominant constellation established in the present or a future stage of
formation of the conditioned reflex.

Finally, the significance of intersignal responses for conditioned-reflex formation must be considered. Differences in the
dynamics of the types of activity which have been identified in intersignal responses and responses to conditional stimuli may be
summarized by saying that the electrographic pictures characteristic of a future stage of conditioning appear during intersignal
responses in a far higher percentage of cases (20-30) than during
responses to the conditional stimuli themselves (3-7). It is evident that intersignal responses, especially in the first stages of
conditioning, facilitate the appearance of anticipatory excitation
(Anokhin, 1962), facilitating formation of the conditioned reflex.
The results suggest that intersignal responses are important for
formation of the reflex. Although a product of the reflex, intersignal responses apparently train the nervous system and facilitate reflex formation. From this point of view it is easy to understand why there is such a large number of intersignal responses
in the period of active conditioning and a decrease in their number in the period of stabilization. Furthermore, Lat (1960) has

actually demonstrated a direct relationship between the frequency of intersignal responses and the speed of conditioning.

During reflex formation and training, intersignal responses anticipate future events to a diminishing degree, and they increasingly reflect current or, to some extent, past conditioned responses, while in the final stage they cease altogether.

SUMMARY

1. Analysis of the electrical phenomena accompanying intersignal responses during defensive conditioning in dogs demonstrates variations in the nature and mechanisms of the different types of intersignal responses. Responses of the first type are unaccompanied by visible changes in electrical activity and are unconnected with the conditioning mechanism. Intersignal responses of the second type are accompanied by electrographic patterns of activity characteristic of a certain stage of reflex formation. The mechanisms of formation of intersignal responses thus differ at different stages of conditioning and they embody a dominant constellation which is formed in the present or future stage of formation of the conditioned reflex.

2. Three types of this second class of intersignal responses are distinguished. Responses of the first type develop against a background of changes in electrical activity characteristic of conditioned responses in the actual stage of reflex formation. They constitute 30–70% of all intersignal responses. The second type is accompanied by changes in electrical activity characteristic of responses to conditional stimuli in a future stage of conditioning. Responses of this type account for 20–30% of the total number of intersignal responses. Finally, the third type of intersignal responses arise against a background of phenomena analogous to those accompanying conditioned responses at a preceding stage of conditioning. They were observed in 5–20% of the cases.

3. The difference in the dynamics of these types of activity distinguished during intersignal and conditioned responses is that the electrographic patterns of activity characteristic of a future stage of conditioning appear in the course of intersignal responses in a much higher percentage of cases (20–30) than during conditioned responses (3–7).

4. During conditioning the relative proportions of types of intersignal responses change: the proportion of intersignal responses of the first and third type relative to the total number in that particular stage increases: the first from 32–37 to 64%, the third from 4–9 to 16%.

5. Intersignal responses of the second type reflect processes in a subsequent stage, and are most numerous (55% of the total number of intersignal responses in that particular stage) in the first two stages of reflex formation (the stages of generalization), and as the reflex is formed and stabilized they gradually diminish to 13%.

Changes in electrical phenomena characteristic of the conditioned response in a future stage occasionally appear also on traces of intersignal responses.

6. The results may be regarded from the standpoint of the significance of intersignal responses for conditioning: although a product of the reflex, intersignal responses apparently train the nervous system and thus facilitate reflex formation, by assisting manifestation of the forthcoming excitation.

Effect of Visual Afferent Impulses on Formation of Cortical Rhythms [*]

L. A. Novikova

Electrophysiological Laboratory, Research Institute of Defectology
Academy of Pediatric Sciences of the RSFSR
Moscow, USSR

It has repeatedly been shown by morphologists that interruption of the flow of afferent impulses disturbs brain structure. Nearly all investigators who have studied the morphology of the brain after destruction of receptors have found degenerative changes in the subcortical nuclei of the damaged analyzers. Less consistent results have been obtained with respect to morphological changes in the cortex after destruction of receptors.

Sechenov applied the concept of the reflex activity of the central nervous system extensively to the analysis of electrical phenomena in the brain. He not only discovered the continuous rhythmic activity of nerve cells but also drew attention to changes in the rhythmic electrical potentials under the influence of afferent stimuli (1882).

The view that cortical rhythms are dependent on the inflow of afferent impulses has been extensively developed by Bremer, the first of whose articles was published in 1935, and it has been reflected in many of the electrophysiological investigations conducted more recently. However, some workers (Roger et al., 1967; Batini et al., 1959; Arduini and Hirao, 1959) have concluded in various preparations that visual, auditory, and olfactory deafferentation causes no significant changes in the electrocorticogram. When the results of these investigations are assessed, it must be remembered that these workers performed their deafferentation under acute experimental conditions on different pre-

[*] Pages 200-212 in the Russian edition.

parations. However, to demonstrate the role of volleys of afferent impulses in maintaining cortical tone and forming cortical rhythms, it is particularly important to investigate the EEG after exclusion of the various analyzers under chronic experimental conditions.

For this reason a study was made of the EEG of blind persons who had been without a functioning visual afferent system for several years. The EEG was investigated in 236 subjects with blindness of different degrees. The EEG was recorded by bipolar and monopolar methods on ink-writing multichannel electroencephalographs of different types. For bipolar recording, the two electrodes were placed in one region of the brain. The EEG of the occipital, central (Rolandic fissure), temporal, and frontal regions of the brain was recorded.

Depending on the character of their disease the subjects were subdivided as follows: diseases of the optic neural apparatus 43, vascular diseases 13, diseases of the refractory media 50, congenital developmental anomalies of the organs of vision 30, atrophy and subatrophy of the eyeballs 32, glaucoma 17, extensive trauma of the eye and sympathetic inflammation after trauma 12, anomalies of refraction 49. The loss of vision in 142 subjects was congenital. In 36 the loss of vision occurred before three years of age, and only in 58 subjects after four years of age. The time from loss of vision until the investigation varied from 3 to 50 years. In most subjects not less than 10 years had elapsed from the time of losing their sight, and in nearly all subjects it was not less than five years. Most of the subjects were between 8 and 50 years old (226 persons).

Analysis of the results revealed several types of EEG depending on visual acuity. The subjects as a whole could be subdivided accordingly into the following groups: group 1, visual acuity 0; group 2, visual acuity, distinguishes between light and darkness; group 3, 0.01-0.04; group 4, 0.05-0.09; and group 5, 0.1-0.3 (the visual acuity stated is for the better eye).

In the majority of cases of absolute blindness (83%) no alpha-rhythm is present in the EEG (Table 1). An alpha-rhythm was recorded in the EEG of only seven of 57 blind persons examined, and it was distinguished by a low index, amplitude, and frequency. All these subjects became blind after birth, most commonly at the age of 3-7 years.

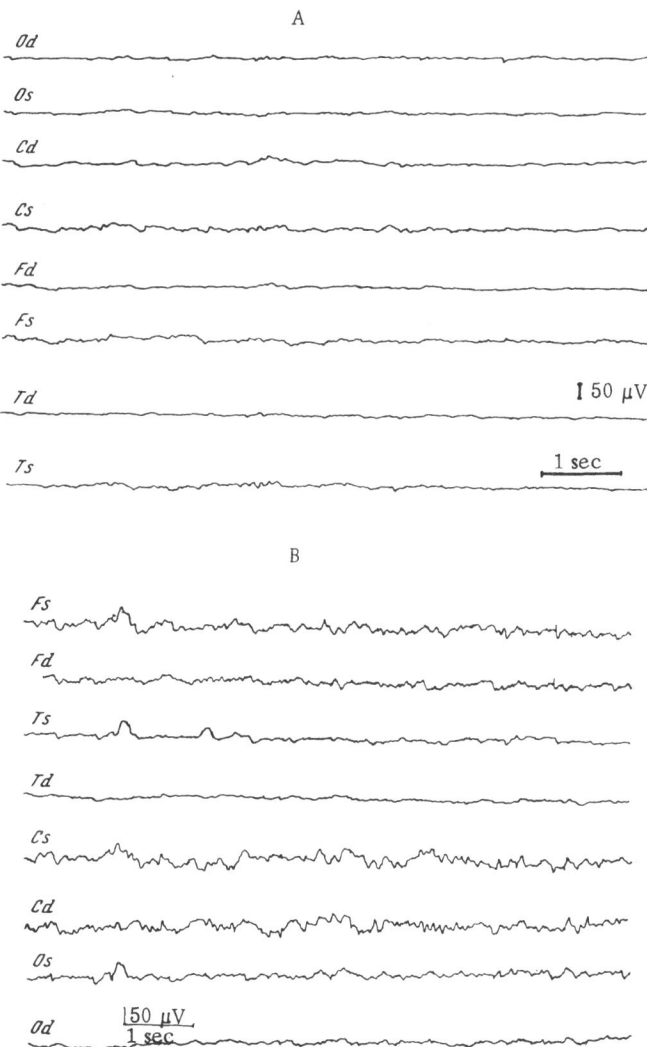

Fig. 1. EEG of the blind. A) EEG of the blind man V-ov, aged 20 years, who had lost his sight because of congenital glaucoma. Visual acuity 0, alpha-rhythm absent; amplitude of waves considerably reduced; B) EEG of blind man S-ov, aged 22 years, blind as a result of congenital microphthalmia. Visual acuity 0, alpha-rhythm absent; focus of maximal electrical activity localized in central regions of cortex. O) occipital region; C) central region; F) frontal region; T) temporal region; d) right side; s) left side.

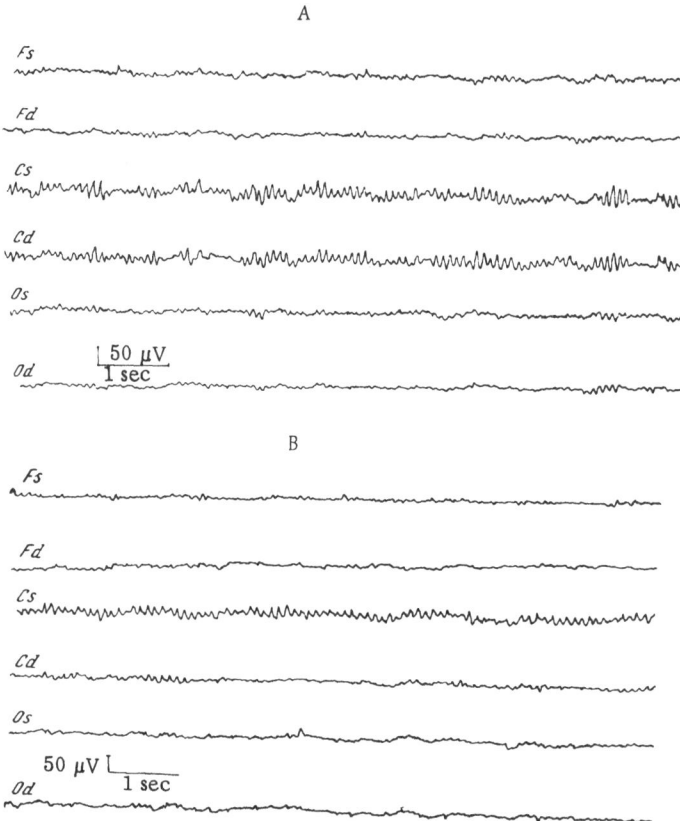

Fig. 2. Magnitude of Rolandic rhythm in EEG of blind subject.
A) EEG of blind girl S-ova, aged 15 years, blinded as a result of injury to the eyes at the age of 4 years. Visual acuity 0, alpha-
rhythm absent. Rolandic rhythm recorded in central region of cortex and spreading to other parts of the brain; B) Rolandic rhythm
most marked in left hemisphere of a blind person.

Examples of EEGs characteristic of absolutely blind persons
are given in Fig. 1A and B. The EEG shown in Fig. 1 was taken in
the blind subject V-v, a man aged 20 years. His blindness resulted
from congenital glaucoma. As Fig. 1A shows, the alpha-rhythm is
absent on the EEG; the amplitude of all the waves is lowered and
the trace is flattened.

The EEG in Fig. 1B also shows absence of the alpha-rhythm,
but the trace in this case is less flattened. Low-amplitude slow

waves and a beta-rhythm predominate in the EEG; the focus of maximal electrical activity is localized in central regions of the cortex.

In many cases clearly outlined, sinusoidal waves resembling the alpha-rhythm in shape and frequency (Fig. 2A), were recorded in the central cortical regions of the blind persons. The facts given below show that these waves belong to the Rolandic rhythm m described initially by Gastaut (1952), and later by others (Lelord, 1957; Klass and Bickford, 1957; Roger, Sokolov, and Voronin, 1958; Chatrian et al., 1959).

1. Sinusoidal waves recorded on the EEG of blind persons are localized in central cortical regions.

2. These waves are frequently monophasic.

3. Sinusoidal waves recorded in the central cortical regions of blind persons, like the Rolandic rhythm, are particularly sensitive to proprioceptive and tactile stimulation.

A Rolandic rhythm is found in the EEG of blind persons much more often than in that of persons with normal vision. In subjects with visual acuity 0 or able to distinguish between light and darkness a Rolandic rhythm is found in 42% of cases, compared with 3-13% of cases, according to different authorities, in persons with normal vision.

Investigation of the Rolandic rhythm in the EEG of the blind subjects showed that it is more marked in the left hemisphere (Fig. 2B).

It is clear from Tables 1, 2, and 3 that the EEG of blind persons capable of distinguishing between light and darkness only, as also in the case of absolutely blind persons, is characterized by absence of alpha-rhythm, lowered amplitude of the waves, and displacement of the focus of maximal electrical activity into the central cortical region. A Rolandic rhythm is found in blind subjects capable of distinguishing between light and darkness more frequently than in absolute blindness.

The reason for the less marked Rolandic rhythm in absolute blindness may be the great depression of cortical rhythms in persons completely without visual afferent impulses. In blind persons with a visual acuity of 0, the traces recorded were extremely flat-

TABLE 1. Magnitude of Alpha-Rhythm in EEG of Subjects with Different Degrees of Visual Acuity

Group No.	Visual acuity	Number of subjects	Magnitude of alpha-rhythm		
			alpha-rhythm absent, %	distinct alpha-rhythm, %	single waves of an indistinct alpha-rhythm, %
1	0	57	83	14	13
2	distinguished light and dark	42	80	10	10
3	0.01 - 0.04	45	48	38	14
4	0.05 - 0.09	40	27	65	8
5	0.1 - 0.3	52	21	77	2

TABLE 2. Amplitude Characteristics of EEG of Subjects with Different Degrees of Visual Acuity

Group No.	Visual acuity	Number of subjects	Type of trace	
			depressed. Amplitude of maximal waves below 30 μV, %	extremely flattened, %
1	0	57	77	39
2	distinguished light and dark	42	62	12
3	0.01 - 0.04	45	60	20
4	0.05 - 0.09	40	53	7
5	0.1 - 0.3	52	27	—

tened in 39% of the cases. Against the background of this marked decrease in amplitude, the Rolandic rhythm could not be distinguished.

The EEG of the third group of subjects (visual acuity 0.01-0.04) differed from that of the two previous groups. In 38% of these subjects an alpha-rhythm was recorded. However, as a rule, it was characterized by a low index, amplitude, and frequency. In the majority of cases (60%) the EEG of this group of subjects was

TABLE 3. Distribution of Maximal Electrical Activity in
EEG of Subjects with Different Degrees of Visual Acuity

Group No.	Visual acuity	Number of subjects	Location of focus of maximal electrical activity		
			focus in occipital region, %	focus in central region, %	no definite focus of maximal electrical activity, %
1	0	57	14	46	40
2	distinguished light and dark	42	10	59	31
3	0.01 - 0.04	45	24	35	41
4	0.05 - 0.09	40	47.5	32.5	20
5	0.1 - 0.3	52	74	17	9

still depressed in character. The number of traces with the focus of maximal electrical activity located in the occipital cortex was higher than in the preceding groups.

In the group with visual acuity 0.05-0.09 an alpha-rhythm was recorded on the EEG in 65% of the cases; the alpha-rhythm was now characterized by a high index (in 11 of the 26 subjects the alpha-index exceeded 50%). In a high proportion of cases (53%), the amplitude of the EEG remained low. However, the focus of maximal electrical activity was localized more commonly in the occipital than in the central region of the cortex (47.5 and 32.5% of the cases).

In the group with a visual acuity of 0.1-0.3, a distinct alpha-rhythm was recorded in 77% of the cases. It was marked by stability and a normal amplitude and frequency. An unstable alpha-rhythm of low amplitude was observed in the group with visual acuity exceeding 0.1 in cases when the flow of visual afferent impulses was depressed from the second eye.

In the subjects of this group changes in the alpha-rhythm were observed which were never found when the visual acuity was below 0.04, and only occasionally when it was 0.05-0.09. In ten subjects the alpha-rhythm was pointed in shape and intermingled with epileptoid bursts. An alpha-rhythm of this type evidently re-

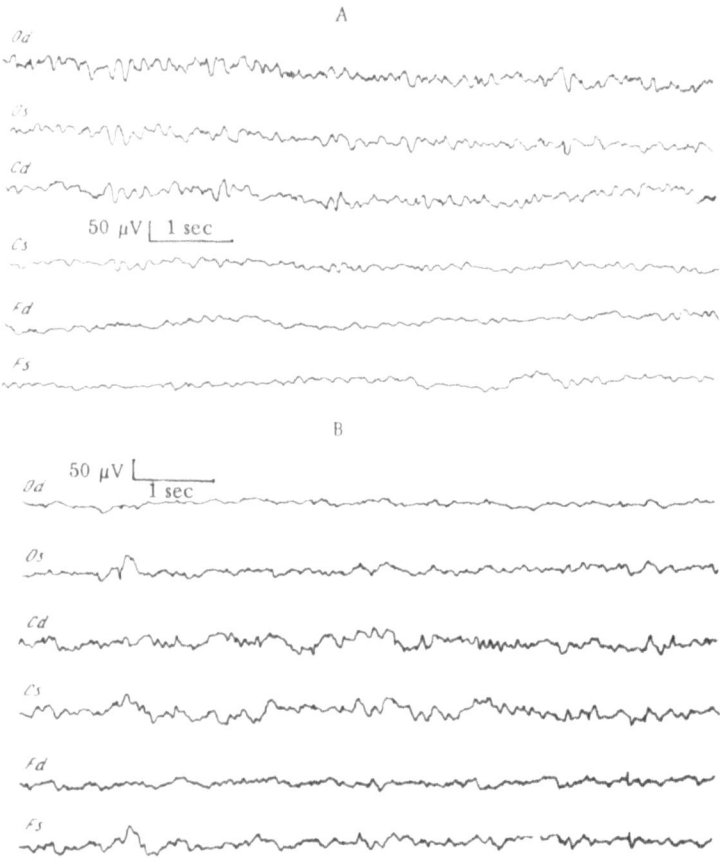

Fig. 3. EEG of subject S-an, blind, deaf, and dumb. A) EEG recorded at 8 years of age, 3 years after loss of vision to the level of distinguishing between light and darkness as a result of optic atrophy. Alpha-rhythm of low voltage recorded in occipital cortex; focus of maximal electrical activity localized in occipital cortex; B) EEG recorded 8 years after loss of sight. Visual acuity had fallen to 0. Alpha-rhythm absent. Focus of maximal electrical activity shifted into central region of cortex.

flects functional changes taking place in the visual cortex when vision is considerably disturbed but not completely absent.

Hence, this investigation showed that in the presence of various diseases of the eye the character of the EEG changes depends on the degree of loss of sight. As the visual acuity increases the alpha-rhythm becomes more distinct and its amplitude and frequency increase. The relationship between the magnitude of the alpha-rhythm and the degree of visual acuity is confirmed by statistical analysis of the results. The statistical analysis revealed a correlation between visual acuity and the alpha-index, the coefficient of correlation being 0.64.

With an increase in visual acuity, not only was the alpha-index increased, but so also was the voltage of the EEG waves (Table 2). The amplitude of these waves increased as the visual acuity rose from 0 to ability to distinguish between light and darkness; a second sharp rise in the voltage of the waves on the EEG was observed in subjects with a visual acuity higher than 0.1.

The increase in amplitude of the waves observed as the visual acuity rose was accompanied by displacement of the focus of maximal electrical activity from the central to the occipital cortex (Table 3).

An important factor in the study of the mechanisms of the changes in cortical rhythms in blindness would be the discovery of a relationship between the degree of the EEG changes and the time of visual deafferentation. Investigations on 12 subjects soon after the onset of blindness showed that for between 0.5 and two months after the onset of blindness the alpha-rhythm may persist. However, in these cases it has low amplitude and a low index. Later, two to six months after loss of sight, the EEG became similar to that characteristic of persons with long-standing blindness.

To discover the role of the time factor in the changes in cortical rhythms during blindness, it is interesting to investigate the EEG of the same persons over a long period.

In Fig. 3, A and B, the EEG of the boy Serezha S., taken 3 and 8 years after becoming blind, is shown. This boy is blind, deaf, and dumb.

Comparison of the two EEGs recorded at different times after injury to the visual and auditory analyzers shows that not only were the waves more depressed than previously, but also that the focus of maximal electrical activity was displaced: in the EEG taken three years after the illness, when afferent impulses were still present, the maximal amplitude of the EEG waves was recorded in the occipital cortex. After complete loss of all traces

of residual eyesight, the focus of maximal electrical activity
shifted to the central regions of the cortex, where single waves of
a Rolandic rhythm were recorded. In the occipital cortex, after
prolonged visual deafferentation, the amplitude of the waves was
sharply reduced and the alpha-rhythm abolished.

Hence, it follows from the example given that as the time
after the onset of blindness increases, depression of the cortical
rhythm is deepened, the alpha-rhythm is destroyed, and the focus
of maximal electrical activity moves from the occipital cortex to
the central regions, where a Rolandic rhythm begins to form.

DISCUSSION

Investigation of the EEG of blind persons shows the importance of visual afferent impulses for formation of the principal
human EEG rhythm, the alpha-rhythm. It was found that an alpha-
rhythm is absent in 80% of cases from the EEG of blind persons.
In those few cases when an alpha-rhythm is recorded on the EEG
of blind persons, the blindness always developed after birth, and
usually after three years of age. Hence, in the absence of visual
afferent impulses, no alpha-rhythm is formed in the human EEG.

The importance of visual afferent impulses for the formation
of cortical rhythms is shown especially clearly if the EEGs of
blind and deaf persons are compared. A stable alpha-rhythm is
recorded in deaf persons, in contrast to the EEG of blind persons
(Fig. 4A).

The role of visual afferent impulses in the formation of the
alpha-rhythm is also clearly demonstrated by the study of the EEG
in cases of homonymous hemianopia, when the flow of visual afferent impulses is interrupted into one hemisphere but preserved in
the other.

The EEG was recorded from 38 patients with homonymous
hemianopia due to organic lesions of the brain. In agreement with
observations made by others (Engel et al., 1945; Blinkov and
Rusinov, 1949; Shmel'kin, 1949; Pampiglione, 1952; Tron and
Presman, 1955; Bergman, 1957), in nearly all cases of homonymous hemianopia a distinct asymmetry of the alpha-rhythm was
found in the occipital cortex. Asymmetry of the alpha-rhythm

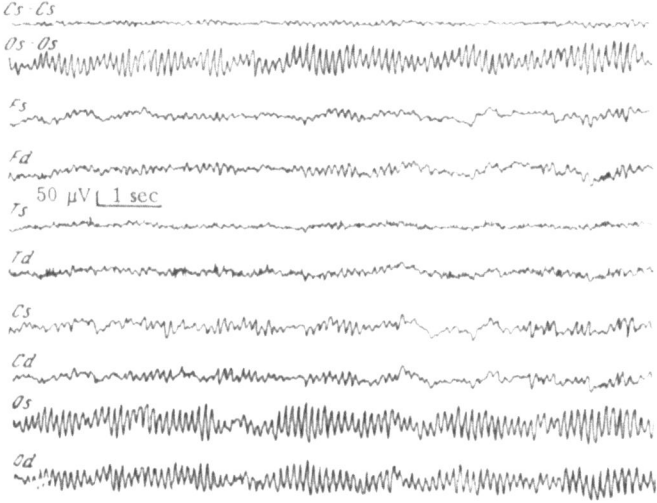

Fig. 4. EEG of a deaf boy K-ov, aged 14 years. A distinct alpha-
rhythm is present mainly in the occipital cortex.

was marked in the presence of lesions of the optic pathways at all
levels.

The following fact clearly emerges from the analysis of
EEGs of blind persons and persons with weak vision: the alpha-
rhythm is disturbed when vision is completely absent and also when
the subject is still able to distinguish between light and darkness.
In a large proportion (about 40%) of cases with a visual acuity of
0.01-0.04, an alpha-rhythm can be recorded. Appearance of the
alpha-rhythm thus coincides with an increase in visual acuity up to
a level at which the shape of an object can be distinguished. Consequently, for an alpha-rhythm to develop, vision must be good
enough to discriminate between shapes, or in other words, the
alpha-rhythm appears on the human EEG only when the central
nervous system receives clues to the nature of objects from photic
stimulation. Similar conclusions are found in a recently published
paper by Cohen et al. (1961).

As was stated above, the EEGs of the blind are characterized
by a lowered voltage of all waves as well as by abolition of the focus of the alpha-rhythm. The decrease in amplitude is found mainly in the occipital cortex of blind subjects. However, the depres-

sion of cortical rhythms also extends to other parts of the brain. According to the observations of Berger (1900), Kononova (1926), and Pines and Prigonnikov (1936), atrophic and degenerative changes develop in layers IV and V of the cortex of the blind. However, these structural changes are limited to the occipital region and do not extend to other parts of the cortex. If the disturbances of the cortical rhythms of the blind were due to transneuronal degeneration, they also would be limited to the occipital region. Generalized changes in cortical rhythms in blind subjects cannot therefore be explained as manifestations of secondary transneuronal degeneration.

What is the cause of the generalized decrease in amplitude of the waves in all parts of the cortex so characteristic of the EEG of the blind? The view has often been expressed in the literature that the depressive character of the EEG in the blind and disappearance of the alpha-rhythm are due to the constant alertness, to the apparently unextinguishable orienting reaction of persons deprived of their sight (Berger, 1935; Loomis et al., 1936; Stepanov, 1955; Zimkina, 1956; Asafov, Zimkina, and Stepanov, 1955). However, as a special series of experiments which I carried out showed, the orienting reactions in the blind are extinguished just as quickly as in persons with normal vision. After complete extinction of the orienting reaction, no alpha-rhythm appeared on the EEG of the blind subjects, and no increase was observed in the voltage of the EEG waves.

The EEG in homonymous hemianopia also indicates that disappearance of the alpha-rhythm and the depressed character of the EEG are associated with interruption of the flow of visual afferent impulses and not to the subject's state of stress. This is shown by the fact that the alpha-rhythm is abolished only in the hemisphere on the side of deafferentation and persists in the intact hemisphere, whereas the "arousal" reaction is bilateral.

Fig. 5. Changes in EEG of rabbits during a long stay in darkness and transfer to an illuminated room. A) Gradual decrease in amplitude and frequency of waves in occipital, sensorimotor, and parietal cortex of a rabbit kept for long periods in darkness: a) background, b, c, d) 14, 22, and 30 days of functional visual deafferentation; B) gradual increase in amplitude and frequency of EEG waves on restoration of visual afferent impulses: a) EEG of rabbit after a stay of 30 days in darkness, b, c, d, e) after a stay of 3 hr, and 10, 18, and 30 days under conditions of normal illumination.

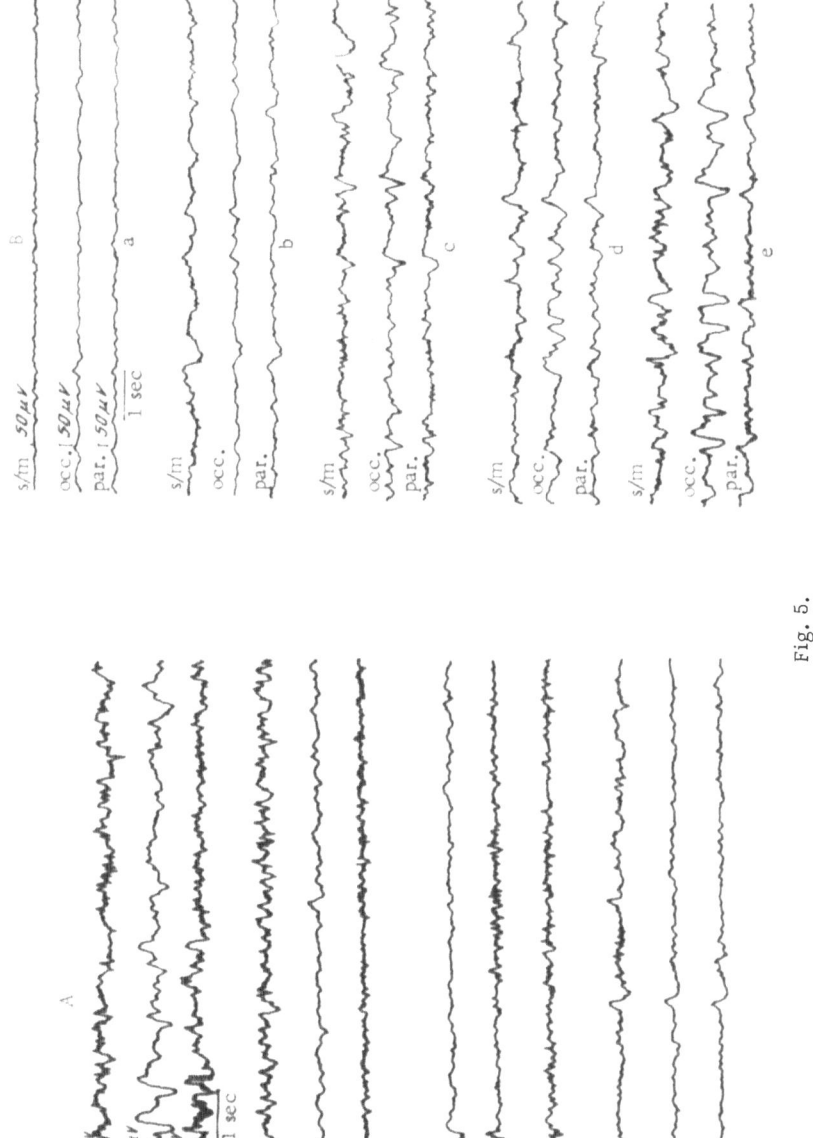

Fig. 5.

To solve the difficult problem of the functional significance of the depressed character of the EEG in the blind, it is important to study its frequency characteristics. It can be seen in the EEG of many blind persons that the decrease in amplitude of the waves is combined with a decrease in the number of fast potentials; in 39% of cases of absolute blindness, extremely flattened traces were recorded.

Hence, it was concluded from the results of all these investigations that the distinctive character of the cortical rhythms in the blind cannot be attributed to the subjects' state of continuous stress, and it cannot, therefore, be associated with increased cortical activation. The depressed character of the cortical rhythms in the blind evidently does not reflect an "arousal" reaction but, on the other hand, it is associated with depression of cortical tone due to the loss of the powerful flow of visual afferent impulses.

This hypothesis, so important for defining the functional state of the central nervous system in the blind, is supported by data in the literature. According to the calculations of Bruesch and Arey (cited by Granit, 1957), about 40% of the total sensory inflow in man arrives along the fibers of the optic nerve, and the loss of such a powerful source of afferent impulses may therefore cause depression of cortical tone.

The activating effect of light on animals has long been described in the biological and ecological literature (Bissonette, 1932; Benoit, 1934; Clark, 1959; Rowan, 1938; Svetozarov and Shtraikh, 1940; Voitkevich, 1944; Karapetyan, 1961). The hypothesis has been expressed that an optico—autonomic or photoenergetic system exists, by means of which the light has a generalized tonic action on the central nervous system (Rowan, 1938; Scharer, 1937; Markelov, 1949; Frey, 1950; Novokhatskii, 1956, and others).

Evidence of the existence of extensive collaterals between the optic pathways and various structures of the brain stem has been obtained by electrophysiological investigations. The work of Gastaut and Hunter (1950), French et al. (1952), and Hunter and Ingvar (1955) has demonstrated the wide irradiation of evoked potentials in subcortical structures during photic stimulation.

The tonic effect of light on the cerebral cortex has been

demonstrated in several investigations of higher nervous activity (Denisov and Kupalov, 1933; Kupalov, 1952; Rozental', 1950; Lobanova, 1954).

The activating action of light on cortical rhythmic activity has been investigated by Volokhov and Davydova (1954) and Kozo-doi and Farber (1936), who described the stepwise increase in voltage of the EEG of animals born blind and acquiring vision.

Important evidence of the tonic action of light on the central nervous system has been found by the work of Chang (1952, 1962), who demonstrated the phenomenon of "potentiation" of the cortex by light.

A number of facts in the literature thus confirm the hypothe-sis of the functional significance of the depressed character of the EEG in the blind. The loss of the powerful flow of visual afferent impulses evidently causes the "depotentiation" of the cortex, re-flected in the depressed EEG.

Indications of the lowered tone of the central nervous system in the blind can be found in the extensive literature dealing with the study of their psychophysiology (Clare, 1941; Hollwich, 1964; Jen-dralski, 1951; Wassner, 1954).

Besides a decrease in amplitude of the EEG in all parts of the cortex in blind subjects, displacement of the focus of maximal electrical activity into the central regions of the cortex takes place. In the EEG of the blind a Rolandic rhythm is seen much more often than in the EEG of persons with normal vision. This rhythm is particularly distinct in cases of congenital blindness or blindness developing early in life. With an increase in visual acu-ity the number of cases in which a Rolandic rhythm is present falls. All these observations suggest that the Rolandic rhythm is connected with a compensatory increase in activity of the motor analyzer. Evidence in favor of this is the fact that the Rolandic rhythm is more distinct on the EEG of the left cerebral hemi-sphere of blind persons, which is more closely connected with the right hand, actively concerned in the process of palpation.

Analysis of the EEG of blind subjects thus reveals two dis-tinct processes: a decrease in amplitude of all the waves, re-flecting lowered cortical tone, and the formation of a Rolandic

rhythm, associated with a compensatory increase in activity of
the motor analyzer.

To confirm the conclusions based on investigation of the
EEG of blind persons, and also to study the physiological mechanisms lying at the basis of changes in electrical activity of the
brain during disturbances of vision, experiments were carried
out on animals. They showed that in the majority of cases after
destruction of the visual receptor (bilateral enucleation) a decrease in amplitude and frequency of the cortical waves is observed. These changes occurred in waves: they were ill-defined
during the first weeks after deafferentation, and became more
marked with time. By the third or fourth month the decrease in
total energy of the potentials had reached 70-95% of the initial
background in many rabbits. Similar, although not absolutely
identical facts were obtained by Sarkisov (1938), Claes (1938),
Volokhov and Davydova (1954), Pigareva and Shilyagina (1955),
Batuev (1963), and Dzidzishvili (1963).

Experiments involving functional visual deafferentation are
of the greatest interest to the study of physiological mechanisms
lying at the basis of changes in cortical electrical activity. Under
these experimental conditions, trauma causing injury to the muscles of the eye is absent and, a particularly important feature,
transneuronal degeneration due to division of the optic nerves does
not take place. Because of this, changes caused by interruption of
incoming visual afferent impulses can be distinguished clearly in
these experiments.

Investigations showed that when rabbits were kept in absolute darkness, the amplitude of all waves in the EEG recorded
from the occipital, parietal, and sensorimotor cortex fluctuated
and gradually fell, and the number of fast potentials was reduced
(Fig. 5A). After 1.5-2.5 months of confinement to darkness, the
amplitude of the EEG in the occipital and sensorimotor cortex fell
in the great majority of rabbits by more than 50%, and in many of
them by more than 75%. The results of these experiments are in
agreement with the observations of Dubner and Gerard (1939),
Livanov (1944b), and Danilov (1960), who found a decrease in amplitude of the cortical EEG in experiments of different types in
which visual afferent impulses were diminished.

A particularly interesting aspect of the functional deafferentation experiments is the investigation of cortical electrical activity after restoration of visual afferent impulses. In the first minutes and hours after rabbits were transferred from lightproof boxes into an illuminated room, no marked changes were found in their EEG. After a longer stay under conditions of normal illumination, a gradual increase in amplitude of the EEG was found in all the rabbits. Together with the increase in amplitude, the number of fast potentials also increased (Fig. 5B).

The results of the experiments on animals confirmed that the depressed character of the EEG after visual deafferentation reflects a lowering of cortical tone and is not due to an "arousal" reaction. 1) The depressed character persisted after administration of chlorpromazine; 2) parallel with the decrease in amplitude of the cortical waves of the animals after deafferentation, an increase was found in the threshold of cortical excitability determined by direct stimulation of the cortex with square pulses and recording of the limb movement; 3) against the background of a depressed EEG of animals after deafferentation, the rhythm-driving response to flashes is impaired.

Hence, experiments on animals, like EEG investigations in blind persons, indicate the special importance of visual afferent impulses in the generation of cortical rhythms. This special role in man and animals is evidently determined by the biological significance of light in the life of animals at the higher levels of evolution.

Functional Properties of Dendrites in the Mammalian Brain[*]

D. P. Purpura

Departments of Anatomy and Neurology
College of Physicians and Surgeons
Columbia University
New York, New York

The problem of defining the functional properties of dendrites has been of central importance in the study of the bioelectrical activity of the brain for several decades. In consideration of the many outstanding contributions of Prof. M. N. Livanov to the analysis of electrocortical rhythms, it may be of interest to indicate here some aspects of recent studies which have provided additional information on the properties of dendrites and their role in the functional organization of different neuronal systems.

It will be recalled that the earliest studies of electrical activities in the mammalian brain brought to light several features of spontaneous and evoked electrocortical potentials which established that brain waves were not summations of all-or-none discharges of neurons. Additional investigations involving extracellular and laminar analysis of evoked potentials, observations on the effects of pharmacological agents, and studies of the relationship of unit and multiunit discharges of cortical neurons to different components of evoked potentials further indicated that much of the electrical activity recorded from the brain had its origin in dendrites. Although there was little agreement concerning the nature and mode of production of so-called "dendritic potentials" it came to be widely accepted that these graded potentials were the major components of brain waves. Studies suggesting that "dendritic po-

---

*Pages 213-225 in the Russian edition.

tentials" were complex summations of excitatory and inhibitory postsynaptic potentials (EPSPs and IPSPs) and that cortical dendrites were electrically inexcitable (Purpura and Grundfest, 1956) completed this series of investigations of extracellular activities of cortical neurons (Purpura, 1959).

Recently more detailed investigations of the functional properties of cortical dendrites have been made possible by the application of intracellular recording techniques to the analysis of different types of neurons in the mammalian brain. Unfortunately only the faintest outlines of the patterns of dendritic activities have begun to emerge from these intracellular studies. Nonetheless the data obtained thus far are sufficiently important to warrant a more intense effort in this direction. Limitations of space do not permit more than a brief description of some recent studies from our laboratory in which intracellular recording has been employed to provide data on the properties of dendrites of different cortical and thalamic neuronal organizations.

Comparative Physiology of Dendrites in Mature and Immature Neocortex

Intracellular data obtained in many studies have clearly indicated that the sequence of events leading to impulse initiations in neocortical neurons of adult animals is similar to that described for spinal motoneurons or interneurons (Eccles, 1957; Fuortes et al., 1957). In the vast majority of neocortical neurons of adult animals orthodromic stimulation elicits EPSPs which secure impulse initiation by depolarization of low-threshold sites in the initial axonal segment—soma membrane. In many instances, however, no clear separation between responses of initial axonal segment and soma membrane sites may be detectable in spike potentials recorded from neocortical neurons. This suggests little functional difference in the threshold for all-or-none spike generation in soma-initial segment regions in neocortical neurons under the usual conditions of synaptic activation. Intracellular studies have also disclosed that spike potentials recorded from the soma may be succeeded by a brief phase of "delayed depolarization" which is apparently indicative of a variable invasion of proximal dendritic regions by the spike potential (Granit et al., 1963b). The degree of invasion of proximal dendritic regions may be influenced by concomitant synaptic depolarization of soma −dendritic membrane as observed in spinal motoneurons (Nelson and Frank, 1964).

It has been repeatedly demonstrated in neocortical neurons of adult animals that induced membrane hyperpolarization sufficient to block orthodromically-evoked soma −initial segment spike discharges reveals no additional responses other than PSPs. Fast prepotentials (FPPs) reflecting partial spikes in dendrites (Spencer and Kandel, 1961a) have been observed only in rare instances in neocortical neurons (Klee, 1966; Purpura and Shofer, 1964). Thus axodendritic synaptic excitation of neocortical neurons in adult animals does not initiate spikes in the dendritic system of these elements (Purpura and Shofer, 1964; Purpura et al., 1964b).

The question of whether apical dendrites of neocortical pyramidal neurons are capable of responding to directly applied depolarizing currents has now been satisfactorily answered in studies of the effects of cortical surface polarizing currents on intracellular activities (Purpura and McMurtry, 1965). It has been shown in these investigations that surface anodal polarization of the motor cortex depolarizes the soma −initial axonal segment region and thereby initiates discharges of pyramidal neurons whereas surface cathodal polarization produces soma hyperpolarization and blockade of spike potentials. Evidently cathodal depolarization of apical dendrites does not elicit spike discharges in these elements despite the fact that depolarizing excitatory actions of cathodal currents are clearly demonstrable on interneuronal elements located in superficial regions of cortex. It is of interest that polarizing currents applied to the surface of the motor cortex produce low-threshold effects on nonpyramidal tract neurons with currents that are too weak to influence the soma membrane potential or evoked PSPs of Betz cells. Cathodal or anodal polarizing currents may also produce dramatic inversions or augmentations of different components of cortical surface−evoked potentials without producing changes in transmembrane potentials recorded at the cell body level of pyramidal neurons (Purpura and McMurtry, 1965). This dissociation between effects of weak transcortical polarizing currents on evoked potentials and intracellular activities has been considered direct evidence that the major components of evoked cortical responses represent summated PSPs generated at dendritic sites remote from PSPs recorded by an intracellularly located microelectrode.

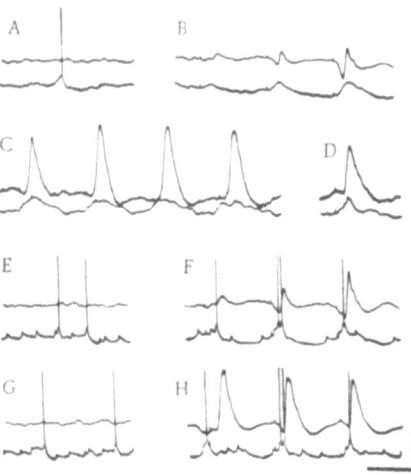

Fig. 1. Development of fast prepotentials or partial spikes presumably of
dendritic origin in a nonpyramidal tract neuron after prolonged weak surface anodal polarization. (A) Control. Spontaneous discharge evolves from
a slow depolarization. (B) First three stimuli of augmenting sequence induced by ventrolateral thalamic stimulation are associated with incrementing EPSPs prior to polarization. (C) Changes in EPSPs during early phases
of anodal polarization. (D) Example of response during late phases of polarization. Note partial spike responses on EPSP. (E) Several seconds after
cessation of anodal polarization. Bursts of partial spikes and occasional
spike discharges are evident. (F) First three stimuli as in B. Note fast prepotentials and partial spikes as well as increased effectiveness of excitatory
synaptic drive. (G and H) 10 sec after E and F. Fast prepotentials and partial spikes persist. Spike amplitude is greater in E through H than in spontaneous discharge (A) prior to polarization. Calibration: 50 mV, 100 msec.
(From Purpura and McMurtry, 1965.)

Perhaps one of the most intriguing findings obtained in intracellular studies of the effects of transcortical polarizing currents
is illustrated in Fig. 1. The neuron impaled at a depth of 0.7 mm
in this experiment exhibited a progressively incrementing EPSP
during repetitive ventrolateral thalamic stimulation (Fig. 1B). A
brief period of weak surface anodal polarization produced a dramatic change in the evoked potential (Fig. 1, C and D) and led to
the development of bursts of rapid depolarizing potentials which
facilitated cell discharge for several seconds after termination of
the anodal polarization (Fig. 1, E–H). Apparently in this case inward currents through distal apical dendritic regions during surface

anodal polarization altered the excitability of proximal dendritic sites sufficiently to induce repetitive partial spikes in these elements. The fact that such partial spikes have never been observed in pyramidal neurons during cathodal polarization of motor cortex may be taken to indicate that when partial spikes occur in neocortical neurons of adult animals these regions do not involve distal portions of apical dendrites. Indeed it is quite likely from the fact that strong membrane hyperpolarization may eventually block partial spikes or fast prepotentials (Purpura and Shofer, 1964) that the latter responses are generated at sites in dendrites that are relatively close to the soma. It must be reemphasized that partial spike generation in dendrites of neocortical neurons in adult animals is an extremely rare event and need not be considered an important feature of the mode of impulse initiation in these cells. Thus the intracellular data obtained in recent years from studies of neocortical neurons of adult animals have amply confirmed the view that spike generation does not occur in dendrites of these cells under ordinary circumstances. It can be shown that this property of neocortical neurons in adult animals is acquired during postnatal development as evidenced by contrasting findings obtained in neocortical neurons of immature animals.

A considerable volume of data has been obtained on the morphophysiological changes observed in feline neocortex during postnatal ontogenesis (Purpura, 1961; Purpura et al., 1964a). It will be recalled that in the early postnatal period pyramidal neurons exhibit well-developed apical dendrites and numerous axodendritic synapses whereas basilar dendrites are poorly developed and axosomatic synapses are rarely encountered (Voeller et al., 1963). Neocortical neurons and synapses as well as evoked cortical potentials acquire characteristics similar to those observed in adult animals by the end of the third postnatal week. However, intracellular recordings from neocortical neurons at this stage have revealed features of "immaturity" which are not obvious from consideration of morphological findings and evoked potential data. The major finding in these intracellular studies of immature neocortical neurons has been the demonstration that under appropriate conditions of synaptic activation impulse initiation may occur in dendrites with all-or-none propagation of the spike potential into the soma (Purpura et al., 1965b).

Observations which have led to the conclusion that spike gen-

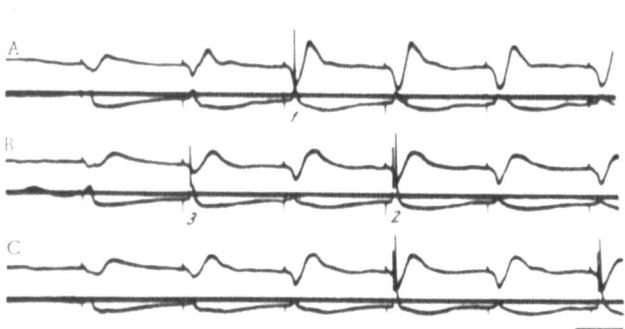

Fig. 2. Changing characteristics of spike potentials during IPSP
summation in a neuron from a 3-week-old kitten. A – C. Three
series of 5/sec ventrolateral thalamic stimulations which elicited
cortical surface augmenting responses. Between each series sufficient time was allowed for membrane potential to return to resting baseline levels. (A) Repetitive stimulation at first elicits a
progressively incrementing EPSP which is superimposed on a synaptically induced hyperpolarization. The third stimulus evokes
an EPSP which triggers an 18-msec latency spike potential (1).
Two different types of spike discharges are illustrated during early
phases of IPSP summation in successive series of stimulations (B
and C). One exhibits two components (2), the other has a prominent shoulder of delayed depolarization (3). Both types evolve
from a level of soma hyperpolarization produced by IPSP summation. Spike potentials in B and C are of shorter latency than the
spike evoked by a prominent EPSP (1). Also note that spikes in B
and C exhibit rapid rise times and lack slow depolarizing prepotentials. Calibration: 50 mV, 100 msec. (From Purpura et al.,
1965b.)

eration and propagation may occur in dendrites of immature neocortical neurons are summarized in Fig. 2. The upper channel
recordings in this figure represent the cortical surface-evoked
responses to 5/sec stimulation in the ventrolateral thalamus of a
21-day-old kitten. The lower channel responses are simultaneous
intracellular recordings obtained from a neuron in the sensorimotor cortex. A third beam trace is superimposed on the baseline
intracellular record. Three periods of 5/sec stimulation are illustrated (Fig. 2, A-C). It is evident that the first stimulus elicits
an intracellular response characterized by a prolonged synaptically evoked membrane hyperpolarization or IPSP. Although the
second and subsequent stimuli in Fig. 2A elicit a progressively

incrementing early EPSP there is some tendency for the prolonged IPSP to exhibit summation. The third stimulus in Fig. 2A
elicited an EPSP which triggered a "normal" spike (1) at a critical level of soma depolarization. In Fig. 2, B and C, it can be
seen that when the IPSP exhibited more pronounced summation,
spike potentials were elicited which exhibited rapid rise times and
were not preceded by a slow depolarizing potential. A spike (2)
in Fig. 2B exhibits a second 50-mV-component on the shoulder of
the first spike whereas the shoulder of delayed depolarization is
seen on another spike (3). Similar spikes without depolarizing
prepotentials are noted in Fig. 2C.

Findings illustrated in Fig. 2 have been interpreted as follows. Although EPSPs ordinarily secure impulse initiation in immature neurons at soma –initial axonal segment sites, the prolonged and powerful IPSPs depress primary impulse trigger sites
but do not affect EPSPs generated in dendrites. Continued stimulation of axodendritic excitatory synaptic pathways results in facilitation of dendritic EPSPs and impulse initiation in dendrites.
Spikes generated in this fashion propagate into the soma with a
relatively high safety factor despite the membrane hyperpolarization associated with the IPSPs. Spikes of dendritic origin recorded from the soma thus lack depolarizing prepotentials and
may exhibit one or more components if spike initiation occurs at
several loci in dendrites. In addition to the spike potentials shown
in Fig. 2 B and C, a relatively large number of immature neocortical neurons have exhibited fast prepotentials or partial dendritic
spikes as well as marked variations in spike potential configuration. The data obtained in intracellular studies of immature neocortical neurons provide the first description in the mammalian
central nervous system of the characteristics of intracellular
spikes of dendritic origin. The fact that similar types of spikes
have not been observed in neocortical neurons of adult animals
suggests that the excitability properties of neocortical dendrites
are dramatically altered during late phases of postnatal ontogenesis. The implication here is that such ontogenetic changes in dendritic properties of neocortical neurons may be as important for
the normal functional development of the cerebral cortex as more
obvious morphological changes in neuronal and nonneural elements. Whether membrane changes in neocortical dendrites resulting from various traumatic insults might lead to the reappear-

ance of spike generating properties in dendrites is a question of
no little importance for the understanding of the development of
abnormal excitability in neurons located in chronic epileptogenic
lesions of neocortex in adult animals.

Spike Generation and Propagation

in Hippocampal Dendrites

The demonstration of orthodromic spikes without depolarizing prepotentials in immature neocortical neurons has prompted
us to reexamine the properties of neurons in other than neocortical locations in the adult brain. In view of the relatively large
proportion of hippocampal neurons which exhibit fast prepotentials
or partial spikes, presumably of dendritic origin, during spontaneous or subiculum evoked activity (Spencer and Kandel, 1961a), it
was considered likely that under appropriate conditions of axodendritic synaptic bombardment impulse initiation might occur in
dendrites of hippocampal neurons with all-or-none propagation into the soma. To test this hypothesis two types of experiments
were carried out in adult encéphale isolé cats. In one series of
preparations stimulation of the subiculum at different frequencies
was carried out during intracellular recording from hippocampal
neurons. When such stimulation was controlled to avoid seizures
characterized by cyclic episodes of soma depolarization and hyperpolarization, alterations in spike potentials were noted such as
those illustrated in Fig. 3. During the early phase of repetitive
stimulation there was an initial increase in EPSPs followed by
marked attenuation, then elimination of EPSPs. Immediately
thereafter EPSPs and "normal" spikes were not observed. With
the build-up of negative seizure waves (in recordings from the
ventricular surface) intracellular spikes without depolarizing prepotentials were noted. These spikes were generally preceded by
brief negative deflections as shown in Fig. 3, E and F. Spike discharges with pre-spike negative deflections were generally of
smaller amplitude than "normal" spikes evoked by EPSPs. Rhythmically recurring spikes during subiculum-induced seizure activities also frequently exhibited one or more inflections on rising
phases and prolonged depolarizing after-potentials. Additional
characteristics of spike potential changes are shown in Fig. 3, G
and I. It will be noted that during early periods of recovery of
normal responsiveness subiculum stimulation elicited a prominent

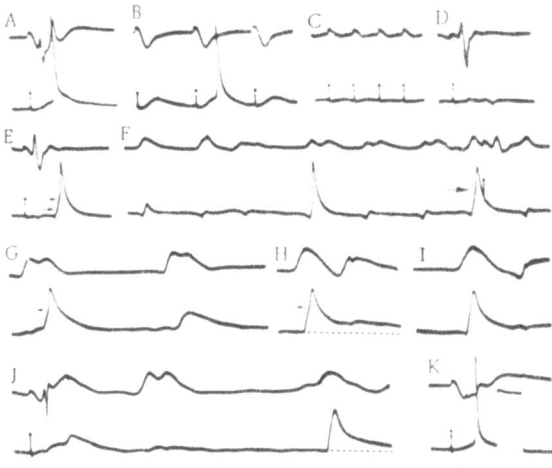

Fig. 3. Reversible changes in EPSPs and spike potentials during seizure activity initiated by repetitive subiculum stimulation. (A) Control response to 1/sec subiculum stimulation. Short-latency EPSP elicits a 75-mV spike potential which is followed by a prominent depolarizing after-potential. The surface evoked response exhibits a multiphasic spike-like component and late negativity which is temporally related to the spike discharge of this element. (B) Stimulus frequency is increased to 20/sec. Early phases of the EPSP are augmented in amplitude but cell discharges are not secured with all EPSPs. The spike potential is decreased in amplitude and increased in duration. Surface evoked positivity has lost all superimposed multiphasic components. (C) During 50/sec subiculum stimulation EPSPs are severely attenuated or eliminated along with cortical surface responses. Note minimal membrane depolarization (4-6 mV) during the period of repetitive stimulation. (D) Immediately after resumption of 1/sec subiculum stimulation an EPSP is not evoked in this element although a cortical surface response occurs. (E) Several seconds later, before appearance of seizure activity in surface recordings, subiculum stimulation elicits a long latency discharge associated with a pre-spike negativity. Several inflections on rising phase are indicated by arrows. Note prolongation of late phases of spike. (F) Rhythmically recurring negative deflections observed in intracellular records are occasionally associated with spike potentials of varying configuration. A partial response is associated with the first negative deflection. Subiculum stimulation after the last spike in this series is ineffective. (G—I) Various types of "spontaneous" depolarizing potentials observed during late phases of seizure activity. (G) Initial component of discharge is triggered by a slow depolarization. (H and I) Discharges with prolonged depolarizations arise directly from the base. (J) Early recovery of subiculum evoked EPSPs which fail to evoke typical spike potentials. "Spontaneous" discharge arising directly from baseline has an extraordinarily prolonged decay phase. (K) 20 sec after J. Although residual alterations in surface-evoked activity are detectable, EPSP and spike potential exhibit complete recovery. No significant change in baseline membrane potential has occurred during the entire sequence of PSP and spike potential changes shown in E-K. Calibration: 50 mV, 20 msec. (From Purpura et al., 1966c.)

compound EPSP which failed to trigger a spike discharge (Fig. 3J).
However, spontaneous depolarizing potentials with prolonged declining phases were observed during this period. These dramatic
changes in spike potentials were entirely reversible (Purpura et
al., 1966b). It has been possible to show in these studies that the
negative deflections preceding the spikes with prominent depolarizing after-potentials are intracellular reflections of large amplitude extracellular negative field responses. The latter result
from the summation of discharges of adjacent units during the
"activated state" induced by repetitive subiculum stimulation. During this "activated state" extracellularly recorded giant negative
spikes of 10-20 mV are frequently observed. The characteristics
and distribution of these giant extracellular negativities indicate
that in the "activated state" induced by synaptic bombardment of
hippocampal dendrites the soma —dendritic membrane of hippocampal neurons is involved in spike generation. Thus spikes arising in remote dendritic sites, probably in the synaptic fields
served by the temporo —ammonic pathway, propagate into the soma
with a high but variable safety factor in the "activated state." It
is not clear at this time what factors are operating to condition
the appearance of spikes without depolarizing prepotentials during
subiculum induced seizures. However, there is some evidence
that depression of excitatory synaptic pathways which generate
EPSPs in or close to the soma is prerequisite for the development
of impulse trigger sites in dendrites (Purpura, 1966). Additional
effects of subiculum stimulation including prolonged soma —dendritic conductance changes may also play a role in facilitating the
development of dendritic spikes.

A second type of study that has been particularly important
in providing additional information on the processes involved in
spike generation in hippocampal dendrites is illustrated in Fig. 4.
In this study intracellular recording from hippocampal neurons
was carried out during single shock subiculum stimulation and
concomitant transhippocampal polarization. Application of polarizing currents to the ventricular surface of the exposed hippocampus in these experiments produced effects on hippocampal neurons
which were opposite to those observed in neocortical pyramidal
neurons following motor cortex polarization (Purpura and McMurtry, 1965). In the case of hippocampal neurons, anodal polarization hyperpolarized the soma and markedly attenuated soma

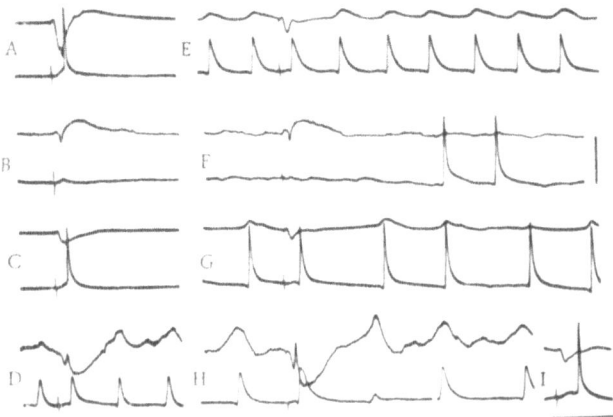

Fig. 4. Reversible effects of transhippocampal polarization on EPSPs and spike potentials. Upper channel responses were monopolarly recorded from the ventricular surface
of the hippocampus (CA_2) close to the site of introduction of a 2 M K^+- citrate filled
micropipette employed for intracellular recording. One polarizing electrode (Ag-
AgCl-saline bridge) was placed a few mm from the surface and intracellular electrodes; the other was located on temporal or nuchal muscles. Negativity signalled by
an upward deflection in upper channel alvear surface responses. Intracellular potentials displayed on an oscilloscope operated on a c mode. (A) Control response elicited
by single shock (1/sec) subiculum stimulation. (B) A few seconds after onset of anodal
polarization (0.8 mA) of the alvear surface. EPSP is markedly attenuated and late surface positivity is eliminated. (C) Immediately after cessation of anodal polarization.
Recovery of evoked EPSP and spike potential is not complete. (D) Two seconds after
onset of cathodal polarization (0.8 mA). (E) Cessation of cathodal polarization and full
development of seizure activity in which negative surface waves are associated with
intracellular spike potentials with prespike negative deflections. (F) Anodal polarization during seizure shown in E interrupts seizure discharges and hyperpolarizes soma as
indicated by increased amplitude of spikes without depolarizing prepotentials. (G) Resumption of seizure discharges immediately after termination of anodal polarization.
(H) Superimposed cathodal polarization in late phase of seizure initiated by first period of cathodal polarization (D). (I) Early recovery of subiculum evoked responses
30 sec after cessation of seizure activity. Calibration: 50 mV, 50 msec. (From Purpura and Malliani, 1966.)

EPSPs whereas cathodal polarization of the ventricular surface
depolarized the soma and basilar dendritic system (Purpura and
Malliani, 1966). Cathodal polarization also triggered repetitive
spikes of low amplitude which exhibited negative prepotentials similar to those observed during subiculum-induced seizures (Fig. 4D).
Such spikes without depolarizing prepotentials were increased in

amplitude following termination of cathodal polarization (Fig. 4E).
Subsequent anodal polarization blocked the rhythmicity of these
discharges and revealed only sporadic spikes of large amplitude.
The latter spikes arose directly from the baseline and exhibited
prolonged depolarizing after-potentials (Fig. 4F). Seizure activity
reappeared following cessation of anodal polarization (Fig. 4G)
and was augmented during another period of cathodal polarization
(Fig. 4H).

The different effects of anodal and cathodal polarization of
the surface of the neocortex and ventricular surface of hippocampus are attributable, in part, to the different geometrical orientation of the cell bodies of pyramidal neurons in these two types of
cortex in respect to the location and direction of the applied currents. Thus anodal polarization of neocortex hyperpolarizes apical dendrites and depolarizes pyramidal cell bodies whereas anodal polarization of the ventricular surface of the hippocampus hyperpolarizes the soma and depolarizes apical dendrites. Cathodal
polarization produces opposite effects but additional complications
are apparently introduced in the case of hippocampal neurons by
the potentiality of dendrites of hippocampal neurons for developing
spikes in response to strong cathodal currents. Once such spikes
are initiated they may be altered in amplitude and frequency depending upon the direction of soma membrane potential change produced by subsequent anodal or cathodal currents. Thus unlike apical dendrites of neocortical pyramidal neurons of adult animals,
hippocampal dendrites respond to steady depolarizing currents
with repetitive discharges. Such spikes recorded from the soma
arise without depolarizing prepotentials and have frequency characteristics similar to those observed during "activated states" induced by repetitive subiculum stimulation. Studies of the effects
of transhippocampal polarization thus indicate that the appearance
of rhythmically recurring orthodromic discharges without depolarizing prepotentials is largely a consequence of prolonged changes
in hippocampal dendritic excitability rather than the development
of complex interactions in intra- and extra-hippocampal synaptic
pathways. It is not unlikely that cathodal polarization is effective
in this respect by inducing "pacemaker-like" potentials in distal
dendritic regions which lead to rhythmically recurring synchronous
discharges of dendrites. Viewed in this fashion the frequency char-

acteristics of orthodromic spikes of dendritic origin would be determined by the time course of the depolarizing process initiating dendritic spikes as well as the prolonged post-spike conductance changes underlying the prominent depolarizing after-potentials of spikes propagating into the soma from dendritic sites.

Properties of Dendrites of Specific
Thalamic Relay Cells

This brief survey of the mode of impulse initiation in neurons of the mammalian brain would be incomplete without mention of recent findings from our laboratory on the properties of thalamic neurons in the ventrobasal (VB) complex (Maekawa and Purpura, 1966). For present purposes it is sufficient to point out that intracellular studies of specific VB cells were primarily initiated to provide information on the manner in which these elements are activated by lemniscal afferents. It is well established that lemniscal afferents are distributed largely in relation to dendrites of VB cells (Bowsher, 1961; Scheibel and Scheibel, 1966). Despite this predominant axodendritic distribution of lemniscal afferents, it has been shown that lemniscal—VB relays are capable of following relatively high-frequency afferent volleys with great fidelity (Poggio and Mountcastle, 1963). Previous studies had indicated that specific evoked EPSPs in relay cells of the ventrolateral thalamic nucleus are of relatively small amplitude and short duration. However, these EPSPs are extremely effective in securing cell discharges, occasionally through an intermediary rapid depolarizing prepotential (Purpura et al., 1966d).

It should be emphasized here that analysis of the mode of impulse initiation in VB cells has been possible only in those cells exhibiting large amplitude (50-60 mV) spike potentials and absence of injury discharges following impalement. Any degree of traumatic depolarization is sufficient to "mask" variations in impulse initiation as a consequence of the induced soma-depolarization which greatly facilitates impulse initiation at "primary" trigger sites in the soma—initial axonal segment. The analysis has also been aided by transmembrane polarization of VB cells in order to reveal contributions of fast prepotentials to orthodromically evoked responses.

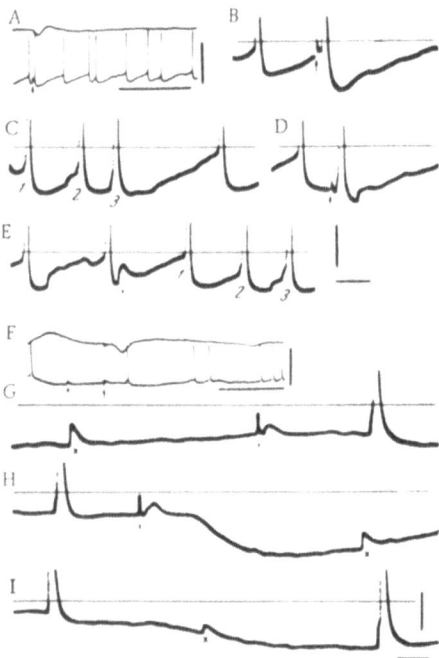

Fig. 5. Variations in firing level and characteristics of fast prepotentials in ventrobasal thalamic relay cells. (A-E) Intracellular recordings from one element. (A) General features of 55-mV spike potentials during lemniscal-evoked discharges and spontaneous activity. Upper channel record is specific evoked response recorded from posterior sigmoid gyrus. (B-E) Photographic enlargements of samples of records taken during various phases of activity. (B) Lemniscal stimulation (at arrow) elicits a spike potential which is preceded by a barely detectable EPSP. (C) Variations in firing level during spontaneous activity. Note increase in amplitude of rapid depolarizing prepotential (1-3). (D) Lemniscal stimulation during post-spike hyperpolarization elicits an EPSP which triggers cell discharge below firing level of spontaneous spike. (E) Another series of spontaneous spikes and an EPSP (at arrowhead) which fails to trigger a discharge. (F-I) Recordings from another cell. Different amplitudes of partial spikes are shown marked with an (X). Note the difference in rise time of these responses compared with lemniscal evoked EPSPs in G and H (at arrow). A small second component is seen superimposed on the evoked EPSP in H. Large depolarizing prepotentials are evident in spontaneous spikes during IPSPs in H and I. Calibration: A, F 50 mV, 100 msec; B-E and G-I 10 mV, 10 msec. [From Maekawa and Purpura (unpublished).]

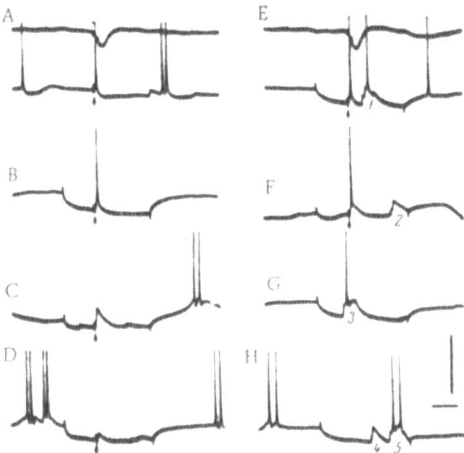

Fig. 6. Intracellular recordings from a VB cell which exhibited marked spontaneous variations in firing level and summations of rapid depolarizing prepotentials or partial spikes. (A) Lemniscal-evoked response (at arrow) is succeeded by an IPSP. (B-H) Responses during induced membrane hyperpolarization as indicated by on-and-off artifacts of current pulse. (B) During hyperpolarization a second component is superimposed on EPSP. This second component or partial spike is seen in the absence of spike discharge in C. (D) The EPSP is shown in isolation. (E) Spike discharge is secured (at arrow) by lemniscal stimulation whereas a spontaneous spike is secured by summation of multiple partial responses (1). (F-H) Additional partial spikes are labeled (2-5). These exhibit variable degrees of summation leading to cell discharge during membrane hyperpolarization. Spontaneous responses are shown in G and H. Anode-break responses are not commonly observed in VB cells unless they are traumatized by impalement or affected by very strong hyperpolarizing currents. However, these currents must produce increases in membrane potential which are well above the reversal potential for IPSPs in order to elicit anode-break responses. Calibration: 50 mV, 20 msec. [From Maekawa and Purpura (unpublished).]

Examination of spontaneous and evoked spikes of VB cells has indicated that under optimum conditions spike potentials may be triggered at different levels of membrane polarization. These marked variations in the firing level of spontaneous discharges are illustrated in Fig. 5 A—E. Spikes might be triggered by slow or rapid depolarizing prepotentials. The latter are particularly prominent during increases in membrane polarization resulting from post-spike undershoots or IPSPs. It is of interest that rapid depolarizing potentials or partial spikes may be of different amplitude in suitable recordings from VB cells (Fig. 5, F—I).

Studies of the effects of induced membrane hyperpolarization on lemniscal evoked EPSPs have established that in the vast majority of cells, specific EPSPs are uninfluenced by inward current injections which increase membrane potential 20–30 mV above resting baseline values. Weak inward currents may reveal isolated small EPSPs (Fig. 6D) or EPSPs with superimposed partial spikes (Fig. 6C). Rapid depolarizing prepotentials of various amplitudes frequently summate to secure impulse initiation during spontaneous activity (Fig. 6E, G and H). Spikes arising without depolarizing prepotentials during IPSPs succeeding evoked spikes have also been observed in this study. These intracellular data from thalamic VB cells have been interpreted as indicating that axodendritic EPSPs are capable of eliciting partial or full spikes in dendrites of VB cells. The finding that the latter elements may respond at high-frequency (200–400/sec) to steady depolarizing currents taken together with observations on the variations in firing level of orthodromic spikes provides a basis for understanding the role of dendritic inputs in securing high-frequency discharges of these relay cells (Maekawa and Purpura, 1966).

Comment

It is not necessary to belabor the point, which is self-evident from the foregoing remarks, that dendrites of neurons in different locations in the mammalian brain exhibit marked differences in their potentiality for spike generation. The vast amount of intracellular data obtained from neocortical neurons of adult animals is consistent with the hypothesis that dendrites of these elements do not ordinarily exhibit spike electrogenesis (Purpura and Grundfest, 1956; Purpura, 1959). However, it is now clear that this hypothesis is not applicable in the case of immature neocortical neurons, hippocampal neurons, and several types of specific thalamic relay cells. To what extent impulse initiation may occur in other types of cells under different conditions of activation or, for that matter, in neocortical neurons of adult animals during abnormal states of activity remains a problem of no little importance.

It has been argued that partial spike generation in dendrites may serve to enhance the efficacy of remote dendritic synapses in securing impulse initiation in soma —initial segment regions (Eccles, 1960; Spencer and Kandel, 1961a). The studies described above go one step further in showing that in the case of immature neocortical neurons (Purpura et al., 1965b) and hippocampal neurons (Purpura, 1966) depression of primary impulse trigger sites in the soma — initial axonal segment regions by IPSPs generated in or close to the soma may selectively inhibit axosomatic excitatory inputs but not axodendritic EPSPs. Thus the capacity for spike initiation and propagation in dendrites may provide a mechanism for securing cell discharge despite selective depression of axosomatic EPSPs.

The question arises whether spike generation and propagation ever occur in dendrites of hippocampal neurons under normal physiological conditions. In the case of immature neocortical neurons spike generation in dendrites may increase the safety factor for axodendritic synaptic activation at a developmental stage in which axosomatic excitatory synapses are poorly developed. However, in the hippocampus, development of propagating dendritic spikes is usually associated with augmented states of hippocampal excitability and seizures. This suggests that factors underlying the "suppression" of impulse initiation in neocortical dendrites during postnatal ontogenesis are also operating in hippocampal neurons, but to a lesser extent. Evidently the maturational "suppression" of impulse initiation in dendrites of neocortical neurons contributes to the normal functional activity of these elements by requiring integration of axosomatic with axodendritic synaptic activities for the latter to be effective in other than a "modulating" capacity. It is of interest that the major afferent projections to both hippocampal neurons and thalamic ventrobasal relay cells distribute largely in relation to dendrites of these elements. The greater potentiality for spike generation in dendrites of hippocampal neurons and VB cells apparently confers additional degrees of freedom for maintaining high-level input–output relations in these axodendritic synaptic pathways.

SUMMARY

Intracellular studies of a variety of neurons in the mammalian brain have been carried out to reexamine the potentiality for dendrites to exhibit partial or all-or-none spikes. Intracellular data obtained from neocortical neurons of adult animals has provided support for the earlier view that dendrites of these elements do not ordinarily exhibit spike electrogenesis. This is in marked contrast to findings in immature neocortical neurons. Under appropriate conditions of synaptic activation depression of "prima-ry" impulse trigger sites in the initial axonal segment—soma re-gion may lead to a shift in the site of impulse initiation into den-drites. Spike potentials recorded from the soma in these in-stances have rapid rise times and lack depolarizing prepotentials. Immature neocortical neurons also exhibit marked variations in configuration and frequently show fast prepotentials or partial spikes of dendritic origin.

Investigations of hippocampal neurons have disclosed that repetitive stimulation of axodendritic synaptic pathways arising in the subiculum may result in the appearance of intracellular spikes without depolarizing prepotentials. These spikes are usu-ally associated with pre-spike negative deflections which are iden-tified as intracellular reflections of giant extracellular negative responses. Orthodromic spikes without depolarizing prepotentials are also elicited in hippocampal neurons during cathodal polariza-tion of the alvear surface of the exposed hippocampus. Data ob-tained in these studies indicate a high potentiality for all-or-none spike generation in hippocampal dendrites of adult animals.

Intracellular recording from thalamic ventrobasal (VB) cells has also been carried out to define the manner in which lemniscal evoked EPSPs effect synaptically secure discharges of VB cells. Analysis of the effects of induced membrane polarization and ob-servations on variations in the mode of impulse initiation indicate that lemniscal evoked EPSPs are generated largely in dendrites and that such EPSPs are effective in securing repetitive dis-charges by virtue of their capacity to elicit partial and full spikes in VB cell dendrites. It is evident from these studies that spike generation in dendrites may be an unusual or abnormal activity in some elements, i.e., neocortical neurons of adult animals, where-as in other cells spike generation in dendrites may be an effective mechanism for maintaining high-level input-output relations es-pecially in those neuronal organizations in which major afferent projection systems are distributed exclusively in relation to den-drites.

Analysis of Information Concerning the Electrical Activity of the Brain*

A. Remond

*Hôpital de la Salpêtrière
Paris, France*

During World War II a group of communications specialists studying conditions of transmission of messages by telegraph, telephone, or radio, succeeded in rationalizing the significance of these messages by defining and measuring their inherent properties, such as their "quantity of information," communication, etc. The information theory which resulted from these investigations has had a profound influence on technology as well as on scientific thought. It has probably been largely responsible for the development and success of large modern computers, which can equally well be called information analysis machines, a term emphasizing both their function and their origin.

Electrical messages arising in the central nervous system differ considerably depending on whether they are obtained from the brain surface, forming what is classically the electroencephalogram, or whether they are recorded directly from the interior of individual nerve cells by means of microelectrodes. It is by means of microelectrodes that neurophysiologists are investigating the nature of elementary brain function. So far as the electroencephalogram is concerned, its use in medicine has developed considerably because of the original diagnostic possibilities thereby introduced into neuropsychiatry.

Elementary neural messages and global EEG messages are so complex that there is now an increasing tendency to deal with them or to simplify their study by means of automatic information analysis.

---

*Pages 226-234 in the Russian edition.

Difficulties of EEG Interpretation

The physician examining an electroencephalographic trace for the first time is surprised by its apparent complexity. His difficulty is all the greater because, when considering the general appearance of the EEG, he automatically compares it with that of the electrocardiogram, whose reassuring monotony greatly simplifies the detection of trivial abnormalities. Instead of the single line or occasionally the three lines recorded simultaneously on the EKG, on the EEG he sees 8 or 16. In their details, the different curves appear to him to be hardly comparable. He finds it difficult to understand how the variations in the shape of the curves with time are organized. Finally, if he compares the electroencephalograms of two individuals chosen at random, he may find them astonishingly different (Fig. 1).

Specialists seeking to derive benefit from these recordings thus find themselves at a considerable disadvantage when attempting to decipher and interpret them. How can they describe or understand the rhythms constituting the basic pattern of the electroencephalogram, which consist of waves of unequal amplitude and duration but with a certain more or less apparent mean value; the irregular bursts into which these waves are grouped; the slowly or quickly changing character of these waves and bursts. This polymorphism, both static and dynamic at the same time, interferes with the definition or diagnosis of the more transient signs of organic brain diseases: vascular, neoplastic, or others capable of modifying the basic picture in widely different ways over a variable, though sometimes discrete, length of the recording. Epilepsy, with its violent but transient graphic storms, usually subclinical in their course, is undoubtedly the disease with the most spectacular electroencephalographic findings, giving the greatest encouragement to those who read these recordings by reassuring them from time to time that their careful inspection is justified.

The Need for Analytical Assistance

The constant search for more typical formulas and wave forms with more precise meaning, as well as for criteria for their definition is still a constant preoccupation, if not a mirage, of specialists in electroencephalography. The qualifications required of those whose work consists of reading electroencephalograms, which

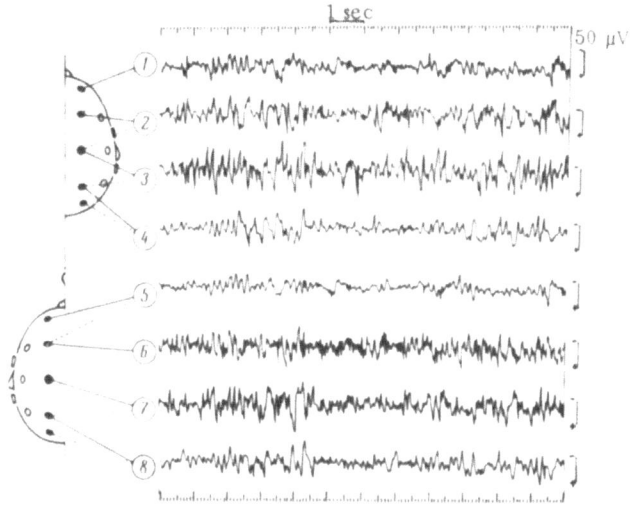

Fig. 1. 8-channel electroencephalogram showing electrical activity of both hemispheres of the human brain taken longitudinally for a period of 12 sec. Note differences in details during this period and between one lead and another. Calibration 50 μV, 1 sec.

include shrewdness, ability to see every line as a statistical mean of its elements and not as the elements themselves, the ability to perceive the ideal form of a characteristic element of the curve through all the distortions produced by the situation of the electrodes or the nature of the leads, explains why so few excel in this field. They also explain the profound wish of many of them to find a method of simplifying, organizing, and controlling the nature of their analytical methods, to take off their shoulders part of the unnecessary load they have to carry.

In this way the need has arisen for replacing visual observation by an automatic type of observation, based on it or complementary to it, so that the elements of a statistical evaluation can be examined without unnecessary fatigue. It is only by applying information analysis methods to the EEG that this requirement can be met. The urge to satisfy this need has recently stimulated research at several centers with the objective of producing automatic machines, based on simple principles, capable of partial decoding of the EEG.

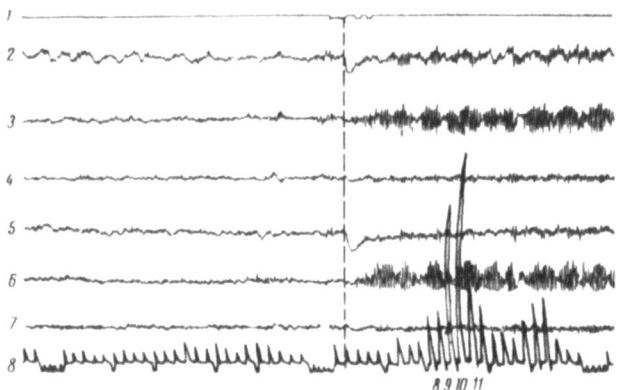

Fig. 2. Frequency analysis of the human EEG. After closing the
eyes, brain activity changes sharply, with appearance of an alpha-
rhythm. Underneath the EEG curves the relative importance of
energy in each frequency band for a period of 10 sec for a series
of successive frequencies from 0.5 to 24 cps is depicted in the
form of a spectrum. On the right half of the figure, the alpha-
rhythm is apparently distributed among frequencies from 8 to 12.
1) Record of blinking; 2) EEG of right frontal cortex; 3) right
occipital; 4) right temporal; 5) left temporal; 6) left occipital;
7) left temporal; 8) trace of frequency analyzer. Broken vertical
line represents moment of closing eyes.

Attempts at Frequency Analysis

The first serious attempt to mechanize the analysis of the
electroencephalogram was made 20 years ago and was based on
frequency analysis.

By means of instruments consisting essentially of a series
of electrical filters, each allowing one particular frequency to
pass, at the end of a period of approximately 10 sec (the "epoch")
the machine was able to determine the distribution of the relative
importance of the frequencies occurring during that time. This
distribution, called the frequency spectrum, was superscribed on
the original EEG (Fig. 2).

Since the EEG frequency spectrum shows little variation
from one epoch to another, its appearance is reasonably typical of
the region of the brain from which the EEG is recorded and also
shows individual characteristics. However, although some prop-

erties of the electroencephalogram were discovered in this way, the method itself was unable to develop because of the complexity of its use and its inconvenience from the practical point of view. It could only be applied to a single EEG trace at a time, and the necessity for frequency shifting from one trace to another made it impossible to compare the activities of different parts of the brain simultaneously.

The appearance of magnetic recorders, capable of reproducing the electroencephalogram in a form capable of easy interpretation, together with the increasing availability of universal digital computers, working to set programs, completely changed the experimental approach to the analysis of electroencephalograms by rendering it more flexible and giving it statistical authority.

The Use of Modern Computers for

Analysis of the Electroencephalogram

When applied to electroencephalography, information analysis in its present form attempts to reveal, by means of a new classification of EEG data, the most demonstrative or hitherto unknown properties, and to clarify the relationship between the EEG and the anatomical, functional, and pathological features on which it is based. From the practical point of view three quite distinct phases in this method of analysis must be considered: obtaining the data, their analysis, and their presentation. In the case being considered, the data are the "amplitude values at a certain moment," the EEG curves merely indicating how these data or amplitude values vary in the course of time.

Obtaining the Data. This phase begins, just as in conventional electroencephalography, by recording the potentials on the scalp by means of electrodes and classical amplifiers. Besides recording on paper, a parallel record is made on magnetic tape, which behaves as a "memory," or in other words it can be reexamined as many times as is necessary, when it reproduces each time the fluctuations of potential similar to those produced by the brain at the time of the initial examination. These recordings on multitrack magnetic tape have undergone considerable improvement, and they form an intermediate memory, capable of multiplying the possibilities of direct analysis. Examination of

the results obtained by the first analysis of a magnetic record
may lead to the decision that different methods of reproducing the
information should be used the next time the tapes are read.

Since the recording of the EEG on magnetic tapes is analog
in type, it cannot be used directly with simple computers of digital type. The "continuous" instantaneous amplitude values contained on the tapes must first be transformed into digits. This
operation is done by a special instrument called an analog-digital
converter. Every millisecond, the voltage values contained on the
tapes are removed simultaneously from each curve of the multichannel recording, and then fed in turn into a coding voltmeter,
which transforms electrical magnitudes into their digital expression with a chosen level of accuracy (for example, with an accuracy of up to three significant figures). This digital expression is
transferred for future use onto punched cards, where the digits
appear as perforations in certain positions on reference columns,
or onto "digital" magnetic tapes, where the digits are expressed
in a system of binary numeration, the system most suitable for
computers.

Analysis of Data. The electroencephalographic data
thus obtained and processed can be fed when required into a computer, which is given a program for their analysis. The program
contains instructions indicating the order in which the introduced
data are to be used, and also the character and order of the operations which the computer must perform when analyzing the EEG
traces to be investigated. On the basis of this program, the computer performs a successive series of operations at tremendous
speed, bringing the data into a situation in which the necessary
arithmetical operations can be carried out with them (for example,
adding them in a certain order, multiplying or dividing them, and
so on), and ultimately it finds and stores the intermediate or final
result of the analysis. When all the operations comprising the
analysis are complete and the results have been collected in the
memory of the computer in the order assigned by the program,
they are ready for delivery to the apparatus presenting the results of analysis.

Presentation of Results. Presentation of the re-
sults consists of the provisional or final transfer of results in-

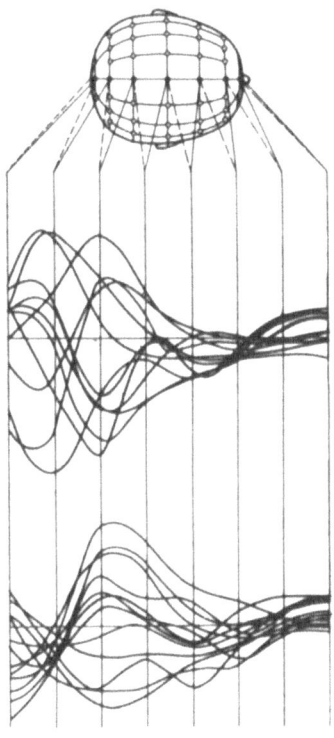

Fig. 3. Series of topograms representing instantaneous values of potentials differences at points located along the midline of the head for an alpha-rhythm. Potentials plotted along vertical lines; localization of electrodes along horizontal lines. A maximum is seen in the parietal region, with a somewhat different configuration in the upper and lower parts of the figure.

scribed in the memory of the output system. This may be a rapid writer, capable of writing from 300 to 1000 lines/min, or an oscilloscope or curve tracer. Depending on the instructions on the program, the results appear in the form of digits or curves obtained from these digits converted into analog magnitudes and expressing the new disposition of the information generated by the chosen program from the data which were introduced into the input of the computer.

The succession of phases given above and of the partial operations may appear complicated, and in fact it consists of a very large number of small elementary manipulations. However, for the user who has no need to bother about all these details, everything happens in a moment because of the fantastic working speed of the electronic devices concerned.

Analysis of Electroencephalographic
Information at La Salpêtrière

Several years ago work began at the laboratory of electroencephalography of the Hôpital de la Salpêtrière on the development of these new methods in the attempt to determine the nature
and forms of the analysis, the logical or mathematical transformations which would enable the largest number of original methods to be applied to the study of electroencephalograms, whether
clinically for diagnostic purposes, or in the field of neurophysiology to investigate the global functional mechanisms of brain activity. It would also be used to continue earlier research into the to-
pology of regional activity.

The classical electroencephalogram at best expresses the
evolution of brain electrical activity as a function of time. On the
other hand, it gives very little information on how the simultaneous spatial relationships between these changes are formed. These
relationships, however, must be of considerable a priori importance, because they indicate the spatial arrangement of the maxima of electrical activity and their origin, they reflect the direction
of displacement of these maxima, and they thereby facilitate their
anatomical or functional localization.

A momentary or very short trace consisting of a series of
discrete linear recordings is very easily carried out by these
methods. To transform a group of instantaneous values obtained
in this way into a continuous curve, which is easy to study, an in-
terpolation calculation is carried out automatically, combining the
values for each lead through a series of hypothetical and most
probable values. The instantaneous topograms thus obtained represent the potential contour at a given moment along the line of
electrodes, and thus give a true "section" through the electrical
field of the head (Fig. 3).

The shape of these instantaneous topograms naturally varies
from one moment to another. To represent many of them in their
correct order and to save space as much as possible, they are
given as a topographic map. Not all the amplitudes of individual
topograms are given. Only those which are regularly spaced on
the height scale of the map are shown. All that remains of each
topogram is the projection of points on the curve having these

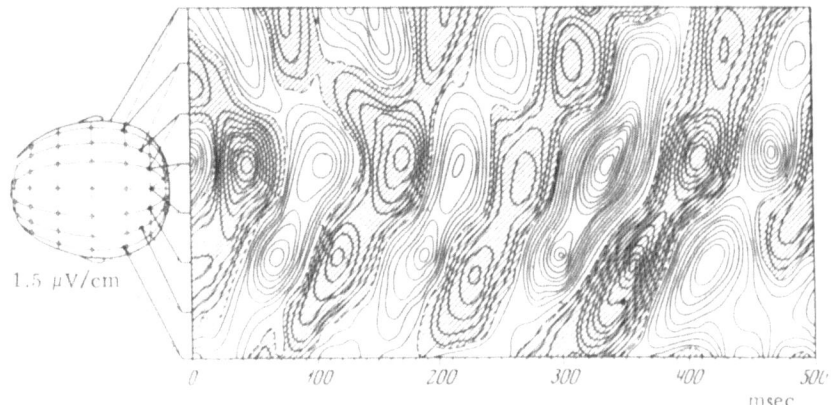

Fig. 4. Spatio-temporal map of the alpha-rhythm. The eight lines of the electroencephalogram are replaced by a map in which the curves represent isopotential levels.
White areas are positive, black areas negative. Levels are organized homogeneously
and the maximum of the alpha-rhythm is displaced from the left hemisphere into the
right.

particular values. The maps ultimately indicate the temporal
evolution of the location of points at the same level for all the
levels chosen. The two coordinates of these maps do not have
the same meaning as in the usual topographic maps. They are
therefore given a different name, being known as "spatio-temporal
maps." In these maps the ordinate represents a geometrical dimension, i.e., the line on the scalp joining the series of electrodes
recorded simultaneously, while the other dimension, the abscissa,
represents time. To recognize positive and negative potentials of
the EEG more easily, the first are traced in black on a white
background and the second in white on a black background (Fig. 4).

The first advantage of this new method of presentation of
electroencephalograms is that it is possible to "read between the
lines" of the ordinary recordings (i.e., to obtain information concerning activity of points located between the recording electrodes). They have thus yielded unforseen information on the organization of brain rhythms. When observed by this means, the
alpha-rhythms, for example, show hitherto unknown properties.
In particular, they appear capable of originating in different regions at different times. During certain periods waves of this
alpha-rhythm may appear simultaneously over the whole of their

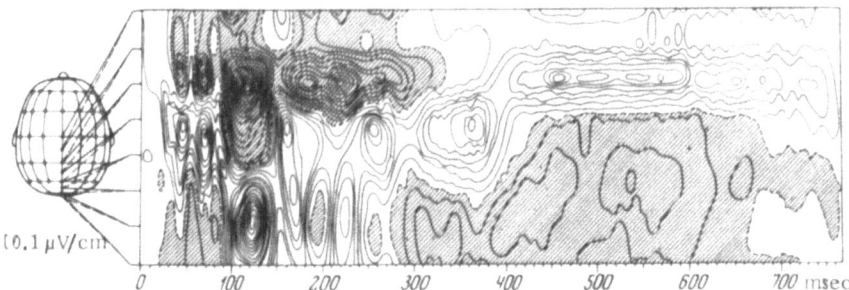

Fig. 5. Spatio-temporal map of averaged evoked potentials representing the mean of 100 successive responses to a flash applied at time 0 (analysis epoch 768 msec). The averaged traces themselves are given in Fig. 6.

territory, affecting both parieto–occipital regions simultaneously, while in other periods they appear to move and to spread over these regions from a certain fixed point of origin, spreading to each point in succession on the surface of their territory. This process does not seem to take place at random. It occurs in definite directions and at apparently definite speeds. It can be influenced, at least in part, by sensory stimulation. This spreading phenomenon may be of great importance to the theory according to which the alpha–rhythm controls or directs in some way or other different cortical activities, which can arise only in certain phases of this process. The spread of the alpha–rhythm thus permits the organization of successive chains of activity from one region to another, without which the proper formation of cerebral mechanisms would perhaps be impossible.

Detection of Latent Phenomena. The electroencephalogram is rightly considered as a mixture resulting from many different phenomena taking place at the same time. Each of them is therefore difficult, if not impossible, to examine in isolation. Information analysis now gives a number of methods for distinguishing some of these aspects of brain activity. For example, this applies to cortical responses to sensory stimulation, usually known as evoked potentials, the amplitude of which may be no more than one-tenth of the amplitude of spontaneous rhythms. Averaging methods, as well as correlation methods, have been used widely to differentiate these responses from masking activity. In the averaging method, for example, use is made of recordings during which the test stimulus is applied many times under identical conditions. The responses to a series of successive short stimuli are then summated, using the time of each stimulation as reference point in time. The larger the number of recorded intervals, the more clearly is revealed what is connected in time with the sensory stimulation. In this way, a typical and reproducible form of the response can be distinguished (Fig. 5). These methods are very sensitive and allow a systematic study of cortical responses to sensory stimulation, for example in the investigation of a lesion affecting their pathways or their relays in the brain.

Reduction of Data. One of the primary aims of analysis of the information in the EEG is to replace the electroencephalogram by an infinitely shorter document, much more characteristic of the functional activity or the anatomical and pathological properties of the brain producing it. The averaged evoked responses discussed above in fact represent a synthetic image highly reduced by statistical methods. Meanwhile, other methods of reduction have been used, based on the topographic presentation of the data, when the activity of a series of leads is averaged in space and in time. Under certain conditions the curves obtained are analogous to the energy contained in the EEG signal, and enable more precise or more general comparisons to be made. These comparisons, when large numbers are involved, may also be performed by computers, which indicate with high precision the existence of a difference or of correlation, and give the actual value of the coefficient of correlation between two responses or a whole series of responses classified in a particular order. Another type of reduction which has also been used in connection with the spontaneous EEG is the power spectrum of the potentials generated along the line of recording electrodes (Fig. 6).

The Future of Information Analysis. Analysis of electroencephalographic information by means of modern computers is only in its infancy. Unfortunately, despite the capacity and enormous possibilities of modern computers, problems arising in this field are still far from solved. All that can be obtained from the computer is the answer to the questions put to it, and therein lies the difficulty. Investigators must show their ingenuity in formulating the fundamental problems which must be solved in new terms capable of translation into a program of calculations likely to lead to statistically significant synthetic conclusions.

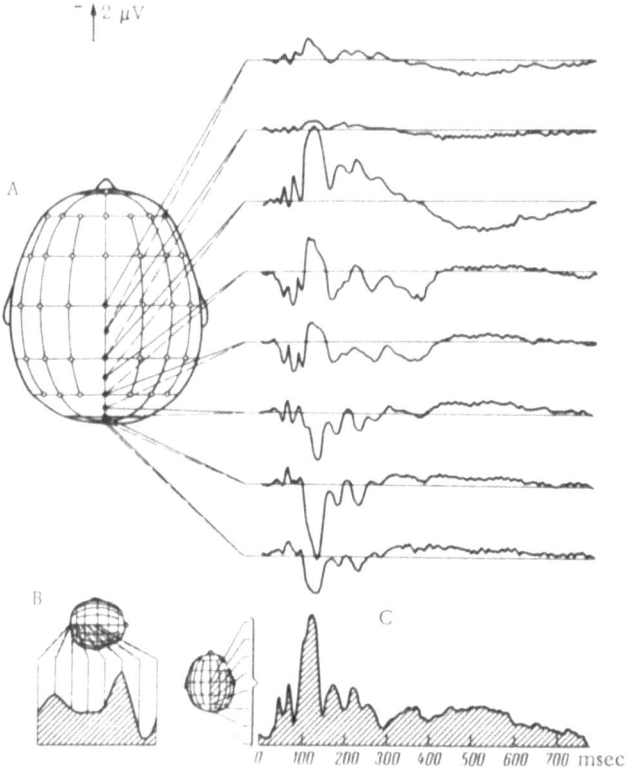

Fig. 6. Averaged evoked potentials used to plot the spatio-
temporal map in Fig. 5. A) Averaged evoked responses presented
in the usual way; B and C) power spectrum averaged responses in
space (B) and in time (C).

Through progress in this field, EEG analysis will become possible,
and will evidently increasingly fulfill its purpose, and the present
methods of describing the electroencephalogram will probably be
considerably modified. These descriptions, in particular, must
become more concise, more accurate as regards localization, and
more effective for diagnosis.

However interesting this progress, its practical importance
may appear to be limited by the high cost of large computers, but
everything is changing rapidly in this field. As the work capacity
of these machines rises, their price falls off rapidly, and the key
to their use in the future may perhaps be for several laboratories
of similar, or even of different type, to group their work. At a

time when electroencephalograms or electrocardiograms can be transmitted without difficulty by telephone or even by radio, the use of a central computer for daily analysis of a large number of electrophysiological documents is merely a question of organization and coordination.

Electrical Excitability of Apical Dendrites[*]

A. I. Roitbak

Institute of Physiology
Academy of Sciences of the Georgian SSR
Tbilisi, USSR

The principal structural elements in the first layer of the cortex
are the ramifications of the apical dendrites, nerve fibers, neuroglia, and small nerve cells. Most fibers of this first layer are
about 0.4–1.0 μ in diameter, although thicker ones are found. Collaterals of the fibers in this first layer terminate synaptically
mainly in the dendritic plexus (Roitbak, 1955, 1964; Okudzhava,
1963). If an electrical stimulus is applied to the cortical surface,
loops of current must spread to all these structures. The electrical response generated around the point of stimulation, a negative
potential 20–30 msec in duration, is interpreted as the postsynaptic potential of the dendrites. It is assumed that impulses arising
in the nerve fibers at the point of stimulation spread along them
and reach the synapses of Gray's type I, where they cause secretion of mediator and generation of the "dendritic potential." In
fact, the response arising several millimeters away from the
point of stimulation must be generated in this way; however, in
the immediate proximity of stimulating electrodes the possibility
cannot be ruled out that other elements may participate in its genesis (Jasper, 1960). The role of the neuroglia in this process is
very doubtful. If it is stimulated directly, stimuli of very high intensity are required to produce depolarization of the neuroglial
cell membrane, and the duration of the response thus produced is
measured in seconds (Hild and Tasaki, 1962). Regarding the presynaptic endings, by analogy with the endings of the neuromuscu-
lar junction it must be assumed that they generate rapid action
potentials of axonal character and that they cannot be recorded
from the surface of the cortex by means of macroelectrodes. So

---

[*]Pages 235-241 in the Russian edition.

far as the dendrites are concerned, it is difficult to determine experimentally whether they possess an electrically excitable membrane or whether they are brought into an active state purely by
the action of a chemical mediator. Evidence in support of both the
first (Eccles, 1964) and the second (Purpura, 1963) hypothesis has
been adduced, but the problem is still at the level of theoretical
argument. During stimulation of the dendrites of a neuron in tissue culture by microelectrodes, excitatory impulses spreading at
a velocity of the order of 0.1 m/sec were recorded (Hild and Tasaki, 1962), but this is still not final proof of the electrical excitability of the dendrites, because under these experimental conditions the dendrites were without synapses. In the present article,
besides this question of electrical excitability of the dendrites, I
shall also examine some properties of the fibers in this first layer.

METHOD

Experiments were performed on adult cats deeply anesthetized with Nembutal to eliminate spike activity of the neurons. The
cortex was exposed and its temperature controlled. Chlorided silver ball electrodes 0.5 mm in diameter were used to record the potentials from the cortical surface. The cortex was stimulated
through a triple electrode consisting of chlorided silver wires
100 μ in diameter cemented together into a triangle whose sides
measured 50-100 μ. Two electrodes were connected to one pole
of the stimulator and the potential at them was regulated by two
potentiometers. The third electrode was connected to the other
pole. Square pulses from a stimulator with radiofrequency output,
with an output resistance of 400 Ω, and with an output for synchronizing the time base of the oscilloscope were used for stimulation.
The first recording electrode was located from 0.5 to 3 mm from
the stimulating electrode, the second more than 10 mm from it.
The cortical potentials were recorded with a type ENO-1 CRO and
a UIPP-2 dc amplifier, or by a Cossor CRO fitted with an amplifier. Recording was carried out under driven sweep conditions.

RESULTS

Response to a Single Stimulus. Recordings obtained 0.75 mm away from the stimulating electrodes and with a

stimulus of increasing voltage are given in Fig. 1A. The amplitude
of the response increases with an increase in the strength of stimulation and its character changes: a positive wave followed by a
second negative wave appears. However, shortening of the latent
period of the response is also observed as the strength of stimulation increases. If the voltage is 6-8 times greater than the
threshold for evoking a negative potential, the response begins to
appear without a perceptible latent period; in this particular experiment the artifact from the stimulus simply interrupted the abscissa for an instant without distorting the subsequent curve. The
minimal synaptic delay, i.e., the time from the apex of the presynaptic impulse to the beginning of the postsynaptic potential, was
0.5 msec. The absence of visible delay in generation of the cortical response to direct stimulation is decisive evidence that the response begins with an immediate response of the stimulated elements, i.e., with a direct response of the dendrites. With a stimulus of low voltage, a response of the dendrites at the recording
point was evidently obtained, arriving there transynaptically. With
an increase in the strength of the stimulus, the loops of current
began to stimulate the dendrites effectively at the point of the recording electrode. This is the probable mechanism of origin of
this phenomenon.

Recordings obtained in the same experiment but when the recording electrode was 2 mm away from the stimulating electrodes
are given in Fig. 1B. The stimuli were of the same intensity. In
this case an increase in strength of stimulation produced only an
increase in amplitude of the response; the latent period did not
diminish even when the stimulating voltage was increased eight-
fold. The absence of a perceptible latent period for cortical responses near the stimulating electrodes is a special phenomenon.
It can evidently be regarded as the result of recording the unbalanced
residue of the stimulation artifact. However, as shown below, even
when this is excluded, a direct cortical response can be shown to
arise.

The cortical response to an electrical stimulus near the
point of stimulation (distance 0.5-0.75 mm) evidently consists of
the sum of the direct response and the postsynaptic potential appearing somewhat later and merging with it. Sometimes the response to stimuli of moderate voltage arising without an appreciable latent period begins to appear after a delay when the strength

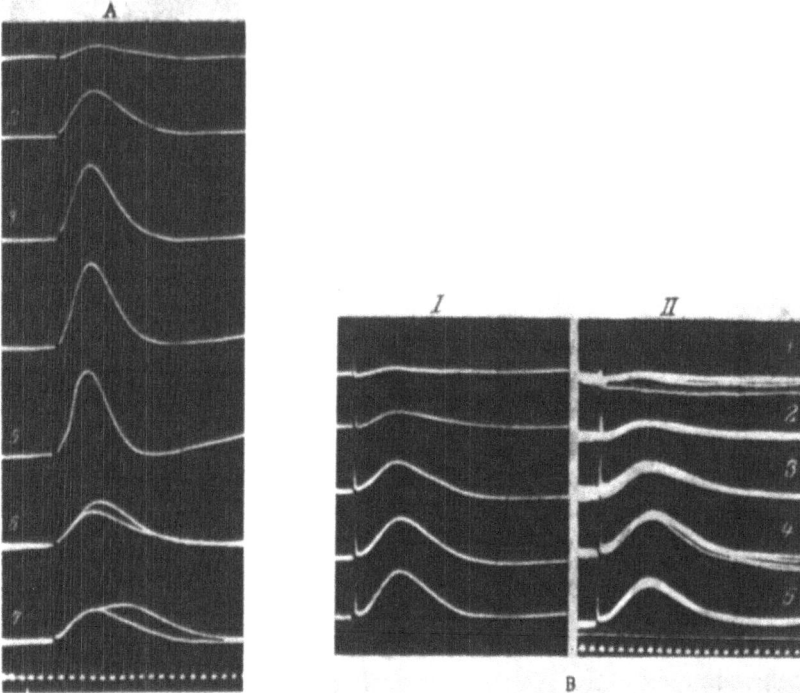

Fig. 1. Changes in cortical electrical response near stimulating electrodes as strength of stimulus is increased. Cat, Nembutal anesthesia
(90 mg/kg); electrodes on suprasylvian gyrus. A) Distance between
stimulating and active recording electrodes 0.75 mm; B) distance between stimulating and recording electrodes 2 mm; I) single response,
II) superposition of five responses. 1) Intensity of stimulus 5 V (0.05 msec);
2) 10 V; 3) 20 V; 4) 30 V; 5) 40 V; 6) Responses to one stimulus (10 V)
and two stimuli at an interval of 2 msec superposed; 7) same, but interval
10 msec. Time marker 2 msec.

of stimulation is increased (Roitbak, 1964). The apparent reason for this paradoxical phenomenon is the anodal action of the
current, blocking the stimulant action of the cathode on the dendrites. Conduction of impulses along the fibers, however, is not
abolished. The considerable latent period of the response to weak
stimuli (Fig. 1A, 1 and 2) could be regarded as indicating the time
required for excitation to spread along the dendrites from the
point of stimulation to the point of recording. The sum of the
length of the two branches in opposite directions may be 0.5–

0.7 mm (Cholokashvili, 1958). There are two objections to this explanation, however: first, the dendrites, according to available information, cannot conduct excitation; second, there are no grounds for considering that excitability of dendrites in the first layer is greater than the excitability of the fibers passing through this layer, some of which are thick and medullated. It must therefore be supposed that when weak stimuli are used, the fibers of the first layer are excited first, and the negative potentials recorded near the stimulating electrodes are postsynaptic. With a considerable increase in the strength of stimulation, loops of current may excite the dendrites directly at the recording point, provided it is not more than 0.5-0.7 mm away. It may also be supposed that excitation of the dendrites takes place at the point of stimulation, but at the point of recording the potential spreads electrotonically along the horizontal branches of the dendrites. The comparatively low excitability of the dendrites relative to electric currents is probably connected with the fact that most of their surface is covered by synapses (Eccles, 1959). There are many grounds for considering that chemical receptor points of the membrane in the region of synapses are not electrically excitable; the receptor point for the mediator with its ion channel cannot be an ion channel for electrical excitation.

Since the dendritic potential disappears after intravenous injection of D-tubocurarine (3-5 mg/kg), it has been concluded not only that this potential (whether far from the point of stimulation or close to it) is postsynaptic, but also that the dendrites are not electrically excitable (Purpura, 1963). This conclusion was made on the basis of disappearance of the dendritic potential during asphyxia caused by stopping artificial respiration (Okudzhava, 1963). The course of the argument is as follows: under the influence of these procedures synaptic transmission is blocked; if the dendrites possessed an electrically excitable membrane, after blocking of synaptic transmission the electric stimulus would still be bound to cause direct excitation of dendrites near the stimulating electrodes.

However, the following objections must be considered. First, great care must be used in interpreting the results of experiments with intravenous injection of D-tubocurarine, because it has been shown that with the doses used, the blood pressure falls, and depression of dendritic potentials was observed in the experiments of Ochs (1962) only when the blood pressure fell below a certain level. Second, during asphyxia, phenomena arise similar to those seen during slowly spreading depression, and in particular, a negative change takes place in the steady potential of the cortex, the dendrites swell, redistribution of ions occurs, and so on, and abolition of the dendritic potential may be due to these other factors. Hence, in the experiments mentioned in which dendritic potentials were abolished, not only the synapses but also the dendrites must have been affected. According to morphological evidence, as cortical function deteriorates, changes take place first in the dendritic branches and synapses in the upper layers of the cortex (Zurabashvili and Naneishvili, 1961).

The facts illustrated in Fig. 1A confirm the view that the dendrites are electrically excitable, and the evidence in favor of this hypothesis is both indirect (Andersen, 1959) and direct. Iwase et al. (1961) observed responses of different character to strong and weak stimuli close to the stimulating electrodes. The response to weak stimuli could be summated if two stimuli were applied at an interval of 3 msec, while the response to the strong stimulus was absolutely "refractory" throughout this time. The response to the weak stimulus did not diminish so suddenly as the recording electrode was moved away as did the response to the strong stimulus. The response to the strong stimulus was regarded as an expression of activity of an electrically excited membrane, the response to the weak stimulus as an expression of activity of a chemically excitable membrane. However, whatever the intervals used, I have never observed complete summation of responses to weak stimuli. On the other hand, responses to strong stimuli were summated to some degree even at intervals of 2 msec or less, in agreement with results obtained by others (Greenbaum and Merlis, 1965).

Response to Paired Stimuli. Responses to a single stimulus and to two stimuli at an interval of 2 msec are shown in Fig. 2. In one experiment (1-3) the stimulating electrodes were bipolar, the artifact was not compensated, and stimuli of high intensity were applied (10 V); in another experiment, tripolar electrodes were used, with a balanced circuit and stimuli of low intensity (3 V). In both cases the response to two stimuli was greater than to one: in the first case by 10%, in the second by 40%, i.e., summation was more marked in the case of weak stimuli. Char-

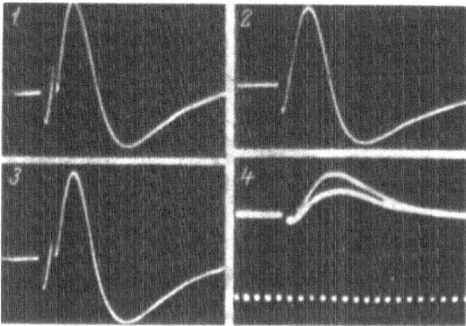

Fig. 2. Cortical electrical responses to single
and paired stimuli near stimulating electrodes.
Electrodes on suprasylvian gyrus: 1-3) Nembutal
75 mg/kg, temperature of cortex 31°, distance
between stimulating and active recording electrodes 0.5 mm; 1) response to two stimuli at interval of 2 msec; 2) response to 1 stimulus; 3)
response to two stimuli at interval of 2 msec;
4) Nembutal 115 mg/kg, triple stimulating electrode, distance between point of stimulation and
point of recording 0.75 mm. Responses to one
stimulus and responses to two stimuli at interval
of 2 msec superposed. Time marker for 4, 2 msec.

acteristically, when shorter interstimulus intervals were used, and
also at the end of the experimental day, this phenomenon disappeared
and could not be obtained with any conditions of stimulation. An
increase in amplitude of the dendritic potential associated with
changes in the conditions of stimulation may be due to an increase
in the number of activated dendrites at the point of recording
(Lloyd, 1959) or to additional excitation of each dendrite. In fact,
the phenomenon of summation of dendritic potentials when the intervals are short has been regarded as the involvement of an additional group of nerve fibers by the second stimulus, the first
stimulus being subliminal. An alternative hypothesis is that of direct
excitation of the dendrites by the second stimulus (Merlis, 1965). In
experiments in which paired stimuli are applied to the cortex, nobody
apparently has taken into consideration the fact that with certain intervals between stimuli the second stimulus will arrive at the point of
stimulation at a time when an electrical response to the first stimulus

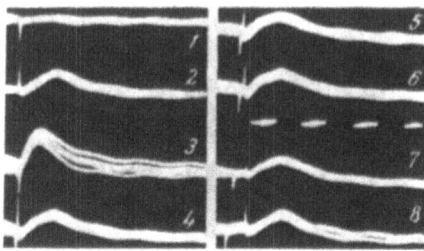

Fig. 3. Electrical responses of cortex to combined stimulation of two cortical points. The
first stimulating electrodes (P_1), second stimulating electrodes (P_2), and recording electrode
(E) were applied to the suprasylvian gyrus in a
straight line; distance P_1-P_2 = 1.5 mm; distance P_2-E = 1.5 mm. Nembutal, 120 mg/kg,
tripolar stimulating electrodes. 1) Response in
E to stimulus through P_1 (10 V, 0.05 msec); 2)
response in E to stimulus through P_2 (5 V, 0.05
msec); 3) response in P_2 to stimulus through P_1;
4) response in E to stimulus through P_2, applied
0.5 msec after stimulus through P_1; 5) interval
1 msec; 6) interval 2 msec; 7) interval 5 msec;
8) interval 10 msec. Time marker 20 msec.

is present at this point. It has been shown that when a dendritic
potential is generated by direct stimulation of the cortex, the impedance of the cortex falls by 3% or more, and the change in impedance coincides in time with the curve of the dendritic potential
(Freygaug and Landau, 1955). With comparatively strong stimulation the dendritic potential reaches its peak after 2 msec (Roitbak, 1955).

It might be supposed that the appearance of a dendritic potential would produce changes in the conditions of stimulation of
the fibers in the first layer or changes in their excitability. Accordingly, special experiments were performed. One of these is
illustrated in Fig. 3. The first stimulating electrode P_1, the second stimulating electrode P_2, and the active recording electrode E
were placed on the surface of the suprasylvian gyrus in one
straight line in such a way that stimulation through P_1 caused a
dendrite potential at the point of the cortex beneath P_2 (Fig. 3, 3),
but did not produce a perceptible response at the points beneath E

(Fig. 3, 1). Stimulation through P_2 was of low intensity, producing a response in E in the form of a negative potential of low amplitude (Fig. 3, 2). The stimulus was applied to P_2 at different time intervals after application of the stimulus to P_1 (Fig. 3, 4-8).

In the experiments no significant change took place in the magnitude of the effect from P_2 in connection with the presence of a dendritic potential at this point.

The fact that a dendritic potential is present at the point of stimulation is, of course, of no significant importance to the origin of the summation phenomenon when the intervals between two stimuli applied at the same cortical point are short. Under these conditions an interval of 2 msec was optimal while with intervals of 1 and 3 msec this phenomenon was less marked. The summation of dendritic potentials occurring when intervals are short is also observed when the first and second stimuli are applied to different points of the cortex and the recording electrode is located between them. Under these circumstances it is observed both when the stimulating electrodes are at a distance of 0.5 mm from the recording electrode and when this distance is 3 mm (Ochs and Book-

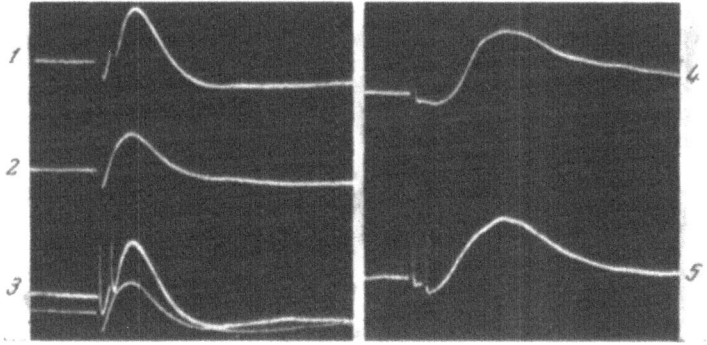

Fig. 4. Cortical responses to single and paired stimuli close to and some distance from stimulating electrodes. The same preparation as in recordings in Fig. 2 (1-3). Distance between stimulating and active recording electrodes 0.5 mm. 1) Response to two stimuli (3 V, 0.05 msec) at an interval of 2 msec; 2) response to one stimulus; 3) responses to one stimulus and response to two stimuli at an interval of 2 msec superposed; 4 and 5) recording electrode moved to a distance of 3 mm away; 6) response to one stimulus (10 V, 0.05 msec); B) Response to two stimuli at interval of 2 msec.

er, 1961). However, if one point is stimulated and the responses are recorded close to it and then some distance away, in the latter case the phenomenon of summation is not found when the intervals are short (Roitbak, 1963). This phenomenon is illustrated in Fig. 4. It shows that the fibers of the first layer are refractory after conducting excitation for 2-3 msec. It also shows that the summation phenomenon for short intervals close to the stimulating electrodes is not associated with excitation of an additional group of fibers by the second stimulus.

The summation phenomenon examined cannot be used directly as an argument in support of the hypothesis of electrical excitability of the dendrites. As mentioned above, it is observed when weak stimuli are used, even those so weak that responses can be found only by averaging with a computer (Merlis, 1965), i.e., using stimuli evoking a postsynaptic dendritic potential, as is shown by the corresponding latent period of the responses (see Fig. 1A, 6 and Fig. 2, 4).

It may be supposed that the second stimulus, arriving during the refractory period of those fibers which became excited and conducted excitation in response to the first stimulus, cannot produce activation of the dendrites innervated by them at distant points. However, presynaptic endings at the point of stimulation which were stimulated below the threshold level by the first stimulus may develop a local potential in response to a second stimulus and electrotonic spread of the current to synapses may give rise to the secretion of an additional number of quanta of mediator, i.e., to the phenomenon of summation.

SUMMARY

If a single stimulus is applied to the cortex, a negative potential is generated near the stimulating electrodes (0.5-0.75 mm away), and with stimuli of comparatively high intensity it arises without a perceptible latent period. The duration of this response is so long that it could not be associated with the electrical response of the fibers of the first layer, and it is too short to be ascribed to a response of the neuroglia. It is evidently a response of the dendrites, a fact which gives evidence that the membrane of the apical dendrites of the cortex is electrically excitable. The comparatively low electrical excitability of the dendrites is evidently the result of the fact that most of their surface is covered by synapses. The cortical response to an electrical stimulus near the point of stimulation probably consists of the sum of the direct response and the postsynaptic potential of the apical dendrites.

In response to two stimuli of any intensity, a summation phenomenon is observed: the response to two stimuli at an interval of 2 msec is greater than the response to a single stimulus. When the distance is 3 mm, summation at this interval is not observed, i.e., the fibers of the first layer are refractory for more than 2 msec. It is postulated that the second stimulus causes a local potential in the presynaptic endings on which the action of the first stimulus is below threshold. Electrotonic spread causes the liberation of a certain amount of mediator in the corresponding synapses.

Correlation Analysis of Central EEG Rhythms of the Healthy Human Cortex[*]

V. S. Rusinov and O. M. Grindel'

Institute of Higher Nervous Activity and Neurophysiology
Academy of Sciences of the USSR
Institute of Neurosurgery, Academy of Medical Sciences of the USSR
Moscow, USSR

Investigations have shown that the electroencephalogram (EEG) reflects the state of brain function. However, the EEG patterns and the generation of its varied rhythms still remain almost as unexplained and conjectural as they were in the first years of their discovery. Visual analysis of the EEG and simple measurement by means of ruler and dividers have proved inadequate to reveal the information presented by the complex pattern of cortical potentials.

In the first years of development of electroencephalography, physiologists attempted the objective quantitative analysis of the EEG by means of mathematical methods. Mathematical analysis of the EEG was first carried out by Livanov (1934). Later, other investigators suggested various systems of automatic analyzers and toposcopes for use in investigating the spectral composition of the EEG and spatial relationships between electrical activity of different parts of the brain. Later, starting with the work of Wiener, the correlation method of EEG analysis was developed. Besides quantitative assessment of the frequency of EEG rhythms, this method can also be used to demonstrate similarity between electrical processes in different regions of the cortex and to express it in temporal relationships (Brazier and Casby, 1952; Brazier and Barlow, 1956; Barlow et al., 1959, and others).

In the present investigation we made an autocorrelation and cross–correlation analysis of electroencephalograms taken from

---
[*]Pages 242-252 in the Russian edition.

different parts of the cortex of 22 healthy human subjects at rest and during afferent stimulation (continuous light, acoustic stimulation) and during muscular exertion (clenching the fingers of both hands into a fist). The occipital, central, and frontal regions of the cortex were investigated, and the cross-correlation method was used to determine the relationship between the EEG of the following pairs of regions: symmetrical points of the occipital and frontal regions, occipital and frontal, occipital and central, and central and frontal regions of the same hemisphere. Monopolar recordings were made, the combined reference electrode being applied to the ear. A two-channel correlograph was used in conjunction with a frequency analyzer with band-pass filters capable of discriminating between delta, theta, alpha, and beta frequency ranges and of assessing their mean amplitudes by parallel integration. Full details of the method used have been described previously (Grindel', 1963, 1964; Grindel' et al., 1964). Auto- and cross-correlation functions were calculated both for the complete EEG and for the frequency band discriminated by the filters.

The correlograms thus obtained were analyzed with respect to the following indices: mean frequency of waves ($\bar{f}$), the ratio between the mean power of the periodic (quasiperiodic) process and the power of the random process (Kp/r); the coefficient of cross-correlation (Kcr), indicating the degree of similarity or connection between two processes; and the time shift of the maximum of the cross-correlation function (TS) reflecting temporal relationships between the processes in the parts of the cortex examined.

Cross-correlation analysis of electrical activity in different regions of the healthy human cortex revealed special patterns of connections between them. As a rule, at rest, considerable agreement is found in the phases of general oscillations in all frequency bands — delta, theta, alpha, and beta — between symmetrical points of the occipital and frontal regions of the right and left hemispheres. The closest correlation is found between recordings taken from anterior portions of the brain: from symmetrical points of the frontal regions, and from the frontal and central regions of the same hemisphere, pointing to their functioning to a large measure as one unit. Moderately close correlation is found between recordings made from the posterior divisions of the hemispheres: symmetrical points of the right and left occipital regions and the central and occipital regions of the same hemisphere (Grindel', 1966).

Relationships between the occipital and frontal EEG and the central EEG are of special interest. The central region of the cortex is a zone of convergence of impulses produced by stimuli of different modalities, the zone of manifestation of the nonspecific response, the projection zone of the nonspecific thalamic pathways.

Distinctive features of electrographic responses in this region of the human cortex to afferent stimuli have been described by a number of workers (Gastaut, 1953; Kats, 1958; Puchinskaya, 1963; Rusinov, 1962). It is interesting to note that it was in this part of the cortex that Walter (1965) discovered a slow potential with both surface and implanted electrodes when recording responses with both surface and implanted electrodes to afferent stimulation. Brazier (1958) found the most intense delta-activity during sleep in healthy subjects also in the central cortex.

The correlation relationships between different parts of the cortex and the central region may be assumed to possess different characteristics reflecting to some extent variations in their cortico-subcortical relationships in healthy persons responsible for typological differences in their higher nervous activity.

Cross-correlation of the occipital and frontal EEG with the central EEG revealed two forms of relationships: 1) dominance of rhythm and its spread from occipital to frontal region and 2) predominance of the random component and the presence of time shifts indicating delay of wave activity in the occipital cortex relative to the central. These two types of cross-correlograms also corresponded to relationships of different character between the occipital and frontal regions of the same hemisphere.

The first type of cross-correlogram of the occipital and frontal EEG with the central EEG was observed in subjects whose EEG had a well-marked alpha-rhythm.

An example of this type of cross-correlogram (CrCG) is given in Fig. 1A. The CrCG of the occipital and central regions of this subject in all frequency bands shows moderate cross-correlation of the order of 0.40-0.50, Kcr amounting to 0.64 for the alpha-rhythm only. The periodic component can be clearly seen in every CrCG, but especially so in the CrCG of the alpha- and beta-rhythm: the value of Kp/r in this case reaches 0.60 and 0.78; correlation occurs relative to the periodic component. In other

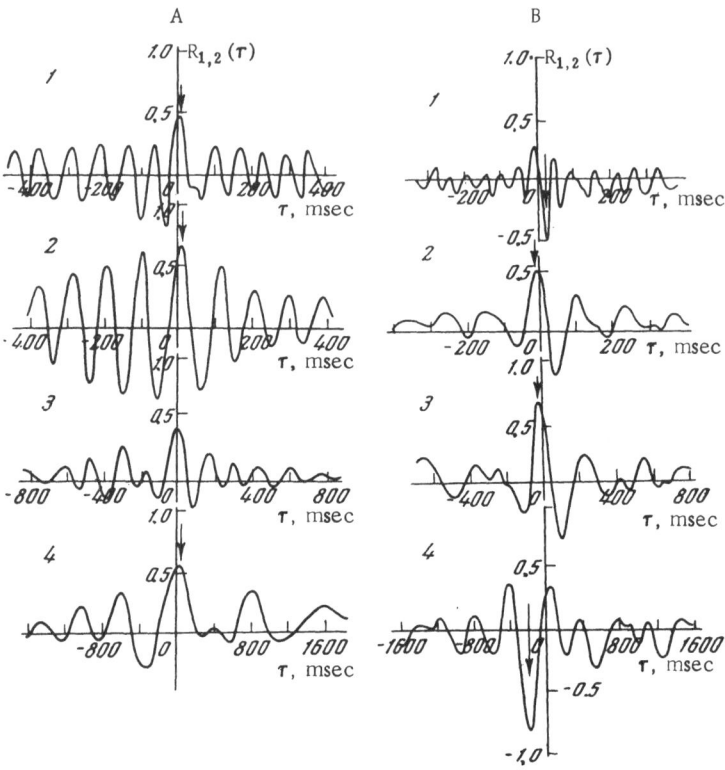

Fig. 1. Cross-correlograms of occipital and central EEGs with discrimination of frequency bands: beta (1), alpha (2), theta (3), and delta (4) in healthy subjects. A) Subject with dominance of alpha-rhythm in EEG; B) subject "without alpha-rhythm" in EEG.

frequency bands the periodic component is weaker ($Kp/r = 0.40$), and the random component is seen more clearly. The time shifts are relatively short — of the order of 0.1 of the period of the investigated waves. Their direction indicates delay of the process in the central cortex.

The CrCG of the occipital and central EEG of another subject whose EEG is characterized by absence of alpha-rhythm differ considerably (Fig. 1B). In this case attention is first drawn to the poorly defined periodic component in all frequency bands, and especially in the band of the alpha-rhythm. Correlation in all bands occurs mainly with respect to the nonperiodic component. Compared with those of the previous subject, the cross-correlation coefficients are greater in absolute value: in the alpha- and beta-bands Kcr = 0.50 and 0.52; in the theta band it is higher still; Kcr = 0.68. In the delta band the absolute value of the cross-correlation coefficient reaches 0.88, but it is negative in sign. A negative Kcr for correlation between delta-activity of the occipital and frontal regions with that of the central region was found to be typical for subjects with absence of dominant alpha-rhythm in the EEG. This fact is interesting in connection with the possibility of special relationships between the specific and nonspecific systems in these subjects. It is interesting to note that the directions of the time shifts in the CrCG of this subject also were different from those in the first case. In the alpha and theta bands the time shifts (TS = -10 and -20 msec) indicate delay of the process in the occipital region. The magnitudes of these shifts are of the same order as in subjects with dominance of alpha-rhythm in the EEG. In the delta band, on the other hand, the time shift discovered is much greater than the shift in the CrCG of the delta-band of the first subject: TS = -200 msec. This shift indicates definite delay of the process in the occipital cortex. The periodic component is weaker on this CrCG (Kp/r = 0.29) than on the CrCG of the subject with an alpha-rhythm, when Kp/r = 0.40, but the frequency of the delta-waves is higher ($\bar{f}$ = 2.43 cps; in the first case $\bar{f}$ = 1.85 cps). Hence, nonperiodic waves and individual delta-waves, opposite in phase in the occipital and central regions and appearing sooner in the central, are common to the central and occipital regions of the subject whose EEG is "without alpha-rhythm."

The CrCG of electrical activity in the central and frontal cortex is less dependent on the character of the EEG. Cross-correlograms of different frequency bands of the central and frontal EEG of the same two healthy subjects — one with dominance of the alpha-rhythm and the other "without alpha-rhythm" on the EEG — are illustrated in Fig. 2. In both bands and in both cases, high values of Kcr are obtained (0.70-0.95). Only with respect to the beta-band in the second subject is the value of Kcr slightly lower (0.48). The time shifts for all bands are small (except for the delta-band in the second subject) — of the order of 5-20 msec, or TS = 0. The rhythm is ill defined. A periodic component is clearly predominant only in the alpha-band of the subject with dominance of alpha-activity.

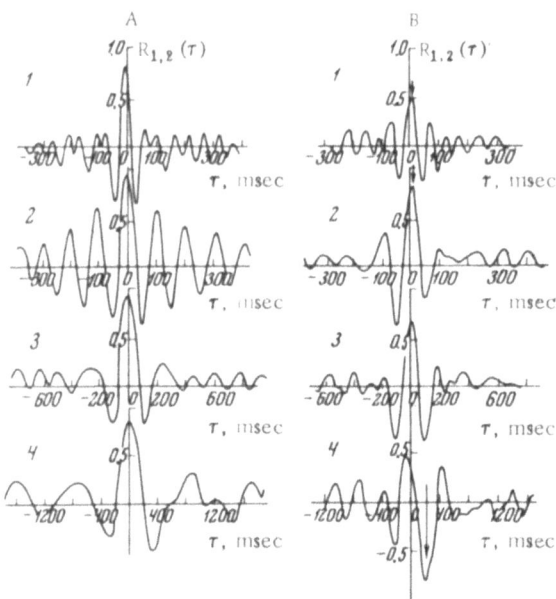

Fig. 2. Cross-correlograms of discriminated EEG frequency
bands: beta (1), alpha (2), theta (3), and delta (4) of the
central and frontal cortex in healthy subjects. Legend as
in Fig. 1.

The CrCG of the delta-rhythm in the second case deserves
special attention. Here, just as on the CrCG of the occipital and
central cortex (Fig. 1B, 4). the value of Kcr is negative, indicating
that the delta waves recorded from these two regions are opposite
in phase. In addition, the magnitude of the time shift is precisely
the same (TS = 200), but it is opposite in direction, indicating earlier generation of the common delta-waves in the central region.

As mentioned above, a negative value of Kcr and a shift of
delta-waves in the CrCG of the central and occipital regions were
observed in some subjects, but this did not always coincide with a
negative value of Kcr on the CrCG of the central and frontal regions: this could be positive.

The fact must be noted that in cases where these special
types of relationship of the central cortex are found in persons
with absence of alpha-rhythm dominance in the CrCG of the occipital and frontal regions in the same band, correlation between the

delta-activity occurs with respect to the nonperiodic component, and TS = 0, a result which we interpret as indicating the synchronous arrival of impulses from subcortical nonspecific structures (Grindel', 1964, 1965, and 1966). The possibility is not ruled out that synchronization of common waves revealed by cross-correlation of the occipital and frontal regions is associated to some extent in these subjects with the presence of waves spreading from the central region to the occipital and frontal cortex. This explanation is confirmed by the fact that the time shifts in the CrCG of the occipital and central and frontal regions in some cases are equal (both amount to 200 msec) and opposite in direction. The presence of a negative value of Kcr in the beta-band may partly explain why the TS is often particularly large during cross-correlation of the occipital and frontal regions in this frequency band. Hence, the temporal relationships between activity of different frequency bands in subjects "without alpha-rhythm" on their EEG are considerably modified by comparison with those for persons with a well marked alpha-rhythm, and the central cortex (the nonspecific zone) plays a particularly important role in this modification of the temporal relationships of electrical activity.

It is interesting to compare the results of frequency and correlation analysis of electrical activity in different regions of the cortex in response to different forms of afferent stimulation. Frequency analysis reveals the characteristics of combined activity or the mean amplitude of the rhythm; correlation analysis gives no ihformation concerning the amplitude of the waves but reveals the degree of periodicity of the process and its mean frequency.

Changes in the alpha-rhythm of the occipital, central, and frontal cortex during photic and acoustic stimulation and muscular exertion (clenching the fingers into a fist) attracted particular attention.

Analysis of the isolated alpha-rhythm of these cortical regions in healthy subjects showed that the amplitude relationships and character of the autocorrelograms at rest depend on the character of each subject's EEG: on whether its alpha-rhythm is dominant or absent.

In subjects with alpha-rhythm dominance in the EEG, maximal integrated activity at rest is found in the occipital cortex; somewhat lower integrator readings are obtained for the central

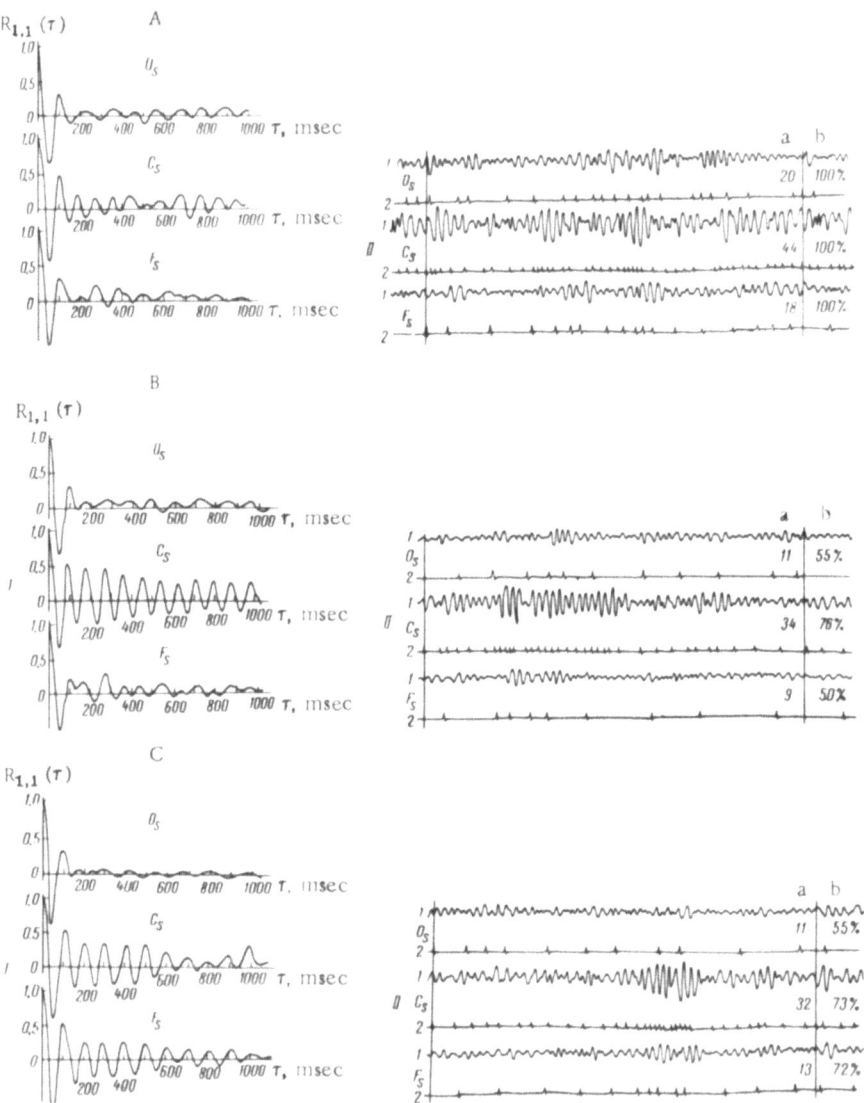

Fig. 3. Changes in alpha-rhythm of left occipital (Os), central (Cs), and frontal (Fs) regions of the cortex under the influence of afferent stimulation in a healthy subject with absence of regular alpha-rhythm in EEG. A) At rest (background); B) during photic stimulation; C) during making fist with fingers. I) ACG of isolated alpha-rhythm of each cortical region; II) results of frequency analysis: 1) alpha-rhythm isolated on analyzer; 2) integrator reading; a) readings of integrator after 5 sec; b) percentages, readings at rest taken as 100%.

and frontal alpha-rhythms. Differences in the frequency of periodic activity and relatively small differences in the values of Kp/r are found on the autocorrelograms (ACGs) of these recordings, but they are smaller than for the frontal ACG.

Different alpha-rhythm correlations are found in subjects with absence of alpha-rhythm dominance in their EEG. The amplitude of the alpha-rhythm is maximal in the central cortex. Integrator readings for this region are several times higher than those for the occipital and frontal alpha-rhythm (Fig. 3). Periodic activity in the background likewise is more marked in the central cortex than in the occipital and frontal: Kp/r for the occipital cortex is 0.08, for the central cortex 0.16, and the frontal 0.10. The highest mean frequency of the periodic activity is also found in the central cortex: $\bar{f}$ central = 10.3 cps; $\bar{f}$ occipital = 9.7 cps; $\bar{f}$ frontal = 9.6 cps.

The ratio between amplitudes of the alpha-rhythm in the occipital, central, and frontal cortex, as revealed by the integrator readings, is similar in the subject with an irregular alpha-rhythm and with a distinct Rolandic rhythm in the background to the ratio between the amplitudes of the alpha-waves in the subject "without alpha-rhythm" in the EEG. Integrator readings for the isolated alpha-band in all three regions demonstrate its low amplitude, but the largest amplitude readings and the highest value of Kp/r are also found in the central region.

During photic stimulation accompanied by a depression response, as revealed clearly by a reduction in the integrator readings in all three investigated cortical regions, the changes in autocorrelation functions of the occipital and frontal alpha-rhythm do not correspond to those of the central alpha-rhythm. In the occipital and frontal regions depression of the alpha-rhythm is accompanied by marked disturbance of periodic activity (by desynchronization); in the central cortex the depression is similar in intensity to that of the occipital cortex but it is not accompanied by a disturbance of periodic activity, i.e., by desynchronization. This pattern of changes in the alpha-rhythm parameters for the various cortical regions during photic stimulation was observed in subjects both with and without alpha-rhythm dominance in the EEG. In the latter case, these relationships affecting the periodicity of the alpha-waves were particularly clear.

The results of frequency analysis and ACG of the isolated
alpha-rhythm of the occipital, central, and frontal regions of the
cortex during photic stimulation in a healthy subject with an EEG
"without alpha-rhythm" are given in Fig. 3B. In all three cortical
regions the amplitude of the alpha-waves is smaller than in the initial background (Fig. 3B). Definite differences are also found in
the ACG; the periodic activity in the occipital and frontal regions,
which was ill-defined before stimulation remains just as ill-defined
during photic stimulation: Kp/r before stimulation and during photic stimulation in the occipital cortex is 0.08, while in the frontal
cortex before stimulation Kp/r = 0.10, and during photic stimulation
Kp/r = 0.07. In the central region, on the other hand, the periodicity of the alpha-waves is definitely increased: Kp/r = 0.16 in the
background, rising to 0.39 under the influence of photic stimulation.
The frequency of the alpha-waves in the central cortex increases
from 10.3 to 11.3 cps, while in the occipital cortex it falls, although
very slightly (from 9.7 in the background to 9.0 cps during photic
stimulation). In the frontal region the mean frequency of the alpha-
rhythm during photic stimulation increases from 9.6 to 11.1 cps.

For the group of subjects as a whole, statistically significant
differences in the values of Kp/r for the alpha-rhythm of the occipital and central (p = 0.05) and central and frontal regions (p = 0.05)
and for the mean frequencies of periodic activity between these
same pairs of cortical regions (p = 0.05; p = 0.01) were established during photic stimulation, the periodicity and frequency of
the alpha-waves being highest in the central region.

Investigation of the alpha-rhythm response during muscular
exertion (squeezing the fingers into a fist) revealed different patterns. Although marked depression of the alpha-rhythm took place,
as revealed by a distinct and significant decrease in the integrator
readings compared with the background, no regular changes in the
values of Kp/r were found either in the occipital and frontal regions or in the central region.

Cross-correlation analysis of the alpha-band of these two
pairs of cortical regions showed that changes in temporal relationships of the activity during photic stimulation were the most regular features.

As mentioned above, the temporal relationships of the alpha-
rhythm in the occipital and central regions of the cortex differed

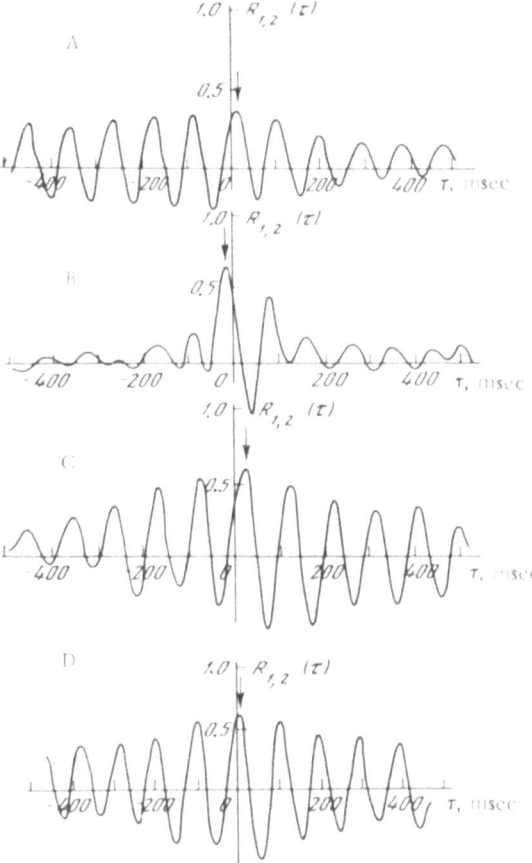

Fig. 4. Cross-correlograms of EEG of right occipital and
central cortex with respect to isolated alpha-rhythm and
their changes during afferent stimulation in a healthy
person with alpha-rhythm dominance in the EEG. Legend
as in Fig. 3; except D) during acoustic stimulation.

in persons with and without dominance of the alpha-rhythm in the
EEG. During the action of photic stimulation, the temporal changes
in the CrCG of the occipital and central cortex of both groups of
subjects assume the same direction, indicating delay of the process
in the occipital region relative to the central. Cross-correlograms
of the isolated alpha-rhythm of the occipital and central cortex of
a healthy subject with alpha-rhythm dominance in the EEG at rest
and during stimulation are illustrated in Fig. 4. They show that

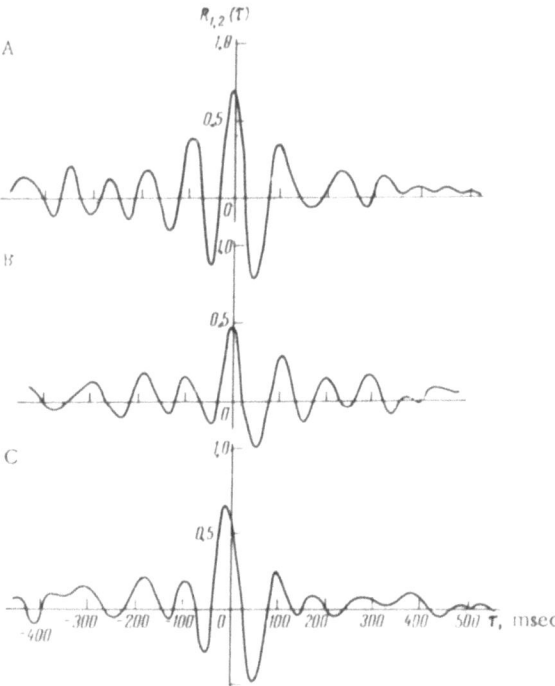

Fig. 5. Changes in cross-correlogram of isolated alpha-
band of EEG of left occipital and central cortical regions in a healthy subject without alpha-rhythm dominance in EEG under the influence of afferent stimulation.
Legend as in Fig. 3.

the time shift in the CrCG at rest is positive (TS = +15 msec), indicating delay of the process in the central region relative to the
occipital. During photic stimulation the periodic activity is considerably reduced, but the time shift becomes negative, indicating that
delay of alpha-activity now occurs in the occipital cortex.

In subjects with absence of dominant alpha-rhythm in the
EEG and with a marked Rolandic rhythm at rest, the time changes
in the CrCG of alpha-activity of the occipital and central regions
indicate the earlier generation of their common rhythm in the central region and its delay in the occipital. During photic stimulation
the direction of the time shifts in the CrCG remains unchanged, and
its magnitude in the subject whose EEG is "without alpha-rhythm"
also remains unchanged, whereas in the other subject (with a

marked Rolandic rhythm) the time shift is increased (Figs. 5 and 6). During acoustic stimulation and making a fist, no regular changes in the time shifts are observed in the CrCG of alpha–activity of these two regions. In the first and second subjects its direction remains as at rest (Figs. 4 and 5), while in the third subject (Fig. 6), with a marked Rolandic rhythm, the time shift changes in direction and now indicates delay of the process in the central region relative to the occipital.

These examples of changes in cross–correlation functions of the alpha–rhythm of different cortical regions in healthy persons under the influence of afferent stimulation show that the spatial-temporal relationships of cortical electrical activity can be considerably modified. When desynchronization of alpha–activity is found in any region, the temporal relationships between regions are disturbed: when a dominant alpha–rhythm is present in the

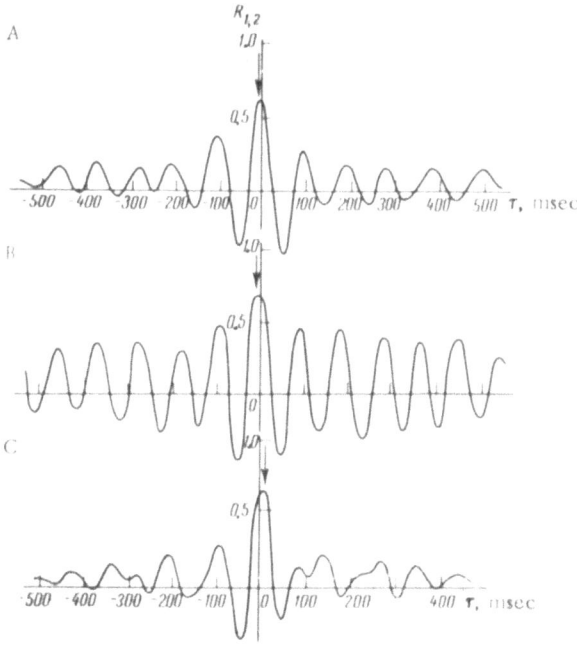

Fig. 6. Changes in cross-correlogram of alpha-rhythm of left occipital and central cortical regions in a healthy subject with dominance of Rolandic rhythm in EEG under the influence of afferent stimulation.

background EEG, photic stimulation, which disturbs periodic activity in the occipital region and synchronizes it in the central, leads
to delay of the alpha-waves in the occipital cortex. If the Rolandic
rhythm is dominant, muscular exertion causes desynchronization
of the process in the central region and the rhythmic waves in the
CrCG change their direction.

The problem of spatial-temporal relationships of cortical
electrical activity is of particular interest to an understanding of
the combined working of different cortical regions in the course of
reflex brain activity.

Special attention has been paid to the study of spatial-temporal relationships of cortical electrical activity and of their
changes in response to different forms of stimulation in the investigations of Livanov and co-workers and of Walter and co-workers
(Livanov, 1962c; Livanov et al., 1964; Walter, 1954; Cooper et
al., 1957; Cooper and Mundy-Castle, 1960, and others).

The investigation of cross-correlation functions of the EEG
in the healthy human subject described above showed that the central region of the cortex plays a special role in establishing definite relationships between the bioelectrical activity of different cortical regions. At rest, if a dominant alpha-rhythm is present in
the EEG, the CrCG from the central cortex is essentially indistinguishable from the CrCG of other pairs of cortical regions.
On the other hand, if no dominant alpha-rhythm is present in
the EEG, in the case of a desynchronized EEG, which has been
shown by special investigations to be characteristic of persons constantly in a state of increased excitability (Gastaut et al.,
1951; Mundy-Castle, 1957; Sokolov, 1960; Thompson and Obrist,
1963; Danilova, 1963; Nebylitsyn, 1963), the correlation functions
of the EEG of the central cortex show certain distinctive properties. In this case waves of opposite phase with the rhythm of delta-
waves are seen in the CrCG of the occipital and central and central and frontal regions. The central region of the cortex is a specific source of delta-activity for these people or the projection
zone of delta-waves from lower levels of the brain, spreading from
it both to the posterior and to the anterior regions of the hemisphere.

Correlation analysis also showed that afferent stimuli, especially photic, evoke a specific response of electrical activity in

the central region of the cortex. In the occipital and frontal regions depressions (decrease in amplitude) during photic stimulation are accompanied by disturbance of the regular rhythm of alpha-waves — by desynchronization of electrical activity, while in the central region the periodicity of the alpha-waves is preserved or even intensified. Meanwhile, the central generator of alpha-activity becomes dominant relative to the occipital generator of alpha-waves, whose working is disturbed by visual impulses arriving along the specific pathways. At the same time, impulses evoked by the same photic stimulation, but reaching the central (nonspecific) region of the cortex, intensify periodic activity there. The differences between the action of the same afferent stimulus on electrical activity in different parts of the healthy human cortex give rise to a constantly changing mosaic of spatio-temporal relationships between cortical regions. The interchange of different types of correlation between the EEG of the cortical regions taking place under the influence of afferent stimuli evidently reflects changes in the state of function of different parts of the central nervous system, changes which occur, in Ukhtomskii's words, "undividedly in time and space," and the complex interaction between constellations of centers which is established in the course of responses of the brain as a single system to external stimulation.

Electrical Responses of the Hippocampus to Stimulation of the Vagus Nerve[*]

F. N. Serkov and N. V. Bratus'

Department of Physiology, Medical Institute
Odessa, USSR

There is now considerable experimental evidence of the existence of hippocampal effects on various autonomic functions (Kaada and Jasper, 1952; MacLean, 1955; Airikyan and Gaske, 1966, and others). By modulating the excitability of the autonomic centers in the hypothalamus, the hippocampus possibly participates in the regulation of visceral activity.

This naturally raises the question of the interoceptive connections of the hippocampus required for obtaining information concerning the state of the viscera. Connections of this type have been demonstrated by evoked potentials (Dunlop, 1958; Ermolaeva and Chernigovskii, 1964; Serkov and Makul'kin, 1966). However, the study of this problem is still incomplete.

The object of the present investigation was to study electrical responses of the hippocampus to electrical stimulation of the vagus nerve trunk and to natural stimulation of the mechanoreceptors of the stomach, and from the results to attempt to explain the character of representation of the visceral afferent systems in the hippocampus and, as far as possible, to define the role of the hippocampus in the analysis of information received from the interoceptors.

METHOD

Experiments were carried out on 27 adult rabbits weighing 1800-2700 g. The preliminary operation was performed under

---

[*] Pages 253-261 in the Russian edition.

ether anesthesiá of short duration. The trunk of the vagus nerve was dissected in the neck, most commonly on the right side, and divided, its central end being placed on buried silver electrodes with an interelectrode distance of 2 mm.

To provide appropriate stimulation of the mechanoreceptors of the stomach, laparotomy was performed, and a rubber balloon connected to an inflating bulb and mercury manometer was inserted into the stomach through an incision in its fundal portion.

The hippocampus was exposed unilaterally or bilaterally. The skull was widely trephined, and the dura and a small piece of brain tissue above the dorsal hippocampus were removed.

A "unipolar" method was used to record the hippocampal potentials. The active electrode, consisting of a thin cotton wick soaked in Ringer's solution, was applied to different points on the surface of the dorsal hippocampus. The reference electrode was fixed to the nasal bones of the skull. Potentials were recorded on a type MPO-2 loop oscillograph with UBP-1 amplifier at its input.

The vagus nerve was stimulated with single pulses (0.5 msec, 2-15 V) The mechanoreceptors of the stomach were stimulated by inflating the balloon inserted into the viscus with air. The pressure was checked from readings of a mercury manometer.

The recording of the electrical responses began as soon as the animal had recovered from the anesthetic.

RESULTS

Electrical stimulation of the vagus nerve trunk regularly causes typical and well marked evoked potentials (EP) in the dorsal hippocampus. The records in Fig. 1 show that the hippocampal EP in response to vagus nerve stimulation as a rule consists basically of two components: an initial electropositive and a subsequent electronegative potential. In this respect, there is complete similarity between the EP of the hippocampus and the EP of the neocortex.

The latent period of the initial electropositive part of the hippocampal EP in response to electrical stimulation of the vagus nerve is longer than that of the primary response of the neocortex to the same stimulus, namely 18-20 msec. The amplitude of this

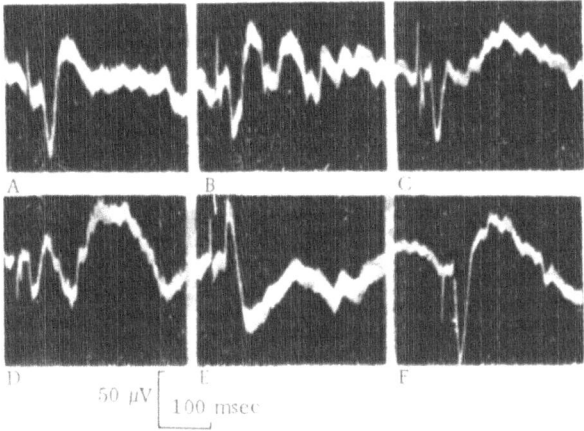

Fig. 1. Variance of EP of hippocampus in response to electrical stimulation of vagus nerve (A, B, C, D, E, and stimulation of sciatic nerve (F).

component varies in different experiments from 50 to 200 μV and its duration from 20 to 30 msec.

Often, the negative component is followed immediately by additional slow waves of varying sign and duration (records in Fig. 1, B, C, and D), as a result of which the total duration of the hippocampal EP may be increased to 300-400 msec. This is much longer than the EP of the neocortex.

In some cases in response to vagal stimulation evoked potentials with an initial negative component were recorded from the hippocampus. The next component was then positive, and the response as a whole appeared as a mirror image of the usual positive-negative complex (Fig. 1E). This distorted form of response was not associated with any particular point on the surface of the hippocampus from which the recording was taken. The main factor influencing the appearance of responses of this type is a deterioration of the functional state of the hippocampus as a result of cooling or drying. Administration of chlorpromazine to the animal also led to the appearance of responses with initial negativity (Fig. 2E).

In some experiments the appearance of the initial positive component of the EP was preceded by one or several comparatively fast low-voltage potentials (Fig. 1, B, C, and D). The first of

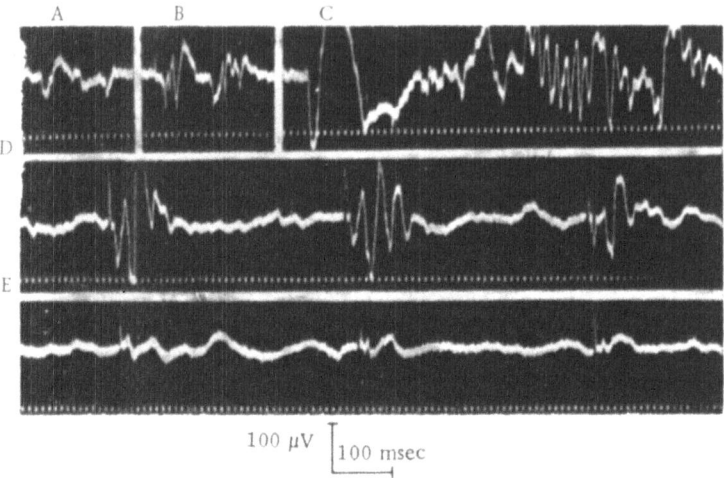

Fig. 2. EP of hippocampus in response to vagal stimulation of varied
strength. A, B, C) 3.5 and 7 V;   D) 15 min after application of 1% strychnine (stimulation 3 V);   E) 35 min after intramuscular injection of chlorpromazine (5 mg/kg) (stimulation 5 V).

these was always positive, while the later ones, if present, could be
either positive or negative. The duration of each such wave was
3-5 msec and the amplitude did not exceed 40 μV. Their latent
period was 6-8 msec, and the time from their appearance to the
beginning of the main components of the EP was 10-12 msec.

To determine the character of representation of the vagal afferent system in the hippocampus and in particular, to identify the
zones of this representation, we recorded evoked potentials from
different points of the dorsal hippocampus. As the records in
Fig. 3 show, although the evoked potentials of different points of
the dorsal hippocampus in response to electrical stimulation of the
vagus nerve are not completely identical, they are nevertheless
well defined at all points shown in the figure. Hence, in the dorsal
hippocampus no area or region could be distinguished which could
be described as a focus of maximal EP activity during vagal stimulation. This shows that the afferent system of the vagus nerve is
represented in the hippocampus not as a separate projection zone,
but diffusely throughout the hippocampus. Other afferent systems
also are diffusely represented in the hippocampus (Serkov and
Makul'kin, 1966). This diffuse type of representation of afferent

systems, characteristic of the hippocampus, differs essentially from the mosaic type characteristic of the neocortex.

Because of this character of representation of afferent systems in the hippocampus, the zones in which afferent fibers of different systems terminate must overlap each other extensively (MacLean, 1959). Experimental evidence of this overlapping was obtained by Serkov and Makul'kin (1966). In confirmation of these findings, in recordings from the same point of the dorsal hippocampus, we also obtained well marked evoked potentials in response to stimulation of both the vagus and sciatic nerves (Fig. 1, A and F).

The representation of afferent systems in the hippocampus is also bilateral. This was confirmed by the fact that in response to stimulation of the right or left vagus nerve evoked potentials appeared in the hippocampus on both sides. They were more

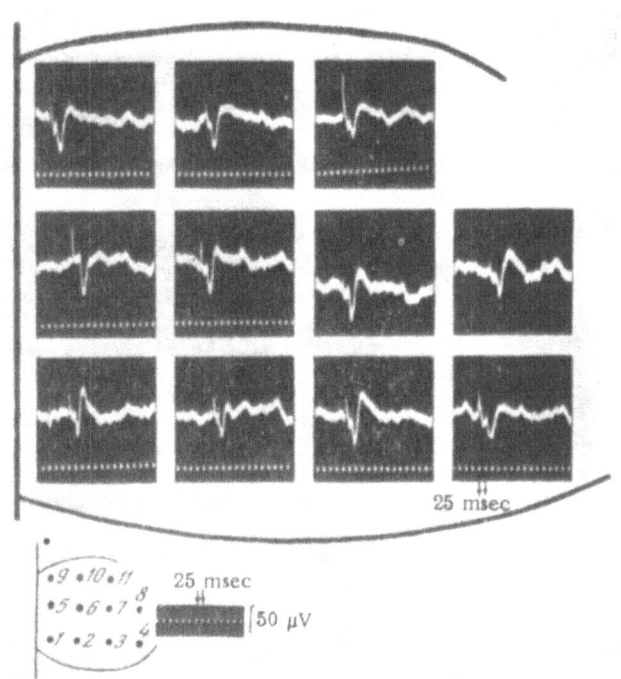

Fig. 3. EP of hippocampus in response to vagal stimulation with recording made from different points of its dorsal surface. Below: diagram showing arrangement of recording points.

clearly defined, however, on the side ipsilateral to the stimulated nerve.

Important factors influencing the principal parameters of the hippocampal EP are the strength and frequency of stimulation. In response to weak threshold stimulation of the vagus nerve, small evoked potentials, mainly only positive, are generated in the hippocampus. With a gradual increase in the strength of stimulation, the amplitude of both the positive and the negative components increases and additional slow components appear. In response to very strong stimulation, evoked potentials with an after–discharge consisting of groups of high–amplitude potentials in a comparatively regular rhythm and with a frequency of about 25/sec may appear (Fig. 2C).

Characteristically the amplitude of the hippocampal EP falls sharply with repetition of the stimulus. This fact was previously noted by Green and Adey (1956) and Serkov and Makul'kin (1966). In our experiments the decrease in amplitude of the main components of the hippocampal EP in response to repeated stimulation of the vagus nerve was found if an interval of 2–3 sec occurred between the first and second stimuli. A more marked decrease was observed with an interval of 1 sec between stimuli. This shows that the first volley of afferent impulses evoking an EP in the hippocampus is followed by a long after–effect, reducing the effectiveness of the next volley. For this reason the hippocampus can react by fully developed evoked potentials only to very infrequent afferent stimuli. According to Green and Adey (1956), the interval between such stimuli must be not less than 10–15 sec.

It is important to note that the amplitude of the different components of the EP is not reduced uniformly. For instance, if the fast, low–voltage initial components which, according to some investigators (Roitbak, 1955; Green and Adey, 1956), reflect presynaptic activity of the afferent fibers, were well developed in the EP, during repetitive stimulation of the vagus nerve they are more stable than the main slow components.

This may indicate that volleys of afferent impulses reach the hippocampus without undergoing any marked weakening in their path, and that consequently, the low frequency with which the hippocampus can react with evoked potentials is dependent on low lability of the hippocampal structures themselves.

This is also confirmed by the results with local application of 1% strychnine solution to the hippocampus. Application of this solution to the hippocampal surface in the region of the recording electrode not only increases the amplitude of the EP, but also makes them more resistant during repetitive stimulation. Evoked potentials of the hippocampus in response to vagal stimulation at 1/sec, 15 min after application of strychnine, are shown in Fig. 2D. Strychnine shortened the latent period of the EP, increased the amplitude of both main components, reduced their duration, and reduced the additional discharge of high-amplitude potentials. The amplitudes of the second and, in particular, the third EP were diminished, but this decrease was not so marked as before the action of strychnine.

We also studied the effect of Nembutal (15-25 mg/kg, intraperitoneally) and chlorpromazine (5 mg/kg, intramuscularly) on the EP. These experiments showed that both Nembutal and chlorpromazine sharply reduce the hippocampal EP in response to vagal stimulation, in some cases blocking them completely.

Because of the specific sensitivity of the reticular formation of the brain stem to these drugs, these results may be interpreted as evidence of its participation in the afferent interoceptive connections of the hippocampus. However, this does not rule out the possibility of a direct action of Nembutal and chlorpromazine on excitation processes in the hippocampus itself.

As well as EPs in response to electrical stimulation of the vagus nerve, we also studied the hippocampal EP in response to appropriate stimulation of gastric mechanoreceptors. In response to rapid distention of the stomach by a balloon under a pressure of 35-40 mm Hg, EPs were recorded in the hippocampus mainly similar to those found during stimulation of the vagus nerve itself (Fig. 4). They consisted either of a single positive phase or of typical biphasic potentials, an initial positive and a subsequent negative phase. Each phase of the response could be complicated by additional components. Just as during stimulation of the vagus nerve, sometimes an EP with initial negativity could be obtained. Because of the nature of the method of stimulation, the latent period was very long and inconstant. Additional waves after the main negative component of the EP were more common in response to natural stimulation of the gastric mechanoreceptors than to stimu-

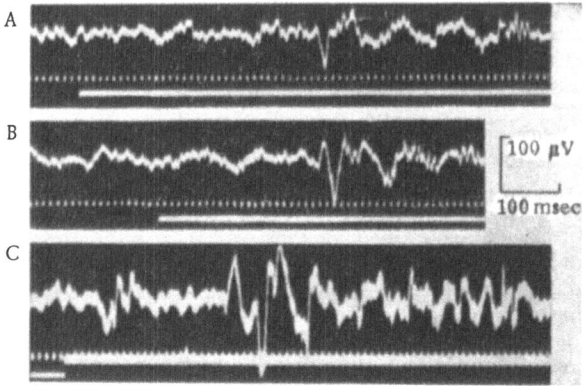

Fig. 4. Variance of hippocampal EP in response to natural
stimulation of gastric mechanoreceptors. Below: distention of stomach.

lation of the vagus nerve trunk. Usually they had the appearance
of theta-waves (Fig. 4A). In some cases, in response to prolonged
stimulation of the gastric mechanoreceptors, the hippocampus generated not a single EP, but a whole series of successive EPs
(Fig. 4C).

Besides the responses described above, stimulation both of
the gastric mechanoreceptors and of the vagus nerve trunk produced changes in the background electrical activity of the hippocampus. These changes took the form most frequently of appearance or intensification of slow electrical activity of the theta-
rhythm type, typical of the electrical response of the hippocampus
to any afferent stimulation (Green and Arduini, 1954).

In agreement with results of others (Airikyan and Gaske,
1965), our experiments showed that the appearance or intensification of a regular theta-rhythm is not the only response of the hippocampus to afferent stimulation. In some cases in response to
afferent, and in particular to strong natural stimulation of the gastric mechanoreceptors, the response took the form of suppression
of the initial theta-activity and a marked increase in electrical activity consisting of fast, low-voltage potentials. This type of response was entirely similar to the desynchronization response in
the neocortex. Mixed types of responses also were observed, in
which synchronization and regularization of the original theta-

rhythm arriving in response to stimulation were followed by de-synchronization or, conversely, stimulation evoked an initial and transient desynchronization, followed by strengthening of the theta-activity.

DISCUSSION

Typical EPs appearing in the hippocampus in response to stimulation of the vagus nerve trunk and to natural stimulation of the gastric mechanoreceptors show conclusively that the volley of afferent impulses thus produced reaches the hippocampus and causes excitation of its neurons. This suggests that afferent fibers of the vagus nerve have their representation in the hippocampus. It is difficult to say how this fact is connected with participation of the hippocampus in regulation of visceral activity, because other afferent systems bearing no relation to interoception are also re-presented in it to a varied degree (Serkov and Makul'kin, 1966). Moreover, EPs arising in the hippocampus in response to stimula-tion of the vagus nerve are almost identical with the EP obtained by stimulation of other afferent systems. In our opinion, this dem-onstrates the uniform character of representation of all these sys-tems in the hippocampus, and also the similarity between physio-logical processes taking place in the hippocampus in connection with the analysis of afferent impulses via these systems.

Analysis of the hippocampal EPs in response to stimulation of the vagus nerve reveals their comparatively long latent period. It is difficult to explain this by assuming that excitation spreads along slowly conducting thin nerve fibers, because hippocampal EPs evoked by acoustic stimulation and stimulation of the sciatic nerve have an equally long latent period (Serkov and Makul'kin, 1966). A more likely explanation is the polysynaptic nature of the pathway from the vagus nerve nucleus to the hippocampal neurons. The possible participation of the reticular formation of the brain stem in the conduction of these impulses was mentioned above. Either directly or through the nuclei of the thalamus and hypotha-lamus, these impulses are transmitted from the reticular forma-tion to the septal nuclei and from thence via the fibers of the for-nix to the hippocampal neurons. This long pathway accounts for their long latent period.

However, the following fact must also be borne in mind. Green and Adey (1956) found that the latent period of the hippocampal EP in response to electrical stimulation of fibers of the fornix and fimbria is 8-10 msec. The path of spread of impulses to the hippocampal neurons in these cases was very short. Judging by the time of appearance of fast potentials in the EP, considered by Green and Adey to reflect excitation of afferent fibers within the hippocampus, the volley of impulses following stimulation of the fornix enters the hippocampus 2-3 msec after stimulation. The main components of the EP, however, do not arise until 6-8 msec after stimulation.

In our experiment the time elapsing between appearance of fast potentials in the hippocampal EP and the time of appearance of the initial electropositive components was 10-12 msec. All this shows that the long duration of the latent period of the main component of the hippocampal EP is due not entirely to the polysynaptic character of the paths of afferent impulses to the hippocampal neuron, but also to certain special features of the conduction of excitation in the hippocampus itself.

All these hypotheses are firmly based only if the fast initial potentials in fact reflect presynaptic excitation of afferent fibers. However, the possibility is not ruled out that during stimulation of the vagus nerve afferent impulses reach the hippocampus along two different pathways. The first has few synaptic relays and the impulses arriving along it enter the deep layers of the hippocampus, excitation of which causes the appearance of comparatively fast surface-positive potentials with short latent period. The second pathway is polysynaptic and the impulses arriving along it enter the hippocampus later, are directed to the pyramidal neurons, and cause the appearance of the main components of the hippocampal EP.

In connection with the problem of genesis of the various components of the EP, the fact that despite the great difference in histological structure of the neocortex and hippocampus, the basis of their EPs in response to afferent stimulation is a stereotyped positive-negative complex is of considerable interest. In the hippocampus, just as in the neocortex, some neurons are so oriented that their long axis is strictly perpendicular to the surface. The presence of identically oriented dipoles in the neocortex and hippocampus possibly also has the result that, during their excitation, the surface potential in both these structures changes identically.

So far as the effect of stimulation of the vagus nerve and the gastric mechanoreceptors on background electrical activity of the hippocampus is concerned, by analogy with the neocortex, changes consisting of synchronization and desynchronization may be regarded as the electrographic expression of elevation or lowering of hippocampal tone. However, it may also be considered that these changes reflect more specific processes. The view that the theta-rhythm arising in the hippocampus under the influence of afferent stimulation is merely an indicator of activation of the hippocampus by the reticular formation of the brain stem, i.e., that it expresses the hippocampal arousal reaction, is now disputed by a number of investigators (Grastyán, 1959). It has been postulated on the basis of fairly solid evidence that this rhythm is connected in some way with the processing, analysis, and memorizing of information entering the hippocampus (Green, 1959; Green et al., 1960; Adey, 1961b; Adey et al., 1961; Magoun, 1965).

Stimulation of the vagus nerve or of its receptor endings in the stomach thus evokes a variety of electrical responses in the hippocampus which indicate that the afferent impulses produced not only reach the hippocampus, but undergo appropriate processing there.

SUMMARY

1. In response to electrical stimulation of the vagus nerve or pressure stimulation of the gastric mechanoreceptors, evoked potentials of the primary response type are produced in the hippocampus. This indicates that representation of afferent fibers of the vagus nerve system exists in the hippocampus.

2. The afferent system of the vagus nerve is represented in the hippocampus not as a separate zone, but diffusely throughout the hippocampus.

3. In their parameters the evoked potentials of the hippocampus in response to stimulation of the vagus nerve are similar to those of the hippocampus in response to stimulation of other afferent systems. They are also similar to the primary responses of

the neocortex. They are formed basically of two components: an initial positive and a subsequent negative component. The latent period of the response is 18–20 msec, the amplitude of the positive component 50–200 μV and its duration 20–30 msec, the amplitude of the negative component 25–200 μV and its duration 50–100 msec. The responses are frequently complicated both by fast potentials preceding the main components and by slow potentials arising after the negative components.

4. The amplitude of the main components of the hippocampal evoked potential falls sharply after administration of Nembutal or chlorpromazine.

5. Stimulation of the vagus nerve or stimulation of the gastric mechanoreceptors may cause the appearance or intensification of a regular theta-rhythm, and also its inhibition.

6. The variety of the electrical responses arising in the hippocampus to stimulation of the vagus nerve indicates its important role in the analysis of interoceptive stimuli.

Investigation of Activity of Cortical Neurons in Early Stages of Conditioning[*]

G. I. Shul'gina

Institute of Higher Nervous Activity and Neurophysiology
Academy of Sciences of the USSR
Moscow, USSR

Many electrophysiological investigations have shown that after the first combinations of conditional and unconditional stimuli, generalized changes in electrical activity take place in various parts of the brain. They consist mainly of increased depression of slow waves in response to the conditional stimulus and the more distinct appearance of fast rhythms (Sakhiulina, Naumova, 1965b; Morrell and Jasper, 1956, and others), the appearance of synchronized rhythms (Livanov and Polyakov, 1945; Shumilina, 1959; Voronin and Kotlyar, 1963, and others), strengthening of distant synchronization of potentials (Dumenko, 1953a; Glivenko et al., 1962b), the appearance of local high-amplitude waves (Livanov, 1962; Shul'gina, 1960; Knipst and Shul'gina, 1966), and changes in the steady potential (Shvets, 1963). However, all these phenomena may be observed to a greater or lesser degree not only during application of combinations of conditional and unconditional stimuli, but also during repeated presentation of new stimuli of adequate intensity (Sokolov, 1960; Vinogradova, 1961; Knipst, 1960; Shul'gina, 1962; Korol'kova and Shvets in this volume, and others).

On this basis the appearance of the above-mentioned changes in electrical activity in the first stages of conditioning is regarded not so much as the reflection of formation of the temporary connection as a manifestation of an orienting reaction. This view is confirmed by the well-known phenomenon of the subsequent weakening, or even complete disappearance of the observed phenomena as the conditioned reflex becomes established.

---

[*]Pages 296-308 in the Russian edition.

On the other hand, results have been obtained by classical Pavlovian techniques to show that the first external manifestation of conditioned reflexes can be seen after 1-10 combinations of conditional and unconditional stimuli (Kulagin, 1955; Shul'gina, 1962, and others). It may therefore be postulated that the changes in electrical activity arising after the first combinations of conditional and unconditional stimuli are more directly related to the process of formation of the temporary connection. However, even a very detailed analysis of overall changes in electrical activity, using different indices, has failed to distinguish the role of processes actually reflecting establishment or strengthening of the orienting response, on the one hand, and the role of processes more directly related to conditioning on the other.

A more detailed interpretation of the conditioning mechanism can be obtained by the study of single unit responses, because they show directly what processes are taking place in the central nervous system after the first combination of conditional and unconditional stimuli.

The few investigations of single units from this aspect have shown that their responses in different structures of the central nervous system can be modified in connection with conditioning (Yoshii and Ogura, 1960; Olds and Olds, 1961; Kamikawa et al., 1964; Burešová and Bureš, 1965; Hori et al., in this volume).

All these investigations were carried out on immobilized animals. Jasper et al. (1962) performed experiments on intact monkeys but give only aggregated data for cortical neuronal responses to light before combination and after establishment of the conditioned reflex, from which it is impossible to judge the dynamics of single unit responses.

The objective of my investigation was to study the activity of cortical neurons of the waking, nonimmobilized animal at various stages of conditioning. It was important to learn whether changes take place in the responses of the neurons to the stimulus which was to become the conditional stimulus, the direction of these changes, and whether changes occur in activity in the intervals between application of stimuli.

By studying the responses of the same neuron before and during application of combinations, I attempted to discover how the new functional system producing the response to the subsequently conditional stimulus is formed on the basis of two functional systems, one of which generated the response to the stimulus hitherto indifferent with respect to formation of a motor defensive response, while the other generated the unconditional response.

METHOD

Experiments were performed on unanesthetized and nonimmobilized rabbits fixed to a frame. Impulse activity was recorded by means of glass electrodes filled with 2.5 M KCl. A micromanipulator designed by Melekhova and Pezhemskii and modified by Diyakonov was fixed to the rabbit's head immediately above a burrhole 2 mm in diameter in the bone. The potentials were amplified by a type UBP 1-01 amplifier and recorded with a cathode-follower and type MPO-2 8-loop oscillograph. A pick-up with variable resistance was used to record the rabbit's respiration and general movements.

The conditional stimulus was a tone of 500 cps from a ZG-10 generator. The unconditional electrical stimulation (US) of the hind limb was generated by a "Physiovar" stimulator (Alvar) and caused slight movement of the limb. The stimulation was repetitive at a frequency of 3/sec and the duration of each pulse was 1 msec.

Experiments were performed on spontaneously active neurons. Responses of the neuron to acoustic stimulation were first recorded. After 5-15 applications, the US was applied against a background of acoustic stimulation acting for 1-2 sec. The combined action of acoustic stimulation and reinforcement also lasted 1-2 sec. The intervals between combinations of stimuli were from 2 to 10 min. If it was possible to make observations on the neuron for long enough and to record its activity during application of 15-20 combinations, the reinforcement was omitted and the unit responses to acoustic stimulation alone were recorded. If, on the other hand, impulse activity stopped earlier (as often happened because of the animal's increasing movement), the responses of the next neuron were recorded during continuing combinations and were compared with responses to subsequent application of acoustic stimulation without reinforcement.

From four to six experiments were performed on each rabbit. Responses of units in the sensorimotor, auditory and, in fewer
experiments, in the visual cortex were recorded.

In the analysis of the results, mean values of the impulse activity were calculated for each successive 100 msec by summating
a series of traces of the unit activity before, during, and after action of the stimulus. The significance of changes in the response
was determined by the use of Wilcoxon's criterion. Distribution
of activity in time was judged from the relationship between the
mean number of impulses during 100 msec ($\bar{X}$) and the dispersion.
The confidence limits for the Poisson distribution were determined
from the table given by Bol'shev and Smirnov (1965).

RESULTS

Unit Responses to Acoustic Stimulation Without Reinforcement and to the US Applied Against a Background of Acoustic Stimulation

As with any other stimulus, all the recorded units can be divided on the basis of their responses to acoustic stimulation into
three main groups (Table 1): neurons not responding, those responding by an increase in frequency, and those responding by a decrease in frequency. As Table 1 shows, most neurons in the sensorimotor cortex respond with an increased frequency, while in the
auditory and visual cortex most neurons responded with a slowing
of impulse activity.

The responses of some neurons were well marked, while
others responded with only a slight change in firing rate from the
background level. Such responses were found only by summation
and averaging of several unit responses. The latent period of the
responses sometimes was stable, but more frequently it varied
from one application of the stimulus to another within limits of
several tens of milliseconds. In most cases it was difficult to determine the latent period precisely because of the considerable
variability of the background impulse activity and the ill-defined
nature of responses to acoustic stimulation.

TABLE 1. Distribution of Cortical Neurons by
Responses to Acoustic Stimulation

Area of cerebral cortex	Number of neurons with:			
	absence of response	increased frequency	slowed frequency	total
Sensorimotor	6	12	8	26
Auditory	4	6	11	21
Visual	3	3	7	13
Total	13	21	26	60

Apart from 10 neurons displaying slower activity, both the strong and weak responses to tone were relatively stable. They varied within a narrow range of intensities, but no clear increase or decrease in the response with repeated acoustic stimulation was observed under these experimental conditions (not more than 5-15 applications). I deliberately did not apply the stimulus many times without reinforcement, assuming that marked extinction of the orienting reaction would prevent conditioning. Judging by the recordings of the animal's motor responses, in most cases this number of unreinforced stimuli was sufficient to prevent generalized movement or significant changes in respiration.

Because of the experimental conditions I was unable to investigate unit responses to the US before conditioning, so as not to cause changes in the animal's functional state which might occur under these circumstances. In addition, since the US was always applied against a background of acoustic stimulation, the responses to the tone as it became the conditional stimulus were compared with responses to combined stimulation.

The distribution of neurons based on their responses to the US applied against a background of acoustic stimulation is shown in Table 2. Compared with responses to tone alone, fewer neurons in this case were nonreactive and more cells reacted by slowing their rate of discharge. The latter were also predominant in the visual and auditory cortex. The response of most neurons to the US was more intensive and prolonged than the response to tone. This was largely due to the fact that the US is a stronger stimulus, but also to the fact that it was repetitive (Fig. 1).

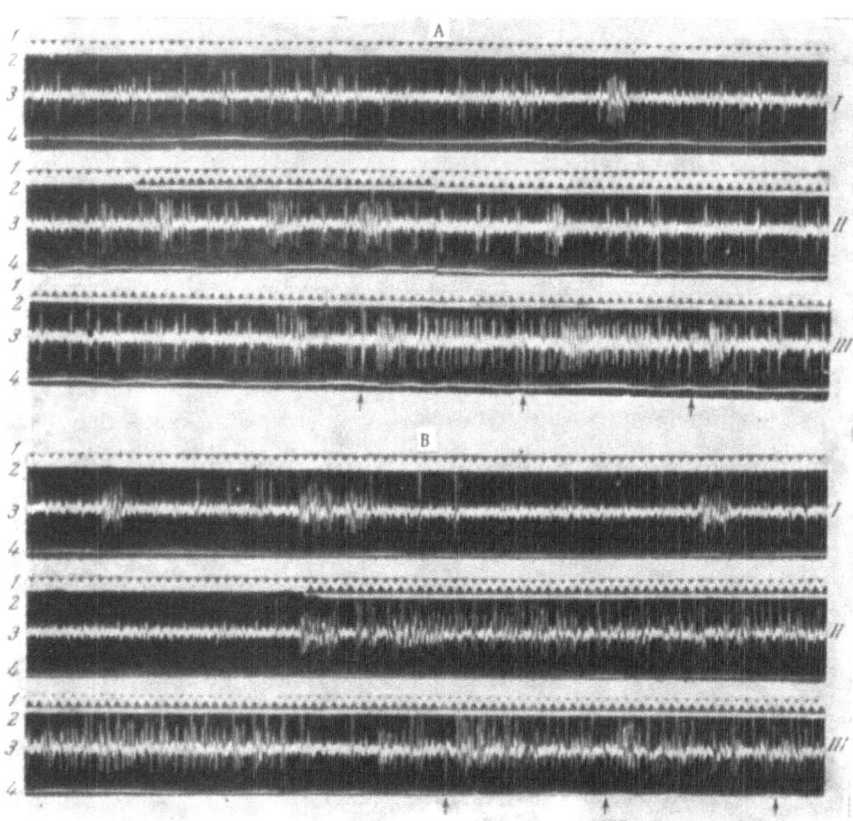

Fig. 1. Increase in activity of a neuron in the sensorimotor cortex in response to acoustic stimulation when combined with US: A, first combination of tone and US; increased spike discharge to tone and to the combined action of tone and cutaneous shock; B) 4th combination of sound with US. Strengthening of response to increased frequency. 1) Time, 50 cps; 2) acoustic stimulation; 3) spike activity (in this case two neurons were recorded simultaneously); 4) respiration and movement of animal. Arrows denote application of US. II and III are the continuation of I.

TABLE 2. Distribution of Cortical Neurons by
Response to Electrical Stimulation of the Limb (US)
Applied Against a Background of Acoustic Stimulation

Area of cortex	Number of neurons showing:			
	absence of response to US	increased discharge	decreased discharge	total
Sensorimotor	5	10	11	26
Auditory	3	4	14	21
Visual	–	3	10	13
Total	8	17	35	60

The latent period of the response of different neurons to the
US varied from 10 to 100 msec. Usually the response to the US
was stable and did not change significantly with successive combinations, although the responses of some neurons to the first application of this stimulus did not coincide with later responses. These
changes took place after 1-4 combinations, usually parallel with
changes in the response to the tone becoming the conditional stimulus, and they will be described below.

Responses of Cortical Neurons

During Conditioning

Recordings of the animal's respiration and movement showed
that after the first combinations of tone with the US, motor responses began to appear to the tone becoming the conditional stimulus. They did not appear after every application of the tone. They
were sometimes observed in response to irrelevant stimuli, and
quickly disappeared when reinforcement was omitted. This character of the motor responses shows that the phenomenon was an
early stage of a generalized but not yet stable conditioned reflex.
In the period of reinforcement of acoustic stimulation by the US
the responses of 50% of the investigated neurons to the tone were
modified, some to a greater degree than others (Table 3).

These changes consisted mainly of an increase in the similarity between the responses to the tone and to the US or to the appearance of a hitherto unobserved similarity. Acceleration of the

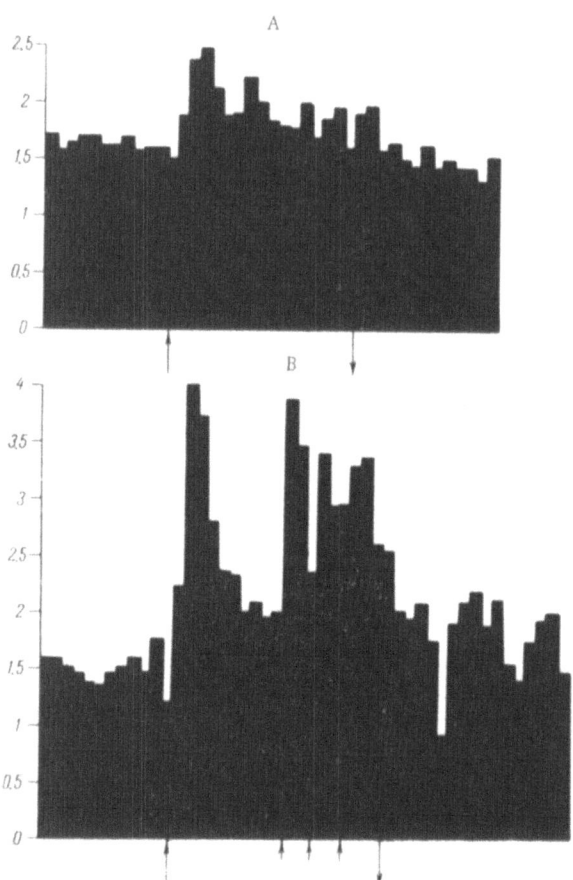

Fig. 2. More marked increase in firing rate in response to
tone when applied in combination with the US: mean data
for 10 neurons. A) Acceleration to tone before conditioning.
Total number of applications of tone 128, B) increase in
intensity of this response to tone when reinforced by US.
Total number of combinations 60. Ordinate: mean number of impulses in 100 msec; abscissa: time from beginning
of recording unit activity. Long arrows denote tone on and
off; short arrows ditto for US.

TABLE 3. Changes in Responses as Tone Becomes the Conditional
Stimulus

Area of cortex	no reaction to tone and US	reaction to tone alone		reaction to US alone		reaction to both tone and US in the same direction		difference	
		+	−	+	−	+	−	+ (to tone)	− (to tone)
Sensorimotor	3	—	2(2↑)	—	3(3−)	11(9↑1↓)	6(2↓2+)	1(1−)	—
Auditory	1(1+)	—	2(2↓)	—	3(3−)	4(3↑)	9(3↓)	2	1(1+)
Visual	—	—	—	—	3(3−)	3(1↓1↑)	7(2↓)		

Figures in parentheses denote number of neurons in which response to tone during conditioning increased ↑, diminished ↓, or a hitherto unobserved acceleration + or
slowing − appeared.

activity was usually more marked, especially in neurons of the
sensorimotor and auditory cortex. It is clear from Fig. 2, in which
the activity of a group of neurons showing a greater increase in
frequency of firing during application of tone − US combinations is
averaged, that under these circumstances, the greatest similarity
is found between periods immediately after application of these
stimuli. This phenomenon can probably be used to explain the fact
that after conditioning has taken place, the simple application of
the conditional stimulus followed by its rapid removal gives rise
to the whole course of the conditioned response previously observed only as the result of prolonged application (Kupalov, 1962).

With the first three neurons of this group, no acceleration
was observed in response to the first application of US. It appeared only in response to the second and third combinations, parallel with more marked acceleration in response to the tone.

In one neuron in the sensorimotor cortex responding neither
to tone nor to the first application of the US, acceleration of impulse activity appeared in response to either stimulus after a few
combinations (Fig. 3).

When reinforcement was omitted, the similarity between the
responses to tone and US sometimes disappeared. For example,
in a neuron of the auditory cortex during pairing, marked acceleration of activity was observed in response to tone and to the US,
followed by slight slowing during applications of the tone without
reinforcement (Fig. 4).

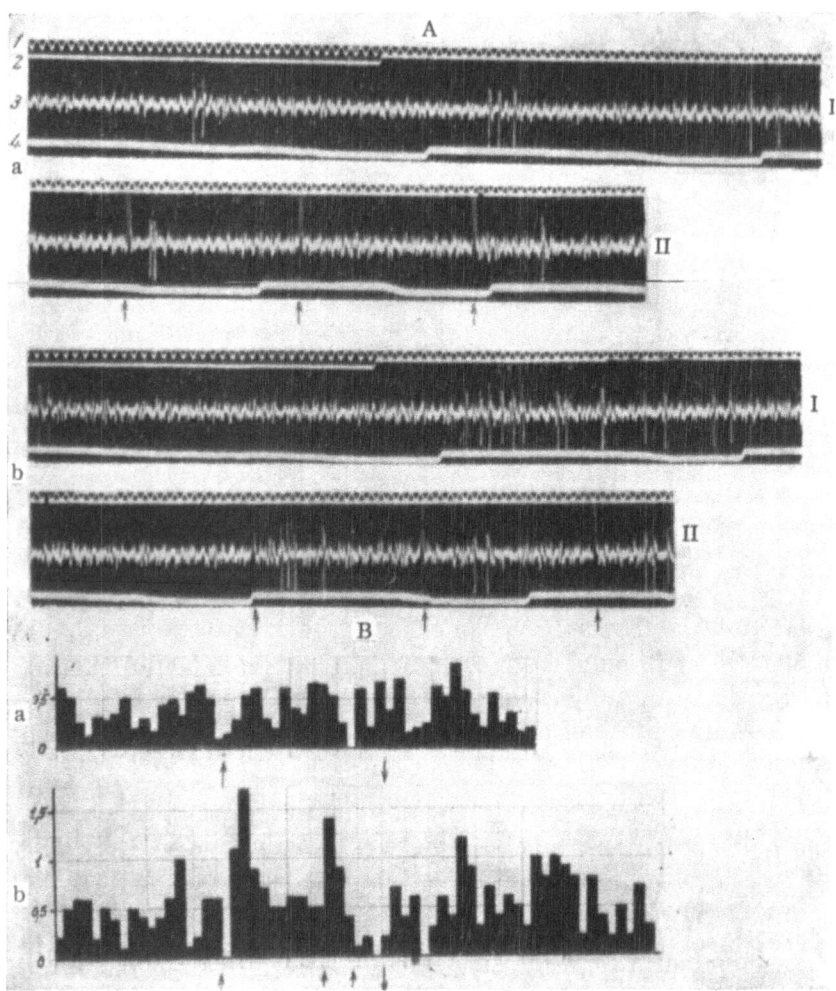

Fig. 3. Appearance of acceleration of impulse activity in response to tone and US when applied in combination to neuron No. 58 of auditory cortex. A: a) 1st combination of tone with US; B: averaged values of impulse activity of this same neuron, a) after 20 applications of tone. Absence of acceleration of impulse activity to tone, b) after 10 combinations of tone with US. Acceleration of impulse activity to tone and to application of US. Legend as in Fig. 1.

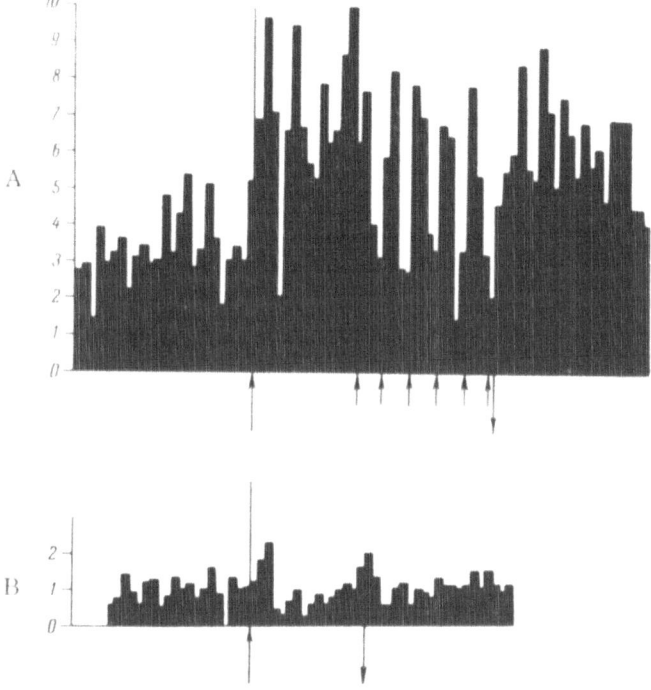

Fig. 4. Change in response of auditory neuron to tone when reinforcement omitted. Mean values of impulse activity. A) After 10
combinations of tone with US. Marked acceleration activity to
tone and to US with prolonged aftereffect; B) after 40 applications
of tone without reinforcement. Slight slowing of impulse activity
to tone. Lowered background activity of neuron.

Slowing of activity in response to the tone, which was also
observed before conditioning in contrast to acceleration, either remained unchanged or became weaker in most neurons during pairing. Weakening was observed in the group of neurons in which this
response was unstable before conditioning and gradually declined
with repeated application of the tone. This suggests that these
neurons belong to the orienting reflex system. This reaction was
not restored after formation of the reflex, but disappeared even
more rapidly than before it. This took place in a parallel manner
to tone and to the US.

A typical example showing the dynamics of the responses is
given in Fig. 5. Marked slowing of activity to the tone was ob-

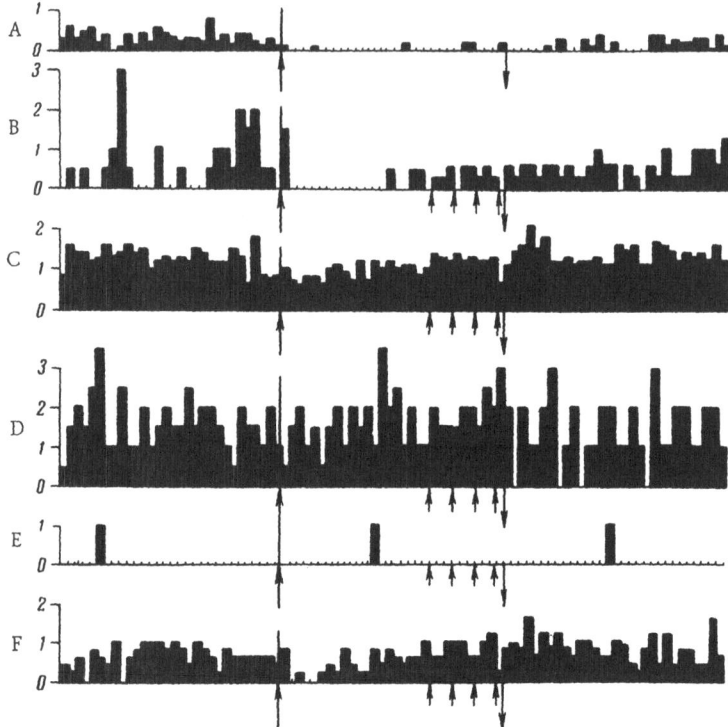

Fig. 5. Responses of a sensorimotor neuron during conditioning. Mean values of impulse activity: A) after 11 applications of tone without reinforcement — marked slowing of impulse activity in response to tone is observed; B) after 1-3 combinations of tone with US — slowing in response to US less marked than to tone; C) after 4-13 combinations of tone with US slowing of impulse activity is less marked both in response to tone and to US; D) after 17 and 18 combinations of tone with US response to combined stimuli is absent; E) after 19 combinations of tone with US background activity of neuron is considerably reduced after application of irrelevant stimulus; F) after 20-24 combinations of tone with US background activity is restored and impulse activity slowed in response to tone. Legend as in Fig. 2.

served in this neuron (sensorimotor area), gradually becoming weaker with repetition (Fig. 5A). Slowing of activity also was observed during the first applications of the US, but was less marked than in response to the tone (Fig. 5B). With subsequent combinations it almost disappeared (Fig. 5C), and at the 17th and 18th combinations it was practically absent (Fig. 5D). After 18 pairings, irrelevant photic stimulation was applied. No response took place to it, but after a few seconds the background unit activity almost completely disappeared (Fig. 5E). Later it recovered gradually, but did not reach its previous high level. In response to the tone, during the next five pairings slight slowing of activity again was observed (Fig. 5F). Instead of an initial slowing, two neurons of this same group responded with an acceleration of activity in a parallel manner to the tone and the US during successive pairings.

If extinction of unit responses to a repetitive stimulus is regarded as a reflection of extinction of the orienting reflex at the neuronal level, as exemplified by this group of neurons, what is observed during the first pairings of the tone and US is not recovery of the extinguished orienting reaction, as might have been expected, but on the contrary, its even more rapid and immediate extinction.

This characteristic of the changes in activity indicates some degree of antagonism between orienting and conditioned responses at the neuronal level in the cortex.

If this fact is confirmed by later investigations, it will be of considerable interest to the elucidation of the relations between these reflexes. In particular, this phenomenon is probably connected to some extent with the fact that a too powerful orienting reflex to any stimulus, like one which has become excessively extinguished, will also prevent the formation of a conditioned reflex.

In another group of neurons not responding to unreinforced acoustic stimulation but responding to the US by slight slowing, under the influence of reinforcement similar slowing was observed in response to the tone alone (Fig. 6).

Comparison of the graphs in Figs. 6 and 2 shows that the latent period of the newly-developed slowing response is much longer than the latent period of the accelerative response, and occurs approximately at the maximum of the latter. This perhaps suggests that the slowing response is secondary, arising as a result of augmented activity of other neurons, for which it probably plays a limiting role.

The changes observed in the response to the tone when reinforced with the US thus consist of the appearance of similarity between the responses to these stimuli. These changes either were determined by the quality of the response to the US or took place in parallel for both tone and US.

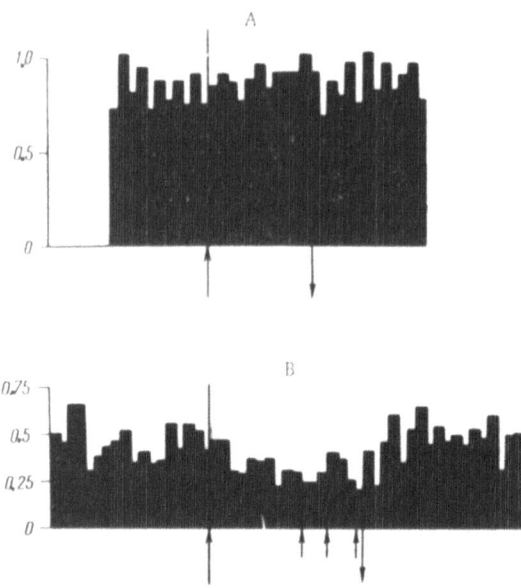

Fig. 6. Appearance of slowing of activity in response to
tone applied in combination with US. Mean data for 8
neurons. A) Response of neurons to unreinforced tone.
Total number of applications of tone 95. Absence of
slowing; B) response of neurons to tone and combined
action of tone and US during conditioning. Total number of combinations of tone with US 73. Appearance of
similar slowing of activity to tone as to US. Lowering
of background activity. Legend as in Fig. 2.

Analysis of the duration of the off-effects shows that in the
case of unreinforced acoustic stimulation the changes in activity
after its discontinuation persisted (or an off-effect reappeared) in
14 of 45 neurons. Usually these changes were of short duration.
In the case of pairing, changes in activity persisted after discontinuation of stimulation in 29 neurons and they were more marked
and more prolonged than in the case of unreinforced acoustic stimulation, due primarily to application of the US (see Fig. 4).
If, however, the off-effects after discontinuing the tone were
intense enough, they also persisted in the period of its reinforcement.

The exception as regards duration of off-changes in activity
is the group of neurons described above showing rapid extinction

of the slowing response under the influence of pairing. After discontinuation of the stimulation, the slowing effect either disappeared much faster during the pairing than before it, or could not be observed at all.

The off-effects in some neurons were so prolonged that this affected the frequency of background activity, increasing it above the background level or decreasing it. An increase in background activity took place more frequently in neurons showing a more marked acceleration of impulse activity or a less marked slowing in response to acoustic stimulation, while background activity was lowered in neurons whose activity was slowed in response to acoustic stimulation subsequently becoming conditioned (Figs. 4, 5, and 6).

The distribution of impulse activity in time remained stable for most neurons (about 80%) before and during conditioning. However, in some neurons it was modified, mainly by the appearance of an increased tendency toward grouped discharges.

These changes in background activity, possibly through a mechanism of posttetanic potentiation (Eccles and Meulders, 1964), probably reinforce traces of excitation and thus facilitate the response to successive stimuli and the implication of new neurons in it. Whether this increased tendency toward grouped discharges is characteristic of the process of formation of the temporary connection or whether it is purely the result of application of repetitive reinforcement is the subject of my next investigation.

DISCUSSION

It is now evident that the response to any stimulus having a peripheral manifestation (movement, secretion, etc.) or without visible external manifestations takes place through the activity of a group of neurons located at different levels of the central nervous system. These neurons, which are interconnected, form a definite functional system. During repetition of a stimulus, some of these neurons gradually drop out of this system (in our experiments these were mainly certain neurons in the cortex responding to the stimulus by inhibition), but most of them, under the same experimental conditions, retain their original response or vary it within slight limits.

If all neurons encountered by the microelectrode in the cortex are recorded and tested without selecting them in accordance with their responses, by the theory of probability the largest number of recorded neurons will be those which are in fact encountered most frequently. Consequently, under these conditions a rough idea can be obtained of the proportion of neurons accelerating or slowing their impulse activity during the formation of a given functional system, and also of the relative number of neurons which participate actively in it or which are nonreactive.

Before pairing of two stimuli, two different functional systems of neurons are formed in response to each of them. These systems differ both in their numerical composition and in the form of their responses. At lower levels, reception of these stimuli is undertaken by different neurons, but the higher the excitation is transmitted in the central nervous system, the more neurons respond to either stimulus. The fact is that at the cortical level most neurons are components of one or more functional systems, and the main differences between them are found not only in the quantitative extent and the structural and spatial arrangement of these neurons, but also in the intensity and form of their responses.

When acoustic stimulation is combined with the US, changes are observed in the quantitative composition of the functional system initially activated by sound. Some initially nonreactive neurons now begin to respond to sound, while others cease to do so. Meanwhile, the form of the responses of those neurons which responded to sound before formation of the reflex also changes.

In most neurons these changes are such that the functional system formed in response to acoustic stimulation becomes increasingly more similar to the functional system responding to the combined action of sound and US.

However, the exceptions to the general rule described above indicate that this is not just a simple replacement of the elements of one system by the elements of another, but in fact the formation of a third system, differing slightly both from the system initially responding to sound and the system initially responding to the US.

The existence of a series of neurons whose response to the first applications of the US against the background of acoustic stimulation differs from responses to subsequent applications indicates that changes in the quantitative composition and form of unit responses take place not only in the system of neurons responding to the conditional stimulus, but also, although admittedly to a far lesser extent, in the system responding to the US.

Only one neuron was observed which was implicated in the response to acoustic stimulation when applied in combination with the US, although not hitherto responding to either of the stimuli. Consequently, neurons which could be regarded as a new intermediate link joining together two hitherto unconnected forms of activity, connecting responses to the US and to a stimulus hitherto indifferent for this particular response, are found very rarely in the projection zones investigated. The reason for this may be that the cortical interneurons which could act as this intermediate link are small in size and recordings can be made from them with much less probability than from the larger pyramidal cells. However, there is another possible explanation, namely that conditioning in fact takes place mainly on account of elements which respond to at least one of the combined stimuli before formation of the conditioned reflex, at least for spontaneously active neurons of the cortical areas investigated. These relationships are probably different for the silent neurons and also for neurons of the cortical association areas.

Hence, changes in cortical neuronal responses show that after the first combinations of conditional and unconditional stimuli, the original responses are not only intensified, but they may also be modified in direction and new responses not hitherto observed may appear. The trend of changes in neuronal responses to the conditional stimulus is largely determined by the quality of these responses to the unconditional stimulus. Evidently as a result of this trend, a new functional system is formed, responsible for the creation of an adequate peripheral response to the stimulus which subsequently becomes conditioned.

What causes the formation of this new functional system? Probably an essential role here is played by changes in the state of each particular neuron and changes in the incoming impulses. The first of these may result from greater activation in the period of action of the unconditional stimulus and from intensification of the off-effects and changes in intensity and distribution in time of the background impulse activity. The strongest evidence of the second of these factors is the change in intensity of unit responses and, consequently, the change in conditions of reaction of other neurons connected with them. Further evidence of a change in the rate of impulses reaching single units may also be given by the existence of a series of neurons in which modified responses to acoustic stimulation took place without significant changes in background impulse activity, and also of others in which new responses appeared concurrently to acoustic stimulation and to the US.

Comparative Aspects of Neural Background Activity[*]

Zh. P. Shuranova

Institute of Higher Nervous Activity and Neurophysiology
Academy of Sciences of the USSR
Moscow, USSR

Interest in the spontaneous or background activity of neural centers, which appeared at the dawn of electrophysiology, has continued throughout the modern period of neurophysiological research. One of the initiators of such research in the USSR is Livanov, whose work has been devoted to the experimental analysis of activity of the brain as a whole as reflected in the EEG, and of single unit activity of cortical neurons (Livanov, 1938, 1940, 1944a, 1965b). At the present time an analysis of the concrete principles governing integration of the activity of neural units is being conducted in Livanov's laboratory. As a part of this analysis I have investigated a relatively simple preparation: the ventral nerve chain of the crayfish. The preliminary results of these investigations are described in this paper and an attempt is made to generalize the results of several authors studying the organization of spontaneous or background activity in neural structures.

Naturally this paper does not claim to describe all the data so far accumulated in the electrophysiological literature on spontaneous activity. Its scope is restricted, first, to the activity of single units only (or, more precisely, to activity of a type which is such that the contribution made by single units can be distinguished); second, to impulse activity only, i.e., to action potentials generated by neurons in accordance with the "all or nothing" law; third, primarily to data obtained from the standpoint of comparative neurophysiology. In other words, integrated activity as seen in the EEG

---

[*] Pages 309-323 in the Russian edition.

is not considered nor are slow local changes in membrane potential in single neurons. In addition, the vast number of facts concerning neuronal activity of the higher levels of the central nervous system of vertebrates are left almost untouched and attention has been confined to the analysis of published data and my own results obtained during the study of invertebrate neurons.

Data Concerning the Spontaneous
Activity in the Invertebrate
Nervous System

The first details of the electrical activity of the invertebrate central nervous system were obtained at the beginning of the modern period of electrophysiology when highly sensitive galvanometers and vacuum-tube amplifiers began to be used. As Adrian (1935, p. 78) wrote in his monograph, "The most remarkable feature of the nervous system of insects is the unceasing active state observed in excised parts of the ganglionic chain. Impulses constantly move up and down this chain . . . they continue for up to 24 hr after dissection of the specimen . . . ." During the same years other investigations were published, carried out on the ventral cord of the crayfish (Prosser, 1934a, 1934b), of the cockroach (Roeder and Roeder, 1939), and so on. These first investigators observed many of the advantages possessed by invertebrates for the study of spontaneous activity: ease of isolation of the central nervous system or its divisions, the fact that its neurons remain capable of functioning well for long periods, and finally, the possibility (at a time when microelectrode techniques were still in their infancy) of analyzing the activity of single units.

Coelenterates. A considerable number of electrophysiological investigations of spontaneous activity in the neural structures of different invertebrates have now been described. It has been shown that the rhythm of the swimming contractions of *Medusa* is neural in origin. It may be assumed that the rhythmic activity of the tentacles of certain polyps of the order *Alcyonaria* is similar in origin (Horridge, 1961). By direct recording of the electrical activity of the polyps *Tubularia crocea* a temporal configuration of action potentials of some complexity was found in the hydranth and dorsal stalk (Josephson, 1962). Action potentials followed one another at constant intervals or were grouped in short volleys separated by periods of silence. Josephson considers that this activity, taking place under relatively constant conditions, is endogenous and due to functioning of the nerve net of the polyp.

Worms. Little information concerning the spontaneous activity of the nervous system of worms is yet available. Some authors (Gray et al., 1938; Koshtoyants, 1957) consider that the background activity in the ventral chain of the leech and earthworm is entirely reflex in origin. Another view, however, is held: the progressive movement of earthworms by means of peristaltic contractions is central in nature, although the spread of peristalsis along the animal's body depends on the tonic influence of tactile receptors or proprioceptors. Progressive movement of *Nereis* by means of parapodia is even more dependent on internal automatism of the nervous system; the swimming movements of the leech evidently are also due to a neurogenic rhythm, and not to chain reflexes, like the stepping movements of an animal (Bullock, 1961).

Most investigations of the background activity of invertebrate neurons have been carried out on arthropods and mollusks.

Crustaceans. Electrical activity in the ventral chain of the crayfish, as mentioned above, was first described by Prosser (1934a, 1934b). Some information on the background activity of the ventral chain of an intact animal (crab) was obtained at roughly the same time by Bonnet (1938), who suggests that the activity described reflects automatism of the work of central neurons and can be compared with the rhythmic activity of the vertebrate brain. Later, Preston and Kennedy (1962), Kennedy (1963), and Biederman (1964) investigated the activity of single fibers composing the interganglionic connectives of the ventral chain of the crab *Procambarus clarkii*. It was shown that about half the investigated fibers possess spontaneous activity. Action potentials either followed each other at strictly constant intervals, or the interval varied within wide limits; the number of these arrhythmic neurons, according to Preston and Kennedy, was about one-third, and according to Biederman about half of the total number of spontaneously active fibers. Kogan and Chorayan (1965), using a microelectrode technique, recorded mostly rhythmic discharges with a frequency of 1-15/sec from the neurons of abdominal ganglion VI of the crab. Arrhythmic discharges were observed in 25-30% of investigated neurons.

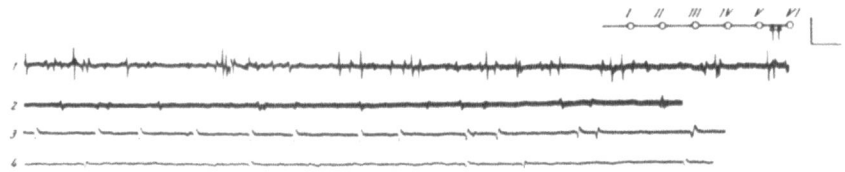

Fig. 1. Background activity in interganglionic connectives of ventral chain of crayfish. 1) On day of isolation of abdominal division of ventral chain; 2) on 3rd day;
3) 6th day; 4) 9th day after isolation. Diagram showing arrangement of recording
electrode given above. Roman numerals represent ganglia of abdominal division of
ventral chain. It is assumed that an action potential spreading in the oral direction
has a downward initial phase. Calibration: 100 µV, 20 msec.

Intense electrical activity in the interganglionic connectives
of the ventral chain was also found in my experiments conducted on
the crayfish *Astacus astacus* L. Action potentials with a complex
configuration may be observed either by taking recordings from
the ventral chain in vivo or from an isolated preparation. Activity
is observed provided that at least one abdominal ganglion is intact.
The character of the activity as a whole was stereotyped during an
experiment lasting several hours. This fact alone is evidence to
show that this activity cannot be a reaction to injury associated
with isolation of the ventral chain or some of its ganglia. Moreover, some neurons remained active during prolonged (up to 10
days) survival of the isolated preparation in van Harreveld's solution (van Harreveld, 1936; Fig. 1). This is obviously an argument
in support of the endogenous origin of the activity described.

Insects. Experiments to study the electrical activity of
the central nervous system of insects were first carried out by
Adrian on the caterpillar and water beetle. In recordings made by
electrodes applied to the isolated ventral chain he found a complex
pattern of action potentials for analysis: "sudden bursts of high
frequency alternate with slow, regular discharges . . . . , which
stop and start again without any visible cause" (Adrian, 1935, p. 79).
According to Roeder and Roeder (1939), the electrical activity in
the ventral chain of the cockroach consists of arrhythmic action
potentials of low amplitude. Sometimes high–amplitude rhythmic
spike discharges were superimposed upon this background. Agents
such as pilocarpine, eserine, nicotine, etc., had a powerful action
on the spontaneous activity. In small doses they caused a reversi-

ble increase in the activity, while in large doses they inhibited it.
Fielden and Hughes (1962), and later Mill (1963) investigated the
background activity of single fibers in the interganglionic connectives and efferent roots of the ventral chain of the dragonfly larva.
They observed great variability in the activity of different fibers:
from action potentials proceeding at a constant rhythm to potentials distributed irregularly in time, and volleys of impulses alter-
nating with periods of absence of activity. In these authors' opinion, grouped activity is very characteristic of this particular preparation. Volleys of impulses are not associated with respiration
or beating of the heart, and are evidently endogenous. The background activity of the metathoracic ganglion of the locust was investigated by Sviderskii (1965). In recordings made from the
ganglion by tungsten microelectrodes he found both regularly discharging neurons and cells generating action potentials at irregular
intervals.

Mollusks.   Spontaneous activity in the central nervous
system of mollusks has been studied mainly by means of intracellular recordings from "giant" neurons. Tauc (1955) described
rhythmic activity in several cells of the abdominal ganglion of
Aplysia; activity persisted for a long time (several hours) in the
neurons of the isolated ganglion. In a series of investigations, Arvanitaki and Chalazonitis (Arvanitaki–Chalazonitis, 1961; Arvanitaki and Chalazonitis, 1961, 1963; Chalazonitis, 1963) investigated
the functioning of different types of neurons of a visceral or parieto–genital ganglion of the same mollusk. They found that a large
proportion of neurons possess long-lasting activity, sometimes
constant in frequency. In addition, they found a group of neurons
generating volleys of impulses; intracellular recordings showed
that each volley is superposed on a slow depolarization wave, a
special form of generator potential. In most cases no signs of
postsynaptic potentials could be found during the phase of de- and
repolarization, as a result of which these workers are inclined to
consider that the slow depolarization waves are truly spontaneous.
Besides these autoactive neurons, activity in the form of volleys of
impulses evoked by synaptic bombardment from other neurons was
observed in many identifiable cells.

Electrical activity of the neurons of certain freshwater mollusks has been studied in the laboratories of Kostyuk and Kogan.
The neurons of these animals are convenient for measuring the

electrical constants of the nerve cell and for investigating the role
of different ions in the functioning of the cell membrane (Gerasimov et al., 1965). According to Karpenko (1965), most intracellularly recordable neurons of the isolated ganglion of the edible
snail show a stable spontaneous activity. Depending on the character of this activity, Karpenko divides the neurons into two main
groups: strictly rhythmic and arrhythmic (in this group he includes neurons giving grouped discharges). During prolonged recordings no sign of transformation of one type of activity into the
other was found. Similar results are described by Kogan and
Chorayan (1965). They point out that rhythmic activity is charac-
terisitc of most neurons of the subesophageal ganglion of the edible snail; arrhythmic discharges of action potentials were generated by 20% of the investigated neurons. Hence, stable background
activity in the nervous system, persisting for a long time in the
absence of special treatment of the animal, is characteristic of the
simplest organized animals (coelenterates). The available data on
this subject, however, are restricted, first, to a group of favorite
preparations (among the mollusks, namely *Aplysia* and certain
freshwater snails; among the arthropods — the crayfish, lobster,
cockroach, and certain others); second, to the fact that investigations have been carried out mainly on the more simply organized
structures, namely the visceral or abdominal ganglion of mollusks,
and the abdominal division of the ventral chain of arthropods. Data
concerning the spontaneous activity of the higher levels of the central nervous system even of these animals are still extremely
fragmentary at the present time.

Temporal Character of Activity of
Central Neurons of Invertebrates

In this paper I have deliberately not mentioned the numerous
data concerning the special features of electrogenesis of invertebrate neurons. What is of most interest is not the physico-chemi-
cal mechanisms of generation of the single action potential, but the
processes responsible for the long persistence of activity in the
single neuron and systems of neurons. From this standpoint, the
investigation of the temporal distribution of action potentials generated by these neurons under relatively constant external environmental conditions is of particular interest.

Some information on the temporal organization of neuronal activity was obtained in my experiments on the abdominal division of the isolated ventral chain of the cockroach. Action potentials were recorded: 1) by applied silver electrodes from the whole interganglionic connective or from a connective split along its length into two symmetrical halves; 2) by capillary microelectrodes inserted into one of the ganglia. Interspike intervals were measured for the neurons retaining their activity for a long time. It was found that the neurons of the isolated ventral chain can discharge: a) rhythmically, i.e., with strictly constant interspike intervals; b) rhythmically, but with a less regular interval; c) essentially arrhythmically; and d) in rare cases, in the form of grouped discharges.

Some of the special features of this activity will now be examined in more detail and compared with those in the literature.

1. Rhythmically discharging neurons. These, of course, were the first type to be met by investigators of the nervous system of various invertebrates. In the illustrations in Adrian's monograph can be seen the regular rhythm of the action potentials generated by a ventral chain neuron of a caterpillar (Adrian, 1935, p. 19). Rhythmically discharging neurons are clearly seen among cells of the isolated ventral chain of the crab (Kennedy, 1963; Biederman, 1964), of the dragonfly larva (Fielden and Hughes, 1962; Mill, 1963), the locust (Sviderskii, 1965), the neurons of the optic lobe of the insect *Calliphora* (Kuiper and Leutscher-Hazelhoff, 1965), many neurons of mollusks (Tauc, 1955; Arvanitaki and Chalazonitis, 1961, 1963; Veprintsev et al., 1964; Karpenko, 1965; Kogan and Chorayan, 1965), and so on.

Hence, the distribution of interspike intervals of many central neurons of invertebrates is symmetrical, and evidently close to normal. The magnitude of scatter, i.e., the degree of spread of the distribution curve of the intervals depends on the value of the mean: the greater the mean, the greater the possible deviations from it, and vice versa. This fact can be clearly demonstrated for the same neuron, whose frequency changes in the course of the experiment for various reasons. Similar data were obtained by Wilson and Davis (1965) for motoneurons innervating the pincer-opening muscle of the crab.

2. It is an essential fact that the activity of rhythmically discharging neurons, although relatively constant over short periods of time (for example, tens of seconds), cannot be constant if examined over longer time intervals (minutes, tens of minutes). The action potentials may remain rhythmic, but the length of the mean interval will be different. The range of values within which the frequency of impulses generated by the same neuron may vary may be fairly wide. For instance, the discharge frequency of the photoreceptor neurons of the crab (which will be discussed below) under my experimental conditions varied from one impulse in several seconds to several tens of impulses per second. The same observations have been described by other authors (Sviderskii, 1965; Kuiper and Leutscher-Hazelhoff, 1965). Hence, the background activity of rhythmically discharging neurons of invertebrates can only be regarded as stationary if it is examined over short periods of time. In this respect, significant differences evidently exist by comparison with the background activity of central neurons of vertebrates, which remains stationary over long time intervals (Kostyuk, 1965).

3. Arrhythmic activity in many invertebrate neurons has also been reported by many investigators (see above). In my experiments, arrhythmically discharging neurons were found in recordings made from the intact interganglionic connective and from half of a connective, and also in microelectrode recordings from intraganglionic neurons. Hence, this fact is firmly established. Unfortunately no published report has appeared on the quantitative analysis of arrhythmic activity of invertebrate neurons. On the basis of my experiments, and also of records from several authors, it can be postulated that this arrhythmia is essentially different from that of most neurons of the vertebrate brain so far studied.

Without going into details, it can be said that most neurons of the vertebrate central nervous system are characterized by irregular background activity, and the distribution of their interspike intervals is evidently near-exponential (Poisson). Such a random distribution of background activity is hardly likely to occur in arrhythmically discharging invertebrate neurons. The arrhythmia in these neurons is evidently a reflection, not of the random predominance of short intervals over long, but of some instability in the value of the mean interval.

4. Grouped activity, i.e., discharges of action potentials subdivided by periods of absence of activity, evidently is not typical of neurons of the isolated ventral chain of the crayfish. Meanwhile, as mentioned above, it is extremely characteristic of the background activity of the ventral chain neurons of the dragonfly larva, and also of several identified neurons in mollusks.

Hence, the information at present available on the temporal organization of background activity of single invertebrate neurons suggests that it consists mainly of more or less regular rhythms or volleys of impulses. In cases when the background activity of the neurons is arrhythmic, it may be postulated that the arrhythmia differs essentially from that typically found in neurons of the vertebrate central nervous system. It should be pointed out, however, that these data are largely preliminary, and the problem of the temporal character of invertebrate neuronal activity is one which must receive specialized statistical investigation.

Possible Sources of Background Activity
of Central Neurons in Invertebrates

Hitherto the concepts of "background" and "spontaneous" activity have been used almost as synonyms. However, it seems desirable to differentiate between these concepts, reserving the term "spontaneous" to mean endogenous activity characteristic of pacemaker neurons, and the term "background" to mean activity of a different genesis characteristic of the nervous system or its components in the background, in the absence of any premeditated action on the investigated structures.

The question of the sources of background activity of central neurons in invertebrates (especially in isolated nervous structures) can now be considered. With respect to some parts of the vertebrate brain the question of whether an intrinsic background activity is present or not is still debated (Burns, 1958), but in the case of invertebrates this question does not arise — activity persists even in isolated structures consisting of a few neurons (for example, in the cardiac ganglion of crustaceans, formed from nine neurons). What is not clear, however, is the way in which this state of steady activity is created. To examine this problem, it is clear that several possibilities must be discussed.

1. Is background activity of these neurons connected with
steady liberation of mediator in the presynaptic ending, resulting
in "miniature potentials" which have been well studied at the neu-
romuscular synapse (Fatt and Katz, 1962)? Although it has been
established that spontaneous fluctuations of potential with parameters close to those of "miniature potentials" (Katz and Miledi,
1963), are present in the motoneurons of the isolated frog's spinal
cord, no experiments of this type have evidently been performed
on invertebrate neuron. At present only a few general observations
suggest that this source does not make any significant contribution
to the background activity of the neurons considered here. If it is
accepted that the critical level of depolarization is 5-15 mV (Veprintsev et al., 1964; Gerasimov et al., 1965, and others), it has
to be postulated that before a single action potential can be produced, synchronized liberation of tens of quanta of mediator must
take place, followed by effective summation of the "miniature potentials" generated by them. In the case of a spontaneous process,
this is evidently improbable, not because the number of synapses
on invertebrate neurons can hardly be very great, but because
synapses on these neurons are distributed along the length of axons
(Kennedy and Mellon, 1964), as a result of which the summation of
"miniature potentials" must be difficult. In my opinion, the following fact suggests that this possibility can hardly take place. It is
well known that the spontaneous liberation of mediator is a purely
random process (Fatt and Katz, 1952); the time distribution of action potentials generated as a result of random summation of "miniature potentials" likewise must obey this rule. Conversely, as
was mentioned above, action potentials generated in the background by invertebrate neurons are not distributed irregularly.

2. An evidently more probable hypothesis is that the background activity of many invertebrate neurons is truly spontaneous
activity of nonsynaptic origin. Various facts are in favor of this
suggestion, especially the possibility of its long persistence in isolated neural structures. Another fact in support of this hypothesis
is the temporal organization of the activity of invertebrate neurons.
It was shown above that they characteristically generate action potentials which follow a more or less regular rhythm. It is well
known, especially in the case of units acting as pacemakers for the
heart, that the autoactivity of pacemaker units is rhythmic (Hoffman and Cranefield, 1962). Finally, further evidence in favor of

this hypothesis is given by the character of appearance of the action potential in the neurons described. It has been shown for certain invertebrate neurons that during background activity an action potential arises as a result of slowly increasing depolarization (the so-called "prepotential") and is accompanied by hyperpolarization, after which the membrane potential again changes toward depolarization (Arvanitaki and Chalazonitis, 1955; Kostyuk, 1965, and others). The most essential feature of this process is evidently the regeneration of slow changes of potential, persisting even when action potentials are absent. At the height of the prepotential, one spike may be formed, in which case the impulses follow each other at a more or less regular rhythm; several spikes may, however, be formed in which case the activity has the appearance of grouped discharges in which the frequency of the impulses first rises and then falls. These are the characteristic features of generation of action potentials by pacemaker elements (Hoffman and Cranefield, 1962). Or, more precisely, they were initially characteristic of these elements, but they can also take place during the functioning of many other excitable structures, starting from muscle fibers and ending with neurons in higher formations of the vertebrate brain, under conditions leading to a shift of membrane potential toward depolarization. It can be postulated that at a level which is slightly more depolarized than the level of the resting potential, the membrane of many (although not of all!) excitable cells becomes unstable and tends to produce constant oscillations, reflected in particular by slow fluctuations in the membrane potential. Many ways are known in which a membrane can be made unstable, so that a state of persistent activity is created in it. In my experiments the sartorius muscle of the frog, vitally stained with photosensitizing dye (eosin) and then exposed to light (Burmistrov and Shuranova, 1965), served as this type of model of generation of "spontaneous" activity. Under these conditions the light evidently evoked persistent depolarization of the nerve endings, as judged by the considerable increase in frequency of the miniature potentials of the end-plate. As a result of this, rhythmic activity appeared in the nerve (and in the muscle). Under certain conditions rhythmic discharges of action potentials were generated in the muscle fibers themselves. Under these circumstances they appeared at the peak of the prepotential, which was absent in the case of action potentials evoked synaptically.

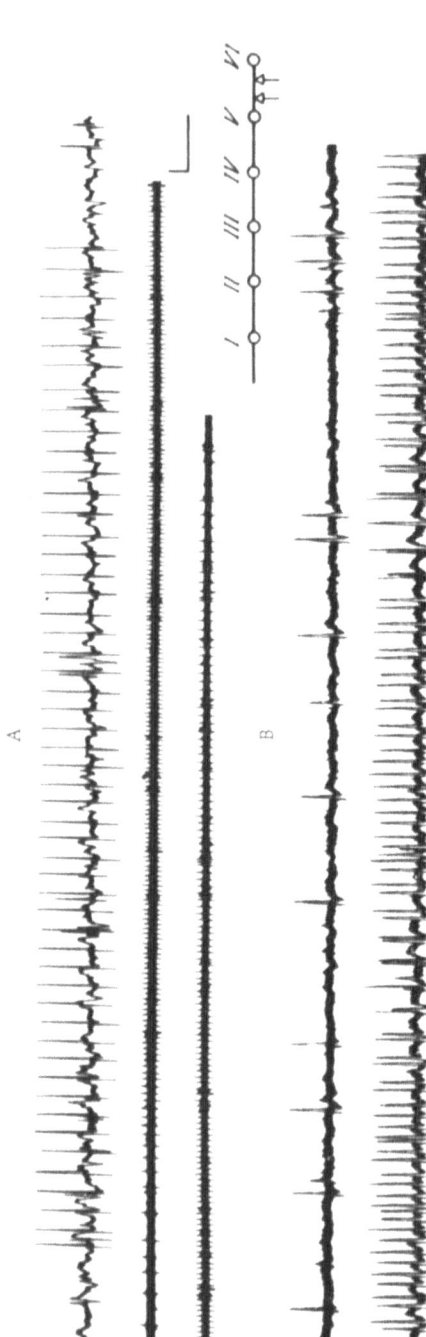

Fig. 2. Effect of certain external factors on background activity of neurons of isolated ventral chain of the crayfish. A) Grouped activity of neurons of isolated ventral chain vitally stained with methylene blue: a, b) different preparation. In b, bottom record is a direct continuation of top. Calibration: for a, 100 μV, 20 msec; for b, 150 μV, 100 msec; B) action potentials of photoreceptor neurons 3 hr after placing preparation in complete darkness (1) and during subsequent exposure to light (2). Calibration: 50 μV, 20 msec. Remainder of legend as in Fig. 1.

On the assumption that the background activity of many invertebrate neurons is connected with their ability to respond by a prolonged discharge of action potentials with no sign of accommodation to a change in membrane potential toward depolarization (compared with the level of the resting potential), the question of the sources capable of producing and maintaining this depolarized state naturally arises. Schematically, the wide variety of these sources may be reduced to two main factors: nonspecific and specific. By "nonspecific" I mean factors acting identically on all neurons, by "specific" those acting mainly or only on some neurons particularly sensitive to them.

It is known, for instance, that the activity of invertebrate neurons (as also of vertebrate neurons) is most strongly dependent on the ionic composition of the medium in which they are found (Prosser and Buehl, 1939), on temperature (Kerkut and Taylor, 1956), on the level of metabolism, and above all on the level of oxido-reductive processes (Prosser and Buehl, 1939; Pangelova and Lyudkovskaya, 1956), and so on. The same observations were made in our experiments: with elevation of the temperature (within certain limits) the number of active cells in the isolated ventral chain of the crayfish and also the frequency of their impulses increased. Using the same preparation, vital staining with methylene blue evidently influenced the level of neuronal metabolism since the background activity in the ventral chain was sharply modified. These changes were expressed not only by an increase in the total number of active neurons, but also by a radical change in their activity, in many neurons becoming grouped (Fig. 2A).

In addition, it is worth noting that many neurons of the central nervous system of invertebrates, including highly organized members such as arthropods and mollusks, possess selective sensitivity to various external environmental factors. Prosser (1934b), for instance, showed many years ago that nerve cells responding to light are to be found in abdominal ganglion VI of the crab. This fact was confirmed by later investigations, when it was found that there are only two such neurons (Kennedy, 1958, 1963). Action potentials of the photoreceptor neurons also were constantly recorded in our own experiments from the interganglionic connectives of the isolated ventral chain of the crayfish. The

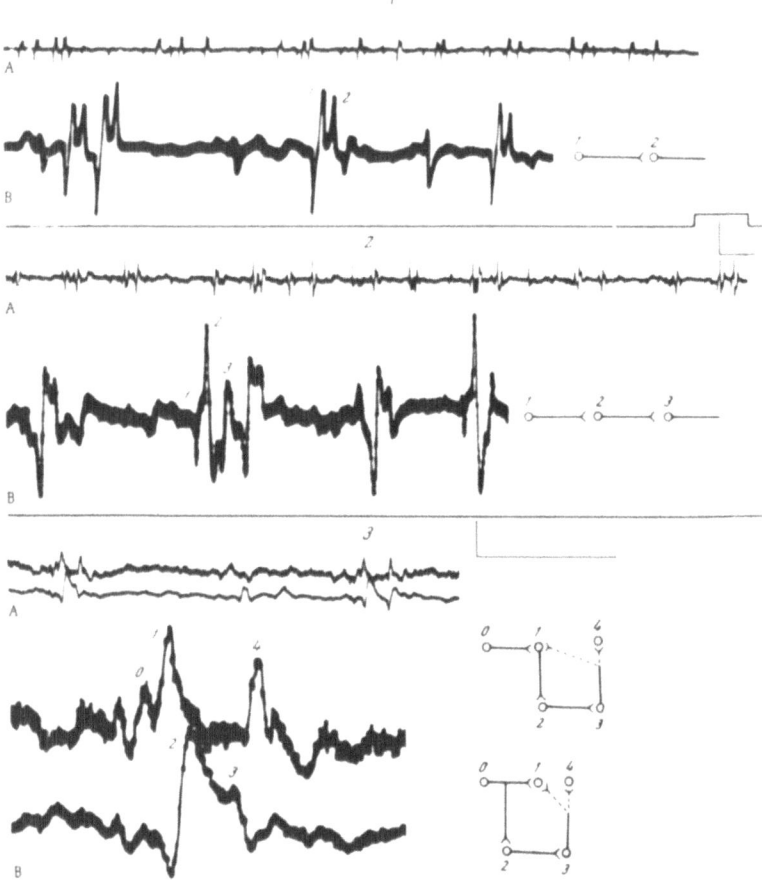

Fig. 3. Reflection of interaction between neurons in pattern of background electrical activity of neurons of isolated ventral cord. 1, 2) recording made from intact interganglionic connective, 3) from symmetrical halves of an interganglionic connective. A) Specimen of original trace; B) part of top record enlarged four times for printing. Numbers on records denote action potentials forming stable configurations. On the right, diagram to show connections between neurons whose activity is shown on records. Calibration: 100 μV, 20 msec.

activity of these neurons, and their ability to respond to light, persisted even after the ventral chain had been isolated for up to eight
days. Depending on several factors, especially the level of illumination, the frequency of the photoreceptor action potentials may
vary considerably: from one per second or less to several tens
per second. The example of these neurons shows clearly that external stimulation (in this case, light) is not the factor directly
"triggering" the pacemaker (the activity does not disappear if the
preparation is kept for a long time in complete darkness —
Fig. 2B), but at the same time, it is evidently this factor which
creates a state of "preparedness" in the neuron for the steady generation of action potentials and which determines their frequency.

Photosensitive neurons are also widespread among mollusks
(Sakharov, 1960; Arvanitaki and Chalazonitis, 1961, 1963, and
others). So-called "cold" and "heat" neurons have also been studied in *Aplysia*. For the former, the frequency of generation of
action potentials reaches its optimum at about 16°, and for the latter at about 31°. Changes in the frequency of action potentials are
directly connected with changes in the membrane potential, which
in the case of the "cold neurons" increases during heating and decreases during cooling (Chalazonitis, 1963). In the same mollusk,
chemoreceptor neurons have been found, the membrane potential
of which changes in one direction or the other with changes in the
partial pressure of oxygen, hydrogen, CO_2, and so on (Chalazonitis,
1963). These central neurons are evidently very little different in
their essentials from a peripheral receptor neuron such as the
crustacean stretch receptor.

3. Finally, the very important question must be examined
as to whether synaptic factors play any part in the organization of
the background activity of invertebrate neurons. This question
does not arise when analyzing the background activity in the central nervous system of vertebrates: the multiplicity of interconnections, the tremendous inflow of impulses from different sources,
the stochastic character of summation of single postsynaptic potentials, and so on — all these factors evidently give the background activity its characteristic features for vertebrate central
neurons. Conversely, as was shown above, nonsynaptic factors
come to the fore when analyzing the background activity of invertebrate neurons. Does this mean that the role of synaptic processes in this phenomenon is reduced to zero?

The first point to be made is that autorhythmic neurons can certainly be injured by physical factors acting synaptically (Perkel et al., 1964). The photoreceptor neurons of ganglion VI of the crab described above, for instance, respond not only to light, but also to tactile stimulation of the animal, the frequency of their discharges increasing with ipsilateral stimulation but decreasing with contralateral stimulation (Kennedy, 1963). In my experiments the following fact was observed. Although the photoreceptor neurons give the most precise and constant response to light, often in recordings made from an interganglionic connective, changes can be observed in the activity of other neurons also; possibly these changes are effected through the photoreceptor neurons, i.e., are evoked synaptically. The possibility of synaptic connections between neurons in the isolated ventral chain is perhaps also demonstrated by the sharp increase in background activity observed in our experiments when a 0.1% solution of nicotine, a substance with selective action on synaptic transmission, was applied to the surface of any ganglion.

Direct evidence is also available to show that interaction between neurons takes place in invertebrates. For instance, the system of connections within the cardiac ganglion of crustaceans have been well investigated (Bullock and Terzuolo, 1957; Hagiwara and Bullock, 1957; Watanabe and Bullock, 1960; Watanabe and Takeda, 1963, and others). It has been shown that of the nine neurons making up the ganglion, the four small neurons are pacemakers and trigger the five larger neurons; reciprocal effects from the large neurons on the small have not been demonstrated. If simultaneous recordings are made from two adjacent neurons of the visceral or parieto-genital ganglion of *Aplysia,* the period of excitation in one neuron may coincide with the period of inhibition of the other, and vice versa. Hence, in this case there are distinct reciprocal relationships (Arvanitaki and Chalazonitis, 1961). The possibility of interneuronal relationships of this type has also been demonstrated for neurons of the edible snail (Karpenko, 1965).

Other evidence of interneuronal relationships may perhaps be given by the arrhythmic activity of several invertebrate neurons described in the literature and observed in our own experiments. Several causes are perhaps responsible for this, including influences of a series of rhythmically discharging cells on a cell initially inactive or possessing its own spontaneous rhythm.

In addition, recording from interganglionic connectives, when simultaneous observations were possible on the activity of several nerve fibers, precise and constant temporal relationships were often observed between the action potentials of two different neurons. This evidently happens because one neuron triggers the other (Fig. 3, 1). As this figure shows, spike 2 follows immediately after spike 1, the interval between them being strictly constant. Similar relationships are found between spikes 1, 2, and 3 in Fig. 3, 2. Fig. 3, 3 shows a more complex case. Action potential 1, generated by a neuron in one half of the ventral chain, evidently triggers action potential 2 belonging to a neuron in the symmetrical half of the chain; after a very short delay, action potential 3 is generated in the same half of the chain, followed soon after by action potential 4 in the other half of the chain. However, another possibility is not ruled out, namely: $1 \rightarrow 2 \rightarrow 3 \rightarrow 1$. If it is remembered that yet another impulse (0) may be found before spike 1, the picture becomes still more complex; an attempt to interpret the generation of this sequence of action potentials is made on the diagrams on the right.

At least three conclusions can be deduced from the facts described above: 1) chains of interconnected neurons are present in the invertebrate nervous system; 2) these chains consist of a limited number of neurons; 3) they are open chains, i.e., they have no feedback. One reason for these special features may be that pacemaker neurons are found at the beginning of the chains, triggering (directly or indirectly through interneurons) effector neurons discharging at the periphery, and thus "tripping" the chain. A feedback mechanism is, of course, found in the operation of the pacemaker neuron — the level of its activity is determined by the degree of tonic excitatory influences exerted on it by various external factors. In the chain itself, on the other hand, no feedbacks have yet been determined experimentally under the conditions of background activity. Indirect evidence in support of this possibility is given by the circulation of excitation in the system of lateral giant axons of the ventral chain of the crayfish during application of a single electrical stimulus to one of the lateral axons, described in the literature (Burmistrov, 1965) and repeatedly observed in my experiments (Fig. 4). The lateral giant axons consist essentially of two chains formed by six serially connected neurons, each of which in turn forms a synaptic contact with the symmetrical element of the other chain. The essence of the phenomenon described

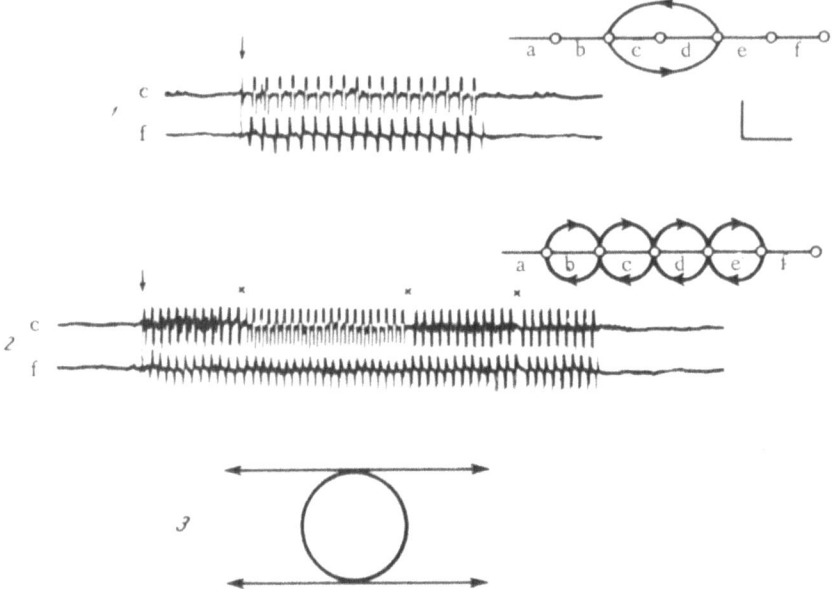

Fig. 4. Circulation of impulses in the system of lateral giant axons of the ventral chain. Above: diagram of abdominal division of ventral chain; letters denote interganglionic connectives; stimulating electrodes placed in b, recording electrodes in f and c. Areas of circulation of the nerve impulse are shown on the diagram. 1) Action potentials circulate around the same path; 2) change in path of circulation of impulse during response. Moments of change are marked by crosses. Calibration: 1 mV, 20 msec; 3) diagram showing a zone of circulation and action potentials leaving it to activate neurons located outside the area of circulation. Time of stimulation shown by artifact marked with an arrow.

is that a nerve impulse, travelling along a directly excited axon, activates the symmetrical lateral axon through one of the intra-ganglionic synapses, and this in turn excites the first axon synaptically (although through a different synapse). Hence, a closed chain of four or more neurons is formed in which excitation can circulate for a long time (for several hundreds of milliseconds; during this period the impulse completes several tens of cycles). The important feature is that (as is clear from the oscillogram and illustrated in the diagram) during each cycle four action potentials move out from the excited area, activating not only those segments of lateral axons outside the zone of circulation, but also other small and medium-sized neurons of the ventral chain. A similar process may perhaps take place in the background activity of the central neurons of invertebrates.

It thus seems that the background activity in the relatively simply organized neural structures of invertebrates (especially arthropods and mollusks) is a reflection: a) of endogenous activity of a series of pacemaker neurons (a fact now reliably established); and b) of neuronal interaction (a soundly based hypothesis, but one which requires experimental verification).

SUMMARY

Facts and considerations concerning the organization of background activity in the central nervous system of invertebrates (mainly in isolated neural structures) are given above. Since it is impossible, for reasons of space, to discuss them in any detail, only a few have been considered.

First, in the nervous system of the most simply organized animals, under relatively constant external conditions a stable background activity is present. An advantage of invertebrates as a test preparation is the possibility of studying this phenomenon in its "purest" form — under conditions of isolation of structures consisting of a limited number of neurons. Many neurons of invertebrates, including those of highly organized forms such as arthropods and mollusks, have been shown to possess true spontaneous activity. This property was evidently inherited by specialized neurons from their "ancestors" — unspecialized excitable elements performing receptor and contractile functions simultaneously (Passano, 1963). With increasing complexity of the nervous system this property has become firmly established, principally in afferent neurons, and more especially in neurons behaving essentially as central receptors.

The presence of spontaneous activity in these neurons is evidently closely connected with their performance of a function similar to that of peripheral receptors (Granit, 1957). In addition, autorhythmic neurons may have a tonic activating influence on spontaneously inactive neural elements of the same or higher levels of the nervous system.

The second category of neurons making their contribution to background activity of neural structures is evidently the association neurons. Compared with the vertebrate brain, the number of these neurons in the invertebrate central nervous system (and especially in those parts of it discussed above) is of course small; nevertheless, in animals such as crustaceans and, more especially insects they can be differentiated histologically at the level of the thoracic nerve-cord (Tsvileneva, 1964). It may be postulated that a certain proportion of the rhythmically discharging neurons are associative in function. These neurons possibly do not possess spontaneous activity in the true sense of the word, for the character of their background impulse activity is directly dependent on the synaptic inflow from other neurons. However, judging from the fact that the activity even of arrhythmically discharging neurons is not irregular, the number of interconnections in the invertebrate neural structures described above must be small, in consequence of which the task of experimental investigation of neuronal interaction is considerably simplified. This conclusion in no way extends to the background activity of the cranial ganglia of these animals, which so far have been very little studied by electrophysiologists.

Neuronal Mechanisms of Habituation[*]

E. N. Sokolov, G. G. Arakelov,
and L. B. Levinson[†]

Laboratory of Analyzers and Department of Cytology and Histology
Lomonosov Moscow State University
Moscow, USSR

Many investigators are at present engaged in the study of "habituation" (diminution and disappearance of the neuronal response as a result of repeated action of the stimulus). This phenomenon is highly selective, it is not merely the peripheral adaptation of receptors, and it can be regarded as the simplest form of "negative learning" (Jouvet, 1961; Thorpe, 1965).

Despite many investigations, the mechanisms of this phenomenon are still unexplained. It is known, however, that habituation is associated with the development of inhibition similar to the development of internal inhibition discovered by Pavlov (1947).

When discussing the neuronal mechanisms of habituation, the main types of inhibitory effects which could be responsible for habituation must first be examined. A detailed study has been made of "reciprocal inhibition," in which collaterals of the axon of the nerve cell excite interneurons forming synapses on the body of the nerve cell, inducing excitation. Interaction between the spinal cord motoneuron and Renshaw cells forming a loop of negative feedback is an example of this type of inhibition (Fig. 1A).

After investigating neurons of the cerebral hemispheres and cerebellum, Eccles (1965a) concluded that there is another type of postsynaptic inhibition, or "afferent collateral inhibition." In this case axons of afferent neurons, besides having a direct excitatory action on a higher neuron, activate interneurons through collaterals.

---

[*] Pages 262-272 in the Russian edition.

[†] Deceased.

These interneurons, ending on a neuron of higher order by inhibitory synapses, form a "direct inhibitory connection" (Fig. 1B).

Eccles (1966) also showed the existence of presynaptic inhibition. The first variant of this inhibition is the influence of interneurons on endings of axons (Fig. 1C). A second variant consists of axo-axonal inhibitory synapses blocking the conduction of excitation as it approaches an activating neuron (Fig. 1D).

The question arises, which of the types of inhibition discussed is responsible for "habituation."

The ganglionic neurons of mollusks provide a convenient preparation for the investigation of habituation. By making intracellular recordings from isolated ganglia of the snail *Helix pomatea,* for instance, Holmgren and Frenk (1961) demonstrated that the electrical responses of a neuron can be extinguished as a result of the repeated application of an orthodromic stimulus.

An important aspect of the study of "parallel inhibition" is its distinction from "reciprocal inhibition." Inhibition in a neuronal network with negative feedback cannot develop without preliminary activation of the neuron exciting the inhibitory interneuron. The threshold of excitation of such a neuron in a neuronal network cannot therefore exceed the threshold of excitation of the inhibitory phase.

In a neuronal network producing "parallel inhibition," on the other hand, the threshold of the inhibitory phase may be lower than that of the excitatory phase or, in other words, inhibition may arise in the neurons without trace of preliminary excitation. Hence, in the case of reciprocal inhibition, as the threshold is approached, omission of the phase of inhibition is to be expected. In the case of "parallel inhibition," on the other hand, the phase of excitation may be omitted if the inhibitory response is shortened to some degree. If it is assumed that posttetanic potentiation of the synapses of interneurons, associated with strengthening of the inhibitory phase, is the predominant feature of neuronal networks with "parallel inhibition," on repetition of the stimulus the excitation will be limited by inhibition, and the phase of inhibition will grow on account of its accumulation (Sokolov, 1965). The effect of "habituation" to a repeatedly applied stimulus is a characteristic feature distinguishing many neuronal networks. Nerve cells in the ganglia of *Limnaea*

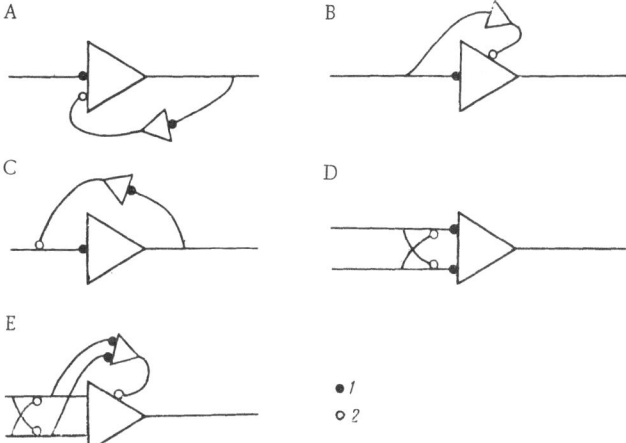

Fig. 1. Different types of inhibitory effects responsible for habituation of a neuron. A) Scheme of reciprocal inhibition; B) scheme of parallel inhibition; C) scheme of presynaptic inhibition through an interneuron; D) Scheme of presynaptic inhibition; E) combined scheme of presynaptic and parallel inhibition. 1) Excitatory synapses; 2) inhibitory synapse. These schemes are conventional, because no synaptic endings are found on bodies of the neurons in *Limnaea stagnalis* .

stagnalis also become "habituated" to repeated orthodromic stimuli in accordance with the rules governing parallel inhibition, similar to those of "habituation" of neurons in some parts of the mammalian brain (Sokolov et al., 1966a, 1966b).

The problem of the participation of presynaptic inhibition meanwhile has remained unsolved. The present paper describes a further stage in our investigation of "habituation" in neurons of *Limnaea stagnalis* to repeated application of a stimulus.

METHOD

The mollusks used in the experiments were collected in the fall and were kept in aquaria at room temperature. The action potentials (APs) of the neurons were recorded by glass microelectrodes with a resistance of 9-15 MΩ, filled with 2.5 M ferric chloride solution. The method of detecting and recording the APs was described earlier (Sokolov et al., 1966a). Neurons were stimulated

through a concentric electrode, about 1 mm in diameter, applied
to nerve fibers joining the ganglion to the body of the mollusk.
Electrical stimuli were applied from a stimulator generating
square pulses from 0.05 to 2.5 msec in duration and from 4 to 32 V
in amplitude.

To test the effect of the spectrum of light from a lamp on the
firing frequency of the neuron, filters SZS-15, cutting out the infrared part of the spectrum, and SZS-12, cutting out the visible
part of the spectrum, were used.

By filling the electrodes with 2.5 M ferric chloride solution,
the neuron could be "labeled" and the position of the tip of the microelectrode in the investigated cell determined. The method consists essentially of introducing Fe^{+++} ions into the investigated
neuron by iontophoresis through the glass microelectrode, and subsequent detection of this iron by histochemical reaction on sections.
To bring about iontophoresis of the iron, a positive voltage from a
dc source (current 5×10^{-8} A) was applied to the recording electrode. The nonpolarizing reference electrode was made negative.
The use of iontophoresis did not affect the physiological state of
the neuron. This method is described fully in an earlier paper.
(Sokolov et al., 1966c).

RESULTS

Functional State of the Neuron

When the membrane of a neuron of the parietal ganglion was
punctured by the electrode, APs with an amplitude of 50 mV and
frequency 3/sec were recorded. At the 35th second of recording
of the APs, a spontaneous cell response was observed, consisting
of a phase of excitation lasting 0.1 sec and containing 6 spikes of
different amplitudes, and a phase of inhibition lasting 0.6 sec. After this phase the background rhythm was restored and the neuron
continued to fire at its initial frequency.

At the 315th second, another excitatory-inhibitory response
appeared, but the phase of excitation contained 8 spikes and lasted
0.8 sec while the stage of inhibition lasted 2.8 sec.

Besides APs with an amplitude of 50 mV, APs of lower amplitude were observed throughout the experiment.

TABLE 1. Determination of Threshold of
Neuronal Response with Change in Amplitude
of Stimulus

Amplitude of stimulating pulse V	Total time of neuronal response, sec	Number of spikes in stage of excitation	Duration of phase of excitation, sec	Duration of phase of inhibition, sec
16	37	9	0.8	36.2
12.8	27	12	1.0	26.0
9.6	—	—	—	—
11.2	—	—	—	—
12.8	8	1	—	7.9
16.0	20	9	1.0	19.9

At the 835th second the firing rate was 2.5/sec. On the appearance of excitatory-inhibitory responses, the body of the mollusk contracted, the electrical response preceding contraction of the mollusk.

Despite repeated application of the filters, alternately cutting out the visible or infrared part of the spectrum, no change in the rhythm of the neuron or amplitude of the AP resulting from exposure to light was observed. Irrespective of the action of light the firing rate gradually diminished while the amplitude of the AP increased. The number of spikes with low amplitude increased at the same time.

Throughout the experiment from time to time spontaneous excitatory-inhibitory responses were recorded in which the duration of the inhibitory phase increased with a decrease in the firing rate, whereas the phase of excitation and number of APs in this phase showed little change.

Neuronal Response to Single

Orthodromic Stimulation

At the 39th minute of recording activity of the neuron, the neural commissures were stimulated electrically with single pulses (32 V, 0.5 msec). The neuron responded to the shock by an excitatory-inhibitory reaction with a total duration of 52 sec. As with the spontaneous excitatory-inhibitory responses, the duration

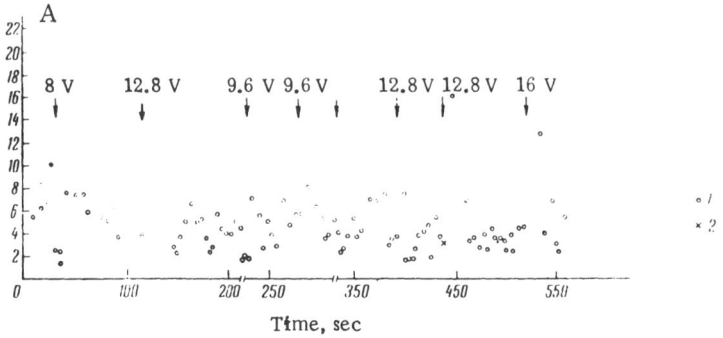

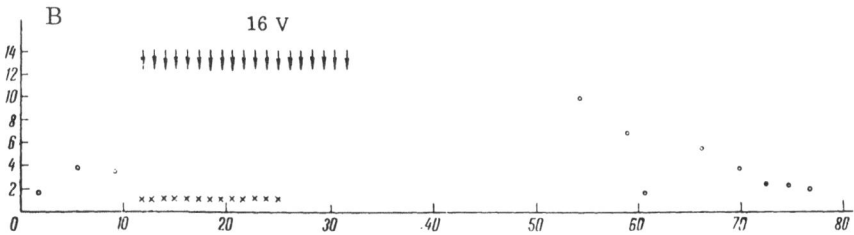

Fig. 2. Neuronal response to orthodromic electrical stimulation. A) Neuronal response to change in amplitude of stimulating pulse denoted by arrow; B) neuronal respose to application of stimulus with amplitude of 16 V and frequency 1/sec. Ordinate: duration of interval between spikes or groups of spikes in seconds. Abscissa: time in seconds. 1) Single pulse; 2) first pulse of volley. The appearance of a grouped discharge to the first stimulating pulse and disappearance of the response at the 14th application of the stimulus can be seen.

of the excitatory phase was 0.8 sec, the number of APs in the stage of excitation was eight, but the stage of inhibition was much longer, namely 51.2 sec. During the response the spikes of low amplitude disappeared. Application of a weaker stimulus (16 V, 0.5 msec) also caused a biphasic response in which the duration of the phase of excitation remained the same (0.8 sec) but the phase of inhibition came to resemble in duration the phase of inhibition in the spontaneous responses.

A further decrease in the amplitude of the stimulating pulse enabled the threshold of the neuronal response to be determined. A decrease in the amplitude of the pulse led to disappearance of the excitatory stage and lowered the activity of the inhibitory phase. This is clear from Table 1.

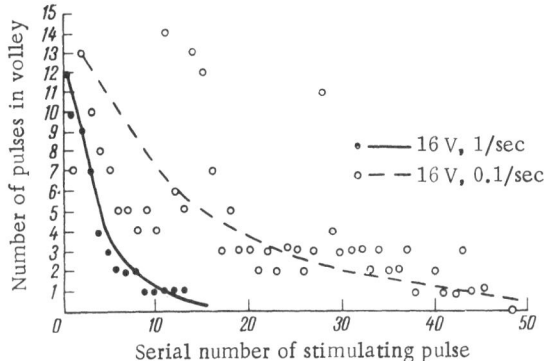

Fig. 3. Number of APs in grouped responses plotted
against serial number of stimulating pulse. Frequency
of stimulation 1 and 0.1/sec.

A pulse of amplitude 16 V evoked a response similar to the
spontaneous excitatory- inhibitory responses. No response was
seen to a pulse of 9.6 V. An increase in amplitude of the stimulating pulse to 11.2 V likewise gave no response. With an increase
in amplitude to the threshold level (12.8 V) the response took the
form of omission of the phase of excitation and shortening of the
phase of inhibition (Fig. 2A).

"Habituation" of Neuron to

Repetitive Orthodromic Stimulation

To elucidate the dynamics of "habituation," neurons were
stimulated orthodromically with pulses of current with a frequency
of 1/sec and an amplitude of 16 V.

The neuronal response to this stimulus is shown in Fig. 2B.
The stimulating pulses caused the appearance of grouped APs and
a subsequent decrease in the number of APs per volley (Fig. 3).
Whereas 10 APs were generated in response to the first pulse,
there were nine to the second and only 1 AP in response to the
tenth. No APs were generated in response to the 14th pulse
(Fig. 2b). Stimulation continued for 19 sec. Discontinuation of
stimulation was accompanied by a long phase of silence lasting
20.4 sec. Subsequently the APs reappeared, the firing rate gradually rising to the background level. In the first second of stimulation at 0.1/sec, groups of APs appeared, just as during stimulation

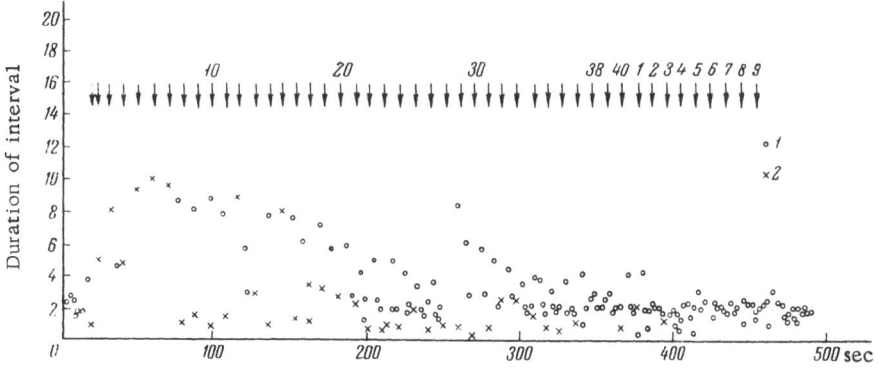

Fig. 4. Neuronal response to repetitive electrical stimulation (16 V, 0.1/sec). Legend as in Fig. 2. Numbers above arrows denote serial numbers of stimuli. Appearance of single AP between volleys and disappearance of grouped response at 44th pulse can be seen.

at 1/sec, followed by cessation of activity. In the interval between the 7th and 8th stimulating pulses, however, besides the grouped response directly connected with stimulation, single APs appeared (Fig. 4). The number of these APs between pulses increased in the course of stimulation. At the 170th second of action of the stimulus, for instance, there were 7 APs in a volley, and one between stimuli, while at the 390th second the corresponding figures were 2 and 5. At the 470th second the neuron did not respond to the stimulating pulse by a group of APs, but APs were generated between pulses. The dynamics of the neuronal response are shown in Fig. 3.

Determination of Threshold of Excitation of Neuron after Electrical Stimulation at 0.1/sec (16 V, 0.5 msec)

At the 10th second after discontinuing repetitive stimulation at a frequency of 0.1/sec, the threshold of excitation of the neuron was measured (Table 2).

Comparison of Tables 1 and 2 shows that the intensity of the neuronal response to a stimulating pulse of 16 V after repetitive stimulation was considerably reduced. Table 1 shows that the neu-

TABLE 2. Changes in Threshold of Excitability
of Neuron after Stimulation at 0.1/sec
(16 V, 0.5 msec)

Amplitude of stimulating pulse, V	Total time of neuronal response, sec	Number of spikes in stage of excitation	Duration of phase of excitation, sec	Duration of phase of inhibition, sec
8	—	—	—	—
16	3.0	—	—	3
12.8	13.0	13	1.0	12.0
16	3.6	2	0.4	3.2
12.8	3.0	1	0.2	2.8

ron responded to the first application of a pulse of 12.8 V with a
distinct excitatory-inhibitory reaction, but at the second application of this same pulse (after an interval of 20 sec) no definite response was observed. This phenomenon was also seen in the neuronal response to a single pulse after exposure to stimulation at
0.1/sec. The duration of all phases of the neuronal response to the
first stimulus of 12.8 V was·13 sec, but to the second it was only
3 sec.

Neuronal Response to Change

in Duration of Stimulating Pulse

The results of measurement of the neuronal response during
a change in the duration of the stimulating pulse from 0.1 to 2.5
msec, its amplitude remaining at 16 V, are shown in Table 3.

It should be noted that the neuronal response to a pulse of
16 V (0.5 msec) was similar in duration to the responses obtained
to an identical stimulus before stimulation at a frequency of 0.1/sec.
However, this recovery took place 74 sec after discontinuing repetitive stimulation.

A decrease in the duration of the pulse to 0.1 msec led to
disappearance of the neuronal responses to the stimulus. A return to the original duration (0.5 msec) once more produced an
excitatory-inhibitory response, but it was shorter than at the first
application.

TABLE 3. Changes in Duration of Neuronal
Response as a Function of Changes in Duration
of Action of Stimulating Pulse with
Amplitude 16 V

Amplitude of stimulating pulse, V	Total time of neuronal response, sec	Number of spikes in stage of excitation	Duration of phase of excitation, sec	Duration of phase of inhibition, sec
0.5	19	11	0.8	18.2
0.1	—	—	—	—
0.5	16	11	0.8	15.2
1.0	27.4	13	1.0	26.4
2.5	16	15	0.8	15.2

A 1-msec pulse lengthened the duration of the response, but this was due to lengthening of the phase of inhibition. When a 2.5-msec pulse was used, the response was shorter than that to a pulse lasting 1 msec.

Neuronal Response to Change

in Frequency of Stimulation

Stimulation at 5/sec (16 V, 0.5 msec) applied for 3 sec caused generation of APs with a high frequency in response to the first pulses and with a sharp decrease in frequency of APs at the end of stimulation. Discontinuation of stimulation was followed by a phase of inhibition lasting 9.9 sec, after which the firing rate was restored.

Stimulation with pulses at 20/sec for 2.2 sec led to diminution and subsequent recovery of the amplitude of the AP. After discontinuing stimulation the neuron generated three spikes of low amplitude, followed by a phase of inhibition. After 20 sec APs of near the initial amplitude were restored.

The response of the neuron to stimulation of 100/sec for 2 sec consisted of the appearance of APs immediately after stimulation began. Later APs were absent, reappearing after stimulation ended, with a silent period lasting 480 sec. APs appearing after the silence were generated with a high firing rate but their amplitude was lower than that of APs generated before stimulation at 100/sec.

Determination of Position

of Microelectrode

After the end of the program of electrical stimulation, Fe^{+++} ions were introduced into the tested neuron through the microelectrode by means of a dc source. The "label" appeared in histological preparations (serial sections) as a blue dot 5 μ in diameter at the border between the nucleus and cytoplasm (Fig. 5). For contrast the section was counterstained with methyl green–pyronine by Unna's method.

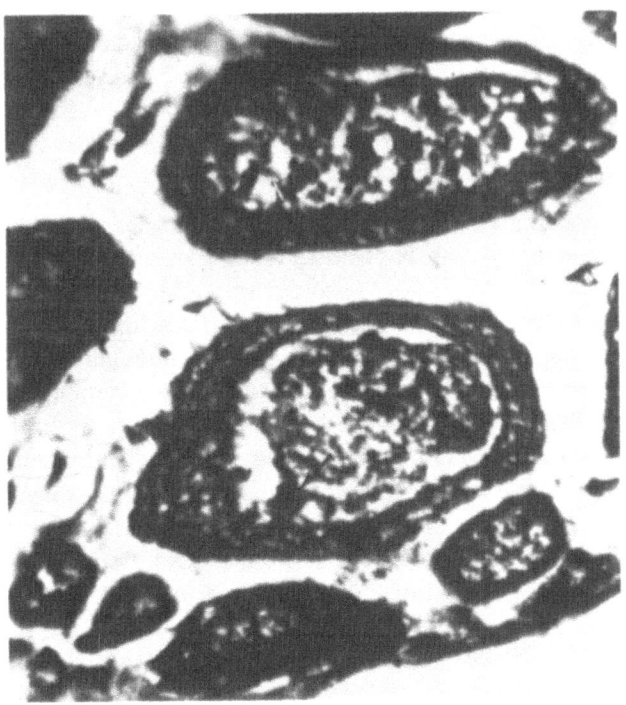

Fig. 5. Mark of tip of microelectrode in neuron of parietal ganglion. Mark denoted by arrow. Objective 90x, occular 7x.

DISCUSSION

Previous investigations (Sokolov et al., 1966a, 1966b) showed that neurons are present in the ganglia of the mollusk, *Limnaea stagnalis*, which possess spontaneous activity and respond to a single orthodromic electrical stimulus by an excitatory-inhibitory reaction. If this stimulus is applied repeatedly, the response gradually becomes extinguished, and the spontaneous activity is restored against the background of stimulation. Another group of neurons is characterized by absence of spontaneous activity. During repeated orthodromic stimulation, in the first stage activation of these neurons is observed, and this is followed by their "habituation" to the stimulus.

Comparison of the possible mechanisms of these excitatory-inhibitory responses favors the scheme with a parallel outlet of inhibition and excitation to the cell (Sokolov et al., 1966b). Proof that parallel inhibition plays a part is given by the fact that the threshold of inhibition of the neuron is lower than its threshold of excitation, and during repeated application of the stimulus inhibition is accumulated and summated, and this is particularly marked in the case of action of weak stimuli.

Independence of the excitatory and inhibitory inputs was also found in the case of this particular cell, confirming participation of parallel inhibition in the "habituation" effect. Meanwhile, the experiment in which the frequency of stimulation was 0.1/sec shows that summation of "parallel inhibition" can account for only the first stage of change in neuronal responses to a repetitive stimulus, when a phase of inhibition of APs is found immediately after the phase of excitation (Fig. 6). If repetitive stimulation is continued, the phase of excitation grows weaker, while the inhibitory phase of the response disappears. The process ends by recovery of the background firing rate, when the neuron does not respond to application of the stimulus. This effect is difficult to explain by the "parallel inhibition" scheme. It may be postulated that at the second stage we have described, presynaptic inhibition comes into play and is strengthened, as a result of which both the response of excitation of the neuron to the arriving pulse and the inhibitory phase of the response disappear, allowing restoration of the spontaneous rhythm. The scheme of the combined action of presynap-

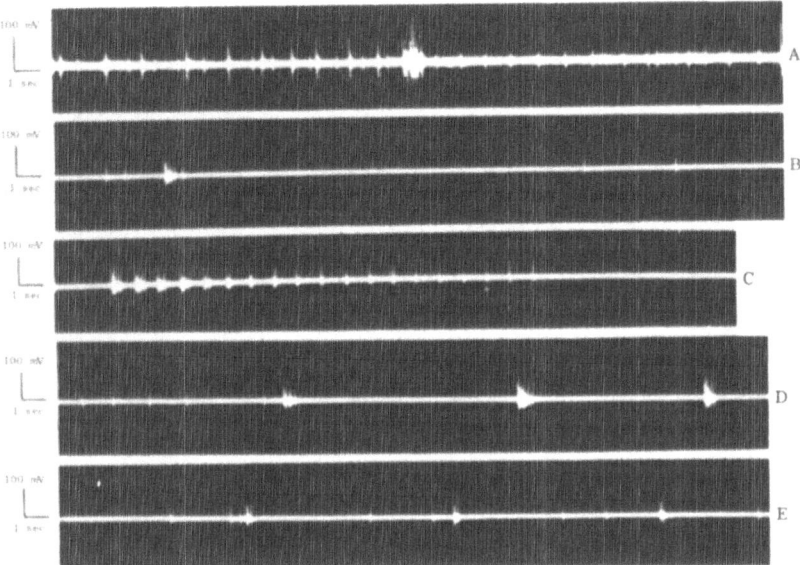

Fig. 6. Types of neuronal electrical activity. A) Background rhythm and spontaneous excitatory-inhibitory response; B) excitatory-inhibitory response of neuron evoked by electrical pulse 16 V in amplitude and 0.5 msec in duration; C) disappearance of response to rhythmic stimulation at 1/sec (16 V, 0.5 msec); D) neuronal response to first stimulating pulses at interval of 10 sec (16 V, 0.5 msec); E) stage of partial extinction of neuronal response to stimulating pulse and recovery of spontaneous activity against a background of stimulation.

tic and "parallel" inhibition just considered is illustrated in Fig. 1E.

At the first stage of "habituation" parallel inhibition comes into action, and later it is joined and strengthened by presynaptic inhibition. The strengthening of inhibition during "habituation" can be explained by the postactivation potentiation of synapses (Eccles, 1966).

In the scheme given above (Fig. 1E), the development of inhibition is delayed relative to the development of excitation in the circuit with both parallel and presynaptic inhibition. This conclusion is confirmed by the response of the neuron to high-frequency stimuli. Since the mechanisms of inhibition cannot be brought into action, the high-frequency volley of stimulating pulses depolarizes the cell membrane, leading to depression of AP generation which is particularly prolonged during the action of stimuli with a frequency of 100/sec.

CONCLUSIONS

1. "Habituation" to orthodromic electrical stimulation takes place in two stages: parallel inhibition and presynaptic inhibition.

2. "Habituation" of a neuron is characterized by selectivity relative to amplitude and duration of the stimulating pulse.

3. Retardation of inhibition relative to excitation has the result that high-frequency orthodromic stimulation depolarizes the cell membrane.

SUMMARY

"Habituation" of neurons of the mollusk, *Limnaea stagnalis*, to orthdromic electrical stimulation was investigated. Using a concentric stimulating electrode applied to the nerve fibers, the neurons were stimulated by pulses between 9.6 and 16 V, 0.5 and 2.5 msec, and with frequencies of 0.1, 1, 5, 20, and 100 cps, and also by single pulses. With a decrease in amplitude of the single pulse, the phase of excitation disappeared while the phase of "inhibition" remained. A similar phenomenon was observed if the duration of the stimulating pulse was shortened. Application of stimuli at the rate of 1/sec led to "habituation" of the neuron to that particular stimulus. A frequency of 0.1/sec also led to "habituation" of the neuron, characterized by restoration of the background rhythm in the second stage of action of the stimulus.

The possible mechanisms of "habituation" of a neuron are examined. Comparison of the results obtained with modern views of mechanisms of inhibition led to the suggestions that "habituation" in certain neurons is due to the combined action of "parallel" and presynaptic inhibition. "Habituation" is selective relative to amplitude of the stimulating pulse.

During the action of high-frequency stimulation depolarization of the neuronal membrane takes place, due to retardation of inhibition relative to excitation of the nerve cell.

The position of the microelectrode in the investigated cell was verified by histochemical detection of Fe^{+++} ions introduced through the microelectrode by means of electrophoresis.

Analysis of Spatial Synchronization of Cortical Potentials in the Rabbit[*]

Z. A. Yanson, L. V. Tishaninova,
A. N. Balashova, and Z. M. Gvozdikova

Institute of Higher Nervous Activity and Neurophysiology
Academy of Sciences of the USSR
Moscow, USSR

The investigation of simultaneous electrical processes over wide areas of the cortex has become possible only with the use of methods of multiple recording of potentials and, in particular, after development of the method of toposcopy (Lilly, 1949; Walter and Shipton, 1951; Livanov and Anan'ev, 1955).

In recent years Livanov and his collaborators have investigated temporal correlations between potentials at separate points of the cortex during different types of activity in order to study the functional significance of spatial synchronization. They have described changes in synchronization of potentials during conditioning. The dynamics of its waxing and waning at different stages of establishment of the temporary connection has been described, and changes in synchronization have been demonstrated in man during mental work, in certain mental diseases, and during the action of pharmacological agents (Livanov, 1962b, 1962c, 1965a; Glivenko et al., 1962b; Livanov et al., 1964; Gavrilova et al., 1963, 1964, 1965; Luchkova, 1965; Korol'kova et al., 1966).

The object of the present investigation was to determine the role of intracortical relationships and influences from subcortical structures on the formation of spatial synchronization.

---

[*] Pages 341-353 in the Russian edition.

METHOD

Experiments were performed on 18 rabbits.

Cortical electrical activity was recorded with removable needle electrodes simultaneously from 46 or 25 points on the surface of the rabbit's left cerebral hemisphere and from the subcortical structures by means of implanted (25 or 28) electrodes made of nichrome wire 70 μ in diameter (Knipst, 1961). The electrodes were inserted in accordance with calculations based on the maps of Sawyer, Everett, and Green. Accuracy of insertion was verified morphologically. The cortical and subcortical potentials were recorded on a type 50 electroencephaloscope.

To assess the degree of correlation between potentials all the leads were compared in pairs (Glivenko et al., 1962a) and this was done separately for the cortex and for all the investigated subcortical structures. The information obtained was analyzed on M-2 and Dnepr electronic computers. For analysis of the experimental data, areas of the cortex with a high degree of correlation only were chosen, i.e., pairs of points whose potentials showed corresponding changes over a period of 1.5 sec for between 75 and 100% of the time. The degree of synchronization of the cortical potentials was expressed in %. The ratio was calculated between the number of pairs of points working with a high degree of correlation at a given moment and all the possible combinations of the investigated pairs of points, taken as 100%. Significance of the changes in synchronization of the cortical potentials was calculated by means of the χ^2 criterion with 5% probability.

In accordance with the aim of the investigation, three series of experiments were carried out. In the first series (acute experiments on 8 rabbits) the effects of repetitive electrical stimulation of the cortex on changes in spatial synchronization of potentials were studied. The motor cortex was stimulated by electrical pulses through the skull (frequency 5 and 100/sec, duration 3 msec, voltage 40-120 V, duration of stimulation 10 sec). Potentials were recorded before stimulation, during the first 5 min after stimulation, and thereafter every 10 min for 1 hr.

In the second series (chronic experiments on three rabbits) electrical stimulation of the hypothalamic region was carried out (100/sec, 1-3 V, 10 sec). Activity was recorded before stimulation

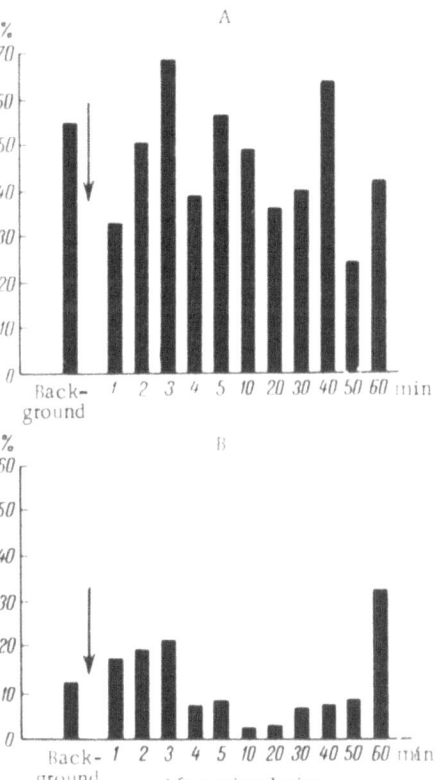

Fig. 1. Changes in synchronization of cortical
potentials in rabbits No. 3 (A) and 4 (B) at
various times after electrical stimulation of
the motor cortex. Abscissa: time before and
after stimulation (in minutes); ordinate: degree of synchronization of potentials (in percent). Arrow denotes electrical stimulation.

and 30 sec and 1, 3, 5, 7, 10, 12, and 15 min after electrical stimulation of the hypothalamus.

In the third series, to determine the pathways of spread of
synchronization (chronic experiments of five rabbits) the cortex
was divided vertically and undercut horizontally. The motor cortex was isolated by an incision about 2 mm in depth, i.e., the
thickness of the gray matter. The occipital cortex was undercut
to separate it horizontally from the subcortical structures by

TABLE 1. Changes in Spatial Synchronization of Cortical Potentials in Rabbits at Different Times after Electrical Stimulation

Rabbit No.	Frequency of stimulation, Hz	Stimulus voltage	before stimulation	Spatial synchronization of cortical potentials, %										
				minutes after stimulation										
				1	2	3	4	5	10	20	30	40	50	60
1	5	100	23.7	39.6	25.9	30.9	70.9	46.5	35.0	—	—	38.2	59.4	37.6
2	100	120	19.3	22.3	18.6	23.3	16.0	20.0	16.5	15.7	27.0	25.4	50.0	46.0
3	100	100	55.0	33.3	49.4	69.4	39.4	56.3	48.8	36.0	39.8	64.5	23.2	47.1
4	100	40	13.0	18.0	19.8	22.2	7.2	7.7	1.8	2.9	6.4	7.5	9.0	32.8
5	100	40	19.0	60.9	14.8	28.9	27.4	39.7	12.5	9.3	6.3	6.5	7.6	10.6
6	5	40	8.0	7.2	8.7	10.8	10.5	10.2	8.7	10.1	8.1	36.9	26.5	27.2
7	5	40	10.0	5.4	9.3	8.3	11.9	—	—	10.5	8.9	6.4	5.6	9.6
8	5	100	14.9	11.6	4.6	7.8	10.0	8.9	9.3	12.4	10.0	9.0	10.0	22.0

means of a scalpel through a burr–hole in the lateral side of the skull. In every case the accuracy of the incisions and undercutting of the cortex were verified morphologically. In these experiments the cortical electrical activity was recorded before the operation, immediately after it, and subsequently for six weeks.

RESULTS

The first series of experiments showed that in 4 of the 8 rabbits there was a decrease (in some cases up to 32%) in spatial synchronization of the potentials in the first minute after cortical electrical stimulation for 10 sec by pulses of different frequencies and strengths (5 and 100/sec, 40–120 V; Table 1). In the other four rabbits synchronization during this period was above its initial level (by 3–42%). As this table shows, no appreciable difference was observed in the changes in synchronized activity when electrical stimuli of different parameters were applied for 10 sec. In most experiments the initial response of the cortical potentials was dependent on their original level. When the background level of synchronization was high, it was lowered; if background synchronization was low to begin with, it rose. Later, starting with the second minute after stimulation, the number of pairs of points working synchronously began to increase in 7 of the 8 rabbits, and after 3–5 min it exceeded the initial level by 2–47%. In rabbit

No. 8 the degree of synchronization also showed some increase at this time, while remaining below the initial level. Later, mainly after the 10th-20th minute, a phase of decrease of spatial synchronization followed: in some experiments down to the initial level, but in others much below it, and this situation continued until 30-40 min after stimulation. A second increase in synchronization of potentials was observed in all cases toward the end of the experiment, i.e., 40-60 min after stimulation. In six of eight rabbits the number of pairs of points working synchronously was 8-30% higher than the initial level at this time. In the other two rabbits the degree of synchronization of the cortical potentials also began to rise, and by the end of the experiment it was close to its initial level.

To show the fluctuations in spatial synchronization after cortical stimulation more clearly, the data for two rabbits are shown graphically in Fig. 1.

These investigations thus showed that repetitive electrical cortical stimulation for 10 sec causes fluctuating changes in the level of synchronization of its potentials. The first phase of increase is observed in the first 2-4 min and the second 40-60 min after stimulation. These changes spread diffusely throughout the cortex.

The question arose: Can the appearance of this diffuse response in the cortex be attributed to influences ascending from the subcortex?

To examine this question experiments were performed to study the effects of electrical stimulation of the hypothalamus on spatial synchronization of the cortical potentials. The second series of experiments, performed on three rabbits, showed that repetitive (100/sec, 1-3 V) stimulation of the hypothalamus for 10 sec in fact causes changes in synchronization of the cortical potentials. The direction of these changes depended on the background existing before stimulation. The dynamics of spatial synchronization in all the experimental animals before and after electrical stimulation is illustrated in Fig. 2. If the initial level of synchronization of potentials was low or average (up to 50%), after stimulation ended the degree of synchronization of the cortical potentials was increased. However, this increase followed a different course in different animals. In rabbit No. 9, immediately after the end of

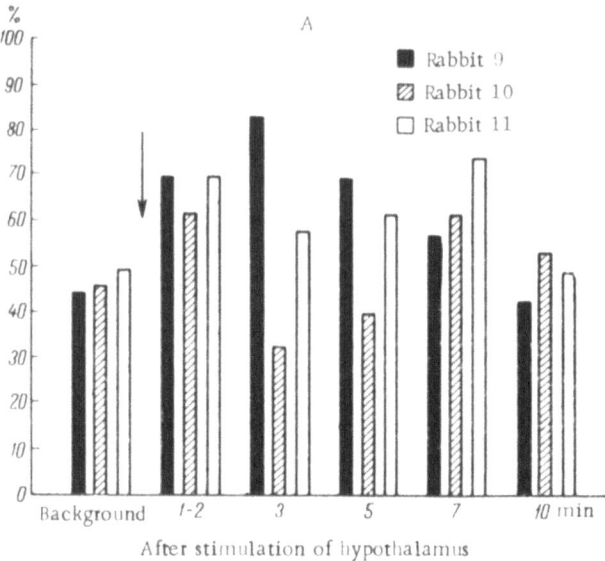

After stimulation of hypothalamus

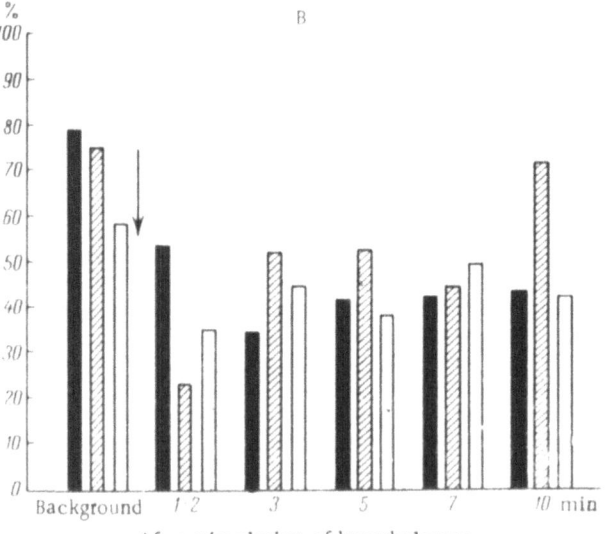

After stimulation of hypothalamus

Fig. 2. Changes in degree of synchronization of cortical potentials
caused by stimulation of the hypothalamus in relation to initial
background. A) Average level of synchronization in background;
B) high level of synchronization in background. Legend as in Fig. 1.

stimulation an increase in the degree of synchronization of cortical electrical activity was observed, reaching its maximum at the third minute. Later, toward the 7th–10th minute, it began to fall gradually to its initial level, which is reached before the end of the experiment. In rabbits Nos. 10 and 11, soon after the initial increase (the first increase) in the number of synchronously working cortical points during the first 2 min after the end of stimulation, a short decrease was observed, followed by a second increase in synchronization at the 7th minute after stimulation (the second increase). Later, toward the 10th–12th minute, the initial background level was regained.

Hence, the results described above show that in some animals an increase in spatial synchronization of the cortical potentials took place immediately after stimulation and lasted for a short time (up to 5 min). In other animals this increase took place in phases; two maxima of synchronization were observed — at the 1st–2nd and 7th minutes after stimulation. These changes in synchronization were obtained from analysis of its level throughout the cortex. Later, a detailed analysis was made of the synchronization of potentials for the visual and parietal areas of the cortex separately and between these two areas. Special calculations of the number of pairs of points working synchronously in these areas showed that their number in the visual cortex increased more than in the parietal (Table 2). In two rabbits the number of "connections" between the two investigated areas showed a particularly great increase. For example, in rabbit No. 11 (experiment 6), their background number was 64 but after stimulation their number reached maxima of 114 at the third minute and 115 at the seventh minute after stimulation of the hypothalamus.

TABLE 2. Number of "Connections" in Parietal and Visual Cortex Individually and between Them (Rabbit No. 11)

Area of cortex	Before stimu-lation	Minutes after stimulation of hypothalamus						
		1	2	3	5	7	10	12
parietal	37	44	33	41	35	44	35	36
visual	45	66	65	63	63	65	52	61
between them	64	99	93	114	70	115	42	45

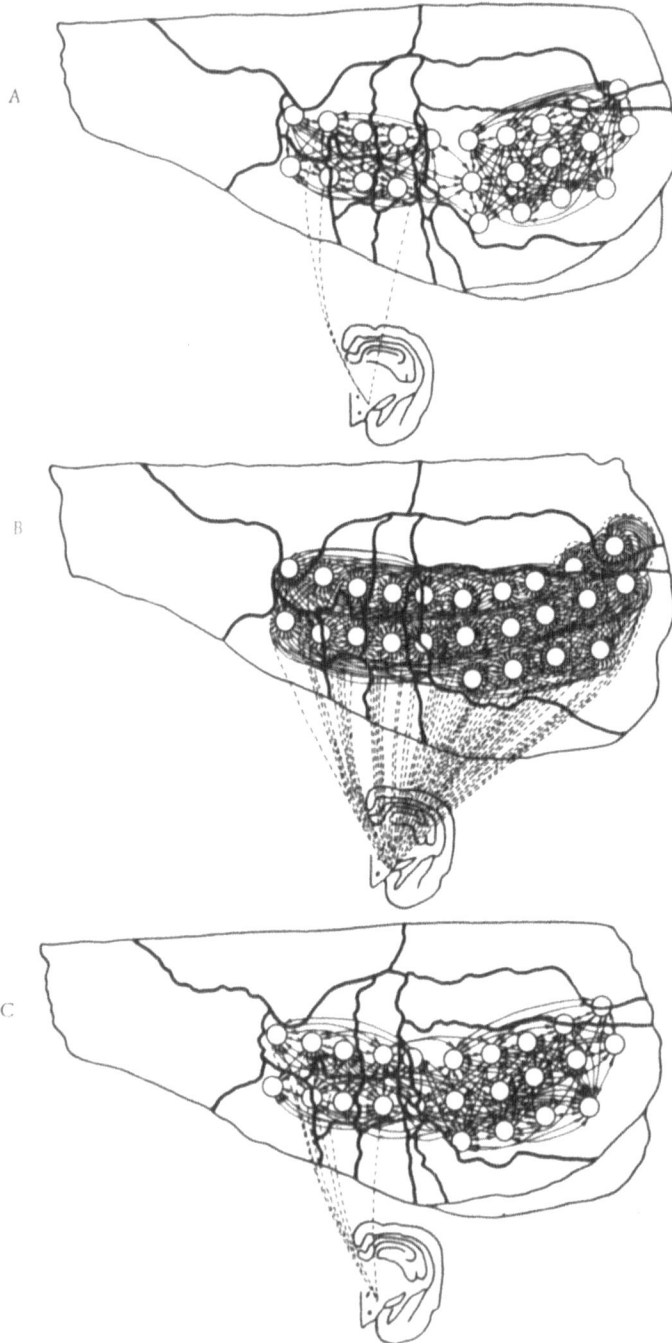

Fig. 3.

The analysis thus showed that spatial synchronization of the investigated areas of the cortex is extensive and spreads diffusely over the cortex.

A correlation likewise was found between the potentials of the stimulated areas of the hypothalamus and of various parts of the cortex. The brain contours drawn in Fig. 3 represent the topographic distribution of points giving a high degree of correlation with each other, and correlations between the stimulated region of the subcortex and separate points of the cortex are also shown. This figure shows that the number of correlations during background activity is small, but under the influence of stimulation their distribution follows a definite pattern. During the first 3 min after stimulation the degree of correlation of potentials between hypothalamus and cortex rose sharply, this being most clearly seen in the visual cortex. By the 5th–7th minute the number of correlations had diminished.

Comparison of the correlation of potentials between hypothalamus and cortex with the change in spatial synchronization of the cortical potentials gave the following result. At the time when cross–correlation of activity between the stimulated part of the hypothalamus and the cortex was greatest, the degree of synchronization had increased and spread diffusely throughout the cortex. The two maxima of the increase in spatial synchronization observed in several animals (Fig. 2) coincided with times of increased correlation of potentials between the cortex and stimulated parts of the hypothalamus.

To obtain a more complete picture of cortico–subcortical relationships, the degree of synchronization was calculated for all investigated subcortical structures (thalamus, hypothalamus, mesencephalic reticular formation) in all the rabbits and the results were compared with the degree of correlations of the cortical potentials. These changes in rabbit No. 9 are shown in Fig. 4 by way of example. In the background, before stimulation, the level

Fig. 3. Changes in degree of synchronization of cortical potentials and number of correlations between hypothalamic region and individual cortical points under the influence of hypothalamic stimulation. A) Before stimulation; B) 3 min after stimulation of hypothalamic region; C) the same, 7 min after stimulation. Arrows denote correlations in cortex. Broken lines represent correlations between hypothalamus and cortex.

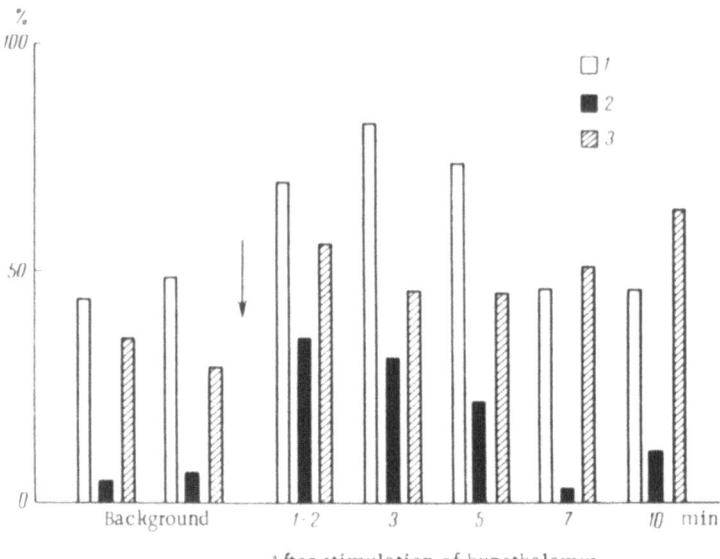

Fig. 4. Changes in degree of synchronization of potentials in the cortex
and subcortex and between them under the influence of hypothalamic stimulation. 1) Synchronization in cortex; 2) synchronization between cortex
and subcortex; 3) synchronization in subcortex. Remainder of legend as in
Fig. 1.

of synchronization of the cortical potentials was slightly higher
(45-50%) than in the subcortex (30-35%). The number of points in
the cortex acting synchronously with parts of the diencephalon and
mesencephalon was only 5-10%. Immediately after electrical stimulation of the posterior hypothalamus ended, the level of synchronization rose considerably both in the cortex and in the subcortex,
and more especially between them — to 40%. Later a gradual decrease in sychronization was observed. All these results demonstrate the considerable influence of the subcortex on the formation
of spatial synchronization in the cortex.

The next subject to be studied was the pathways followed by
influences determining the formation of spatial synchronization in
the rabbit's cortex. For this purpose the third series of experiments was carried out. Experiments were first performed to determine the role of intracortical pathways running within the gray
matter. A vertical incision was made in the region of the motor
cortex of five rabbits. The cortical incision extended from 1 mm

from the midline along the coronal suture to a length 8 mm, depth about 2 mm. Changes in the spatial synchronization of cortical potentials appeared immediately after the operation, and the character of these changes (as in the preceding series) was dependent on the initial background and was generalized. Topographically, the spatial synchronization of background activity in these rabbits consisted of definite "constellations." In some animals there was one large constellation, when nearly all parts of the cortex gave a correlation of potentials of 70% or more. In other animals the zones of synchronously acting points were spatially separated into small constellations. Depending on the initial response to trauma, the animals could be divided into two groups. In those cases (two rabbits) when spatial synchronization in the original state took the form of one large constellation, vertical incision of the cortex caused a decrease in the number of correlations, changes in synchronous activity being observed throughout the cortex (Fig. 5A and B).

Hence, vertical division of the cortex in these rabbits caused weakening of synchronous activity.

If, on the other hand, as was observed in the other three rabbits, spatial synchronization in the initial background was weak (small structures consisting of 3-5 points were predominant), isolation of the cortex led to a marked increase in synchronization. A larger number of cortical points working with a high degree of correlation with each other could be detected (Fig. 5B).

The initial changes in spatial synchronization of potentials were sometimes of short duration and disappeared by the end of the first day of the experiment (one rabbit), but more often they were prolonged and continued for 1-3 days (four rabbits). On subsequent days the changes in synchronous activity of the cortex were fluctuating in character. After one week a marked increase in synchronization was observed, followed by a decrease on the 14th-21st day, and a further increase by the 28th day. These fluctuations of synchronization were observed over the whole hemisphere.

It thus follows from these results that changes in spatial synchronization of cortical potentials take place immediately after the surgical operation and are dependent on the original state of synchronization. It may be postulated that these changes are the

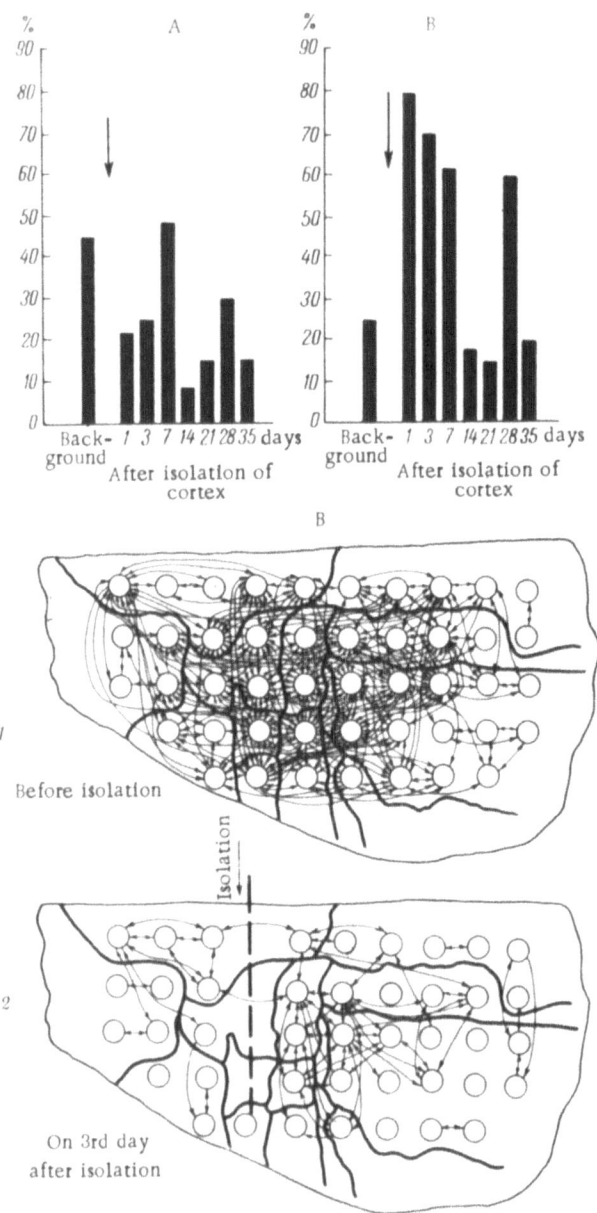

Fig. 5.

result of the response of the cortex as a whole to the operation (probably the combined effects of isolation of the cortex and trauma resulting from the operation).

To assess the effects of surgical division on synchronization of cortical activity, the level of synchronization between separated areas of the cortex was specially investigated. It was found that although synchronization between the separated areas was weakened, it nevertheless continued. To rule out the possibility of ephaptic spread of influences across the incision, special experiments were carried out in which an insulating layer was introduced into the cortical incision. In this case also, as analysis of the experimental results showed, the change in synchronization of potentials was considerably weakened, but did not disappear completely. This may be attributed to incomplete separation of the cortical areas in rabbits: only the dorsal surface of the hemisphere was incised, the medial and lateral surfaces remaining intact. Excitation could evidently enter the isolated areas of the cortex via intact brain tissue, thus by-passing the incision. Persistence of synchronization of potentials after these incisions in the cortex could also be due to influences from subcortical areas.

To determine the role of subcortical structures in the formation of spatial synchronization special experiments were performed. In two rabbits the cortex in the visual area was separated from the subcortex by "horizontal" undercutting incisions.

The experiment showed that, despite differences in the original background of spatial synchronization, separation of the cortex from the subcortex led in both cases to weakening of synchronization of the cortical potentials. In rabbit No. 17, for instance, initially the largest number of points with a high level of correlation between their activity was found in the occipital region, where a definite constellation was formed. Isolation of the cortex from the subcortex led to appreciable changes in synchronization of the

Fig. 5. Changes in spatial synchronization of cortical potentials after isolation by a vertical incision. A) Rabbit No. 12 with high background level of synchronization; B) rabbit No. 14 with low background level of synchronization; C) rabbit No. 12: 1) before isolation, 2) on 3rd day after isolation. Abscissa: time after isolation (in days). Arrow denotes moment of isolation operation. Remainder of legend as in Figs. 1 and 3.

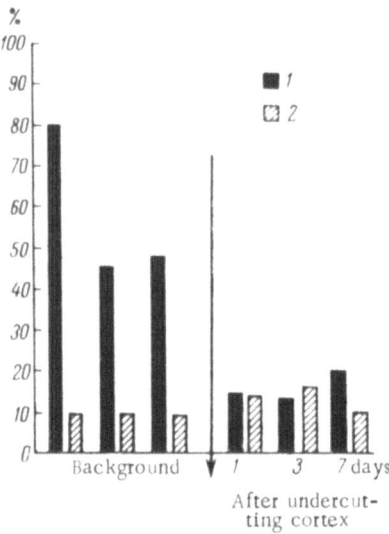

Fig. 6. Changes in spatial synchronization
of cortical potentials after "undercutting"
the visual cortex. 1) Visual cortex; 2) parietal and sensorimotor areas of cortex.
Legend as in Fig. 5.

potentials: a marked decrease in the number of synchronously active points was observed, leading to disturbance of the constellation existing in the visual cortex before undercutting. The study of
synchronization of cortical potentials at different times after the
operation showed that the changes in its level in the undercut and
intact areas were different (Fig. 6). In the undercut area of the
cortex, for instance, synchronization fell sharply immediately after the operation, and did not return to its initial level within one
week. In the intact cortex spatial synchronization underwent very
slight changes, either increasing or decreasing. In the other rabbit (No. 18), before the operation synchronously active points were
uniformly distributed over the surface of the cortex. After isolation of the cortex from the subcortex, differences were found in
spatial synchronization of potentials in the intact and undercut
areas. In the intact area a slight increase in synchronization was
observed, but in the undercut area synchronization was sharply reduced for a long time (one week). These results demonstrate that
the subcortex exerts a considerable influence on the formation of

spatial synchronization of cortical potentials. Although intracortical conduction does play a role in this process, it is less important.

DISCUSSION

The diffuse spread of synchronization of potentials throughout the cortex during electrical stimulation of the motor cortex and hypothalamus evidently indicates that the processes responsible for these phenomena are common in origin. It is known that diffuse cortical responses are produced with the participation of ascending activating influences coming from the mesencephalic reticular formation (Moruzzi and Magoun, 1949; Magoun, 1950, 1952; Bremer, 1953, and others) and thalamus (Jasper, 1954). Murphy and Gellhorn (1945) observed a generalized response similar to that described by Moruzzi and Magoun and other authors after repeated electrical stimulation of the hypothalamus. Several pathways may be responsible for this. Neurophysiological investigations and the technique of strychnine neuronography have demonstrated the existence of many hypothalamo-thalamic and hypothalamo-cortical connections (Murphy and Gellhorn, 1945; Ward et al., 1948; French et al., 1955); direct anatomical pathways also are known (Le Gros Clark and Meyer, 1950; Ruch and Fulton, 1961). More recently results have been obtained indicating the presence of collaterals from the lemniscus into the hypothalamus, by-passing the mesencephalic reticular formation (Shevchenko, 1966).

These observations indicate that in our experiments in which the hypothalamic region was stimulated excitation evidently passed along the hypothalamo-cortical connections into the cortex, evoking a diffuse response there. A response spreading diffusely over the whole cortex also appeared during electrical stimulation of a point in the sensorimotor cortex. How can this be explained? There are reports in the literature of the existence of corticofugal influences on subcortical structures (Niemer and Jiminez-Castellanos, 1950; Jasper et al., 1952; Bremer and Terzuolo, 1953b; Narikashvili et al., 1963, and others).

French et al. (1955) undertook a most careful determination of cortical projections in the reticular formation by the use of

evoked potentials and strychnine neuronography. They showed that
certain subcortical structures receive afferent impulses along
these corticofugal pathways which, in their variety and extent occupy second place after the afferent impulses arriving from peripheral receptors. Segundo et al. (1955) compiled a detailed map
of areas of the monkey's cortex from which a generalized electrographic arousal reaction could be obtained, while stimulation of
other areas gave no effect whatever. Points, stimulation of which
elicited the response, were found on the fronto-orbital surface, in
the frontal projection zones of the oculomotor nerves, in the sensorimotor area, and in the posterior part of the parieto-occipital
region.

Bremer and Terzuolo (1953) put forward the hypothesis that
the arousal reaction evoked by stimulation of certain cortical areas
also develops through the participation of the reticular formation.

A comparison of our findings with those reported in the literature suggests that during electrical stimulation of a point in the
motor area impulses may descend along corticofugal fibers into
the subcortical centers responsible for generation of a response
spreading diffusely over the cortex. These experiments and the
others with electrical stimulation of the hypothalamus show that
the subcortical structures ensure through these influences that
conditions exist which favor a high degree of correlation of cortical potentials. These structures evidently regulate the functional
state at a level at which synchronization is established between individual points in the cortex, i.e., they influence the formation of
spatial synchronization. In this respect, the experiments with horizontal undercutting, limiting the flow of afferent impulses into the
occipital cortex from lower structures, are of special interest.The
number of correlations in the undercut area falls sharply by comparison with the intact, and does not return to its initial level for
a long time.

The results obtained thus indicate that impulses arriving
from subcortical structures have a considerable influence on spatial synchronization of cortical potentials. These results show a
large measure of agreement with those of investigations of the distribution of electrical activity in the brain after certain surgical
operations. Morison and Dempsey (1942, 1943), Penfield (1954),
and Penfield and Jasper (1954), for instance, state that influences

from subcortical centers play an essential role in the generation and distribution of cortical electrical activity.

However, the opposite views have been expressed in the literature, namely that cortical electrical activity spreads mainly via transcortical connections (Kogan, 1958; Monakhov, 1963; Tkachenko, 1963). Our experiments with vertical division of the gray matter of the cortex and subsequent insertion of insulating material into the incision showed that synchronization of potentials between the isolated areas of the cortex is not completely lost. This may perhaps be due to the fact that a vertical incision through the cortex does not isolate the areas on either side completely, but that they remain connected by a dense network of fibers in intact areas of the cortex, the existence of which is shown by the observations of Dzugaeva (1949, 1960). As Fessard (1961) suggests, impulses may travel along these fibers and thus by-pass the incision. This is presumably why the isolated areas of the cortex remain in communication.

It may thus be postulated on the basis of these experimental results that the formation of spatial synchronization of cortical potentials is due both to intracortical influences and, in particular, to influences arriving from subcortical structures which are directly concerned in this process.

SUMMARY

1. Stimulation (electrical stimulation, surgical procedures) of the cortex and hypothalamus influences spatial synchronization of cortical potentials.

2. Changes in synchronous activity of potentials throughout the cortex in response to electrical stimulation and incision of the cortex occur in a series of phases.

3. The direction of the initial changes in spatial synchronization of cortical potentials in response to electrical stimulation or surgical interference is dependent on its original state.

4. The formation of spatial synchronization is due both to intracortical influences and, in particular, to influences arriving from subcortical structures.

Emergence of Synchronization in Electroactive Conducting Media[*]

M. N. Zhadin

Institute of Higher Nervous Activity and Neurophysiology
Academy of Sciences of the USSR
Moscow, USSR

The investigation to be described is concerned with the problem
of the brain as a three–dimensional conductor. ·This problem,
which in nature is more a topic of research in physics than in biology, is nevertheless of great importance to the electrophysiologist studying the distribution of potentials within the brain and in
its surrounding tissues. The brain is a conducting system in
which are scattered a very large number of electrically active
elements. These include the bodies of neurons and glial cells,
axons, and dendrites. Each of these elements generates a current
which flows throughout the volume of the brain and its surrounding
tissues and creates a potential field around its source. The potential at any point of the brain (for example, points A and B in Fig. 1)
is created by the activity of all elements of the brain, however far
from this point they may lie, and at each moment of time it is
equal to the sum of the potentials created at this point by each individual element generating current. Naturally, with an increase
in the distance r_A from the active element to the point of observation A, the potential created by it at this point diminishes. The
number of active elements at a distance r_A from the point of observation increases, however, in proportion to r_A^2. For this reason the activity of these distant elements may make a considerable
comtibution to the potential at the point of observation, which may
actually exceed the contribution of closer elements if the field potential of the source decreases slowly enough with increasing distance.

---

[*] Pages 119-126 in the Russian edition.

The potentials V_A and V_B of the points A and B, lying at a distance a from each other, are created by the activity of the same elementary sources scattered throughout the volume of the brain. Because of the considerable difference between the distances r_A and r_B , active elements located close to the points A and B make essentially different contributions to the potentials V_A and V_B . On the other hand, elements at distances much greater than the distance a between the points A and B will make practically equal contributions to the potentials V_A and V_B. For this reason, if the potential of the elementary source falls slowly enough with increasing distance from the source and the contribution made by elements remote from the points A and B is appreciable, there will be a noticeable similarity in the change of potentials of points A and B, even if all the elementary sources act completely independently.

It thus follows that the factor of conductivity of the media of the brain and surrounding tissues, while undoubtedly useful for the electrophysiologist (for it enables him to record the electroencephalogram), may at the same time lead to the development of a positive correlation between the potentials observed at different points of the brain which have absolutely nothing to do with synchronized activity of individual cells. This effect, which the investigation to be described was intended to study, must be distinguished from synchronization of a biological nature.

To simplify the task as much as possible, the following assumptions are made regarding the medium to be investigated.

1. The brain is a continuous, conducting, electroactive medium, i.e., ignoring the discreteness of brain structure, any conveniently small element of brain volume can be considered as a source of current or as generating a potential field in the surrounding tissue.

2. The distances between the points of measurements are much smaller than the linear dimensions of the brain. It may therefore be considered that the electroactive medium is of infinite size. The electroactive medium as a whole is subdivided into equal elements of volume $\Delta\Omega_i$.

3. The media of the brain and its surrounding tissues are homogeneous and isotropic in conductivity.

4. The conductivity of the medium is purely active in char-
acter, i.e., the potential v_i (ξ, η, ζ, t) of the point $(\xi, \eta, \zeta$ of the
medium, created by an element of volume $\Delta\Omega_i$) bears a linear
relationship on the current strength I_i (t) generated by this element in a period of time t:

$$v_i (\xi, \eta, \zeta, t) = I_i(t) \cdot R [r_i(\xi, \eta, \zeta)], \qquad (1)$$

where r_i (ξ, η, ζ) represents the distance from the point (ξ, η, ζ)
to the element of volume $\Delta\Omega_i$. For frequencies of the order of cycles and tens of cycles per second, which are predominant on the
electroencephalogram, this assumption introduces a negligible
error.

5. Dispersions of currents generated by each individual element of volume are identical, i.e.,

$$\sigma_i^2 = \lim_{T \to \infty} \frac{\int_{-T}^{T} I_i^2(t)\, dt}{2T} = \sigma^2 \quad \text{for all } i. \qquad (2)$$

To distinguish the effect under examination in its pure form,
assume that the coefficient of covariance between currents generated by any two elements of volume of the active medium is equal
to zero:

$$\text{Cov}\, I_i I_j = \lim_{T \to \infty} \frac{\int_{-T}^{T} I_i(t) \cdot I_j(t)\, d_t}{2T} = 0 \quad \text{when } i \neq j. \qquad (3)$$

Since changes of potential in time, without limit of generality are the main concern, a mean value $I_i = 0$ for all values of i
can be assumed.

To measure the potential of the medium, electrodes are introduced into it, i.e., nonelectroactive conducting structures are
inserted into the medium. For this reason potentials generated
in passive conducting structures introduced into an electroactive
medium will be examined.

Let two spherical passive conducting structures of diameter
d be introduced into the electroactive medium at a distance of

$a \gg d$ from each other (Fig. 1).   Let their centers lie at the points A and B. The specific resistance of the introduced struc-tures is assumed to be equal to the specific resistance of the me-dium. Let the symbols r_{A_i} and r_{B_j} represent the distances from point A to the element of volume $\Delta \Omega_i$, and from point B to the element of volume $\Delta \Omega_j$.

According to equation (1), in terms of these symbols the po-tentials of points A and B will have the following form respective-ly:

$$
\left.
\begin{aligned}
V_A(t) &= \sum_i I_i(t) R(r_{A_i}) \\
V_B(t) &= \sum_j I_j(t) R(r_{B_j})
\end{aligned}
\right\} \tag{4}
$$

The coefficient of correlation between the potentials of points A and B is determined by the expression

$$
K = \frac{\mathrm{Cov} V_A V_B}{\sigma_A \cdot \sigma_B} ,
$$

where σ_A and σ_B represent the dispersions of the potentials of points A and B.

According to (4), (3), and (2), the coefficient of covariance is given by

$$
\mathrm{Cov}\, V_A V_B = \sum_{i,j} R(r_{A_i}) R(r_{B_j}) \mathrm{Cov}\, I_i I_j = \sigma^2 \sum_i R(r_{A_i}) R(r_{B_i}).
$$

Bearing in mind that for an infinite medium $\sum_i R^2(r_{A_i}) = \sum_j R^2(r_{B_j})$, in a similar manner we obtain the expression $\sigma_A^2 = \sigma_B^2 = \sigma^2 \sum_i R^2(r_{A_i})$.

Hence,

$$
K = \frac{\sum_i R(r_{A_i}) R(r_{B_i})}{\sum_i R^2(r_{A_i})}. \tag{5}
$$

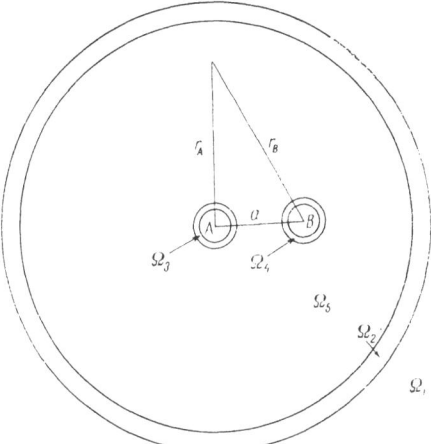

Fig. 1. Subdivision of electroactive medium
into regions of integration.

Multiplying the numerator and denominator by the element
of volume $\Delta\Omega$, and changing from summation to integration relative to volume, we finally obtain:

$$K = \frac{\int\limits_{\Omega} R(r_A)\, R(r_B)\, d\Omega}{\int\limits_{\Omega} R^2(r_A)\, d\Omega},$$

where integration is carried out over the whole volume Ω of the
electroactive medium.

Hence with the assumptions made, the coeeficient of correlation between internal points of passive conducting structures introduced into the medium is determined primarily by a relationship of the type $R(r)$ between the potential of the source and distance from it.

Let the function $R(r) = 1/r^m$, where m represents a certain
positive number. Let the whole medium be subdivided into a series of regions (Fig. 1):

1) Region Ω_1: $r_A > \dfrac{D}{2} \gg a$;

2) Region Ω_2: $\dfrac{D}{2} \gg r_A \geqslant \dfrac{D_0}{2} \gg a$;

3) Region Ω_3: $\dfrac{d}{2} \leqslant r_A \leqslant \dfrac{d_0}{2} \ll a$;

4) Region Ω_4: $\dfrac{d}{2} \leqslant r_B \leqslant \dfrac{d_0}{2} \ll a$;

5) Region Ω_5: $\dfrac{d_0}{2} < r_A < \dfrac{D_0}{2}$, $r_B > \dfrac{d_0}{2}$.

In this case a represents the distance between the centers of the introduced structures.

It will easily be seen that if the function $R(r)$ behaves in this manner, the function $R(r_A) \cdot R(r_B)$ will behave in the regions Ω_1 and Ω_2 as $1/r_A^{2m} \approx 1/r_B^{2m}$, in region Ω_3 as $1/a^m r_A^m$, in region Ω_4 as $1/a^m r_B^m$, and in region Ω_5 as finite and continuous.

In that case

$$\int_\Omega R(r_A)\,R(r_B)\,d\Omega = \lim_{D\to\infty} \int_{\Omega_2} \frac{d\Omega}{r_A^{2m}} + \frac{1}{a^m}\int_{\Omega_3}\frac{d\Omega}{r_A^m} + \frac{1}{a^m}\int_{\Omega_4}\frac{d\Omega}{r_B^m} + \int_{\Omega_5}\frac{d\Omega}{r_A^m r_B^m} =$$

$$= \lim_{D\to\infty} 4\pi \int_{D_0/2}^{D/2}\frac{dr}{r^{2m-2}} + \frac{8\pi}{a^m}\int_{d/2}^{d_0/2}\frac{dr}{r^{m-2}} + C,$$

where $C = \int_{\Omega_5}\dfrac{d\Omega}{r_A^m r_B^m}$ — a finite value.

For the denominator (5) in these same regions we obtain:

$$\int_\Omega R^2(r_A)\,d\Omega = \lim_{D\to\infty} 4\pi \int_{D_0/2}^{D/2}\frac{dr}{r^{2m-2}} + 4\pi\int_{d/2}^{d_0/2}\frac{dr}{r^{2m-2}} + 4\pi\int_{d_0/2}^{D_0/2}\frac{dr}{r^{2m-2}} - 4\pi\int_0^{d/2}\frac{dr}{a^{2m-2}}.$$

If m < 1.5, as $D \to \infty$ the integrals of $4\pi \displaystyle\int_{D_0/2}^{D/2}\frac{dr}{r^{2m-2}}$ in the numerator and denominator of expression (5) increase indefinitely, whereas the other integrals are finite. In the case when 1.5< m< 1.3, with a decrease in the diameter of the introduced structure d, the numerator will tend toward a finite value, and the integral

$\displaystyle\int_{d/2}^{d_0/2}\frac{dr}{r^{2m-2}}$ in the denominator will increase indefinitely. If, on the

other hand, m ≥ 3 the integral $\dfrac{8\pi}{a^m}\displaystyle\int_{d/2}^{d_0/2}\frac{dr}{r^{m-2}}$ in the numerator will increase indefinitely as $1/d^{m-3}$, while the integrals of $4\pi \int\limits_{d/2}^{d_0/2} \dfrac{dr}{r^{2m-2}}$

in the denominator will increase as $1/d^{2m-3}$; the other integrals, however, will be finite.

Hence, when m < 1.5 the coefficient of correlation between the potentials inside the passive conducting structures introduced into the medium is equal to unity for all values of the diameter of the introduced structures d $\ll a$, i.e., a completely synchronized process is established throughout the medium. For m > 1.5, the coefficient of correlation between these potentials differs from zero, is positive, and diminishes with a decrease in the diameter of the included structures: at the limit when d $\rightarrow$ 0 the coefficient of correlation tends toward zero. In other words, when m > 1.5 the process of partial synchronization arises only inside the non-electroactive conducting structures introduced into the medium, and inside the electroactive medium itself synchronization is absent. Clearly, for introduced structures of small diameters d $\ll a$, the following approximate relationships are valid:

$$
\begin{aligned}
K &= B_1\,(m,\,a)\,d^{m} && \text{when } m \geqslant 3, \\
K &= B_2\,(m,\,a)\,d^{2m-3} && \text{when } 1.5 < m < 3, \\
K &\equiv 1 && \text{when } m < 1.5,
\end{aligned} \qquad (6)
$$

where $B_1(m_1 a)$ and $B_2(m_1 a)$ represent certain functions of m and a diminishing monotonously with an increase in a.

In this investigation it was assumed that the introduced structure possesses the same specific resistance as the electroactive medium. In the Appendix the question of the potential of an introduced structure with any specific resistance is examined. It is shown there that for investigation of the potential of a spherical electrode with low specific resistance, present inside an electroactive medium, the electrode may be regarded as a passive inclusion with specific resistance equal to the specific resistance of the medium, and the potential in the center of this inclusion can be studied. Distortion of the field introduced by the other electrode can be disregarded, because the distance between the electrodes is many times greater than their diameter. Hence, the investigation now described can be extended without change to electrodes of low specific resistance.

Such a picture arises in cases when the electroactive medium is in contact with a passive conducting layer. Let a passive conducting layer perpendicular to the z axis cover a semi-infinite electroactive medium occupying a space $z > 0$, and possessing the properties described above. It is assumed that the distance between the points of observation and the thickness of the layer are much smaller than the linear dimensions of the active medium. Just as in the preceding case, we deduce an expression for the coefficient of correlation between the potentials of two points of observation:

$$K = \frac{\int\limits_\Omega R\left[\rho_A(x, y), z - \zeta\right]R[\rho_B(x, y), z - \zeta]\, d\Omega}{\int\limits_\Omega R^2\left[\rho_A(x, y), z - \zeta\right]\, d\Omega} \, .$$

In this case $\rho_A(x, y)$ and $\rho_B(x, y)$ represent the distances from the points of observation A and B to the straight line drawn parallel to the z axis through the point (x, y, z) of the active medium respectively, and ζ represents the coordinate of the points A and B along the z axis. Integration is carried out for the whole volume of the active medium.

Just as was described above, it is easy to show that if the function $R[\rho, z - \zeta]$ changes more slowly than $1/r^{1.5}$ at high values of $r = \sqrt{\rho^2 + (z - \zeta)^2}$ and low values of $r = \sqrt{\rho^2 + (z - \zeta)^2}$, a completely synchronized process will be established throughout the volume of the active medium and the passive layer. If, on the other hand, the function $R[\rho, z - \zeta]$ changes more rapidly than $1/r^{1.5}$ in these regions, synchronization will be absent within the active medium, but partial synchronization will be present in the passive layer. The coefficient of correlation between the potentials of two points of the passive layer in this case will decrease with an increase in distance between them, and as they move closer to the surface of the active medium. With small distances ζ of the points of measurement from the surface of the active medium, the relationship between the coefficient of correlation and ζ will be similar in character to the relationship (6), in which ζ must be substituted for d.

Hence, the factor of conductivity of the electroactive medium and of conducting layers in contact with it and structures inserted into it may lead to the occurrence of partial, or even com-

plete, synchronization between the potentials measured during un-
correlated working of the elementary sources constituting the active medium. The coefficient of correlation between the measured
potentials depends essentially on the nature of the elementary
sources and the distance between the electrodes. However, as is
known, complete synchronization is observed relatively rarely between encephalograms taken even at closely situated points (of the
order of 2 mm apart for the rabbit) separated by a distance much
smaller than the linear dimensions of the brain. It is doubtful,
therefore, if the brain cells are sources of a field of point type
(m = 1). The emergence of physiological synchronization in the
work of cells would lead only to an increase in the coefficient of
correlation between potentials of different points. Many physiologists (Lorente de Nó, 1947; Howland et al., 1955; Bureš et al.,
1962) consider that the nerve cells are dipole sources (m = 2). If
this view is accepted, or the cell field is regarded as a field of a
more complex multipolar structure, it can be concluded that the
synchronization effect now being studied arises only in nonelectroactive conducting tissues (electrodes, cerebrospinal fluid, menin-
ges, bones) in contact with the active brain medium.

A characteristic feature of this effect is the monotonous decrease in the coefficient of correlation between the potentials of
the electrodes when the distance between them increases. Any
deviation from this rule must inevitably go beyond the bounds of
the phenomenon being studied. To reduce the synchronization effect described, the diameter of the introduced electrodes must be
reduced, and when the electroencephalograph is recorded, the
electrodes must be brought closer to the brain surface. The
whole surface of the electrodes except the tip must be insulated
so that synchronized potentials of the upper layers of the nonelectroactive tissue do not influence their potentials.

In conclusion, I wish to express my gratitude to corresponding member of the Academy of Sciences of the USSR, Professor
M. N. Livanov and to Candidate of Biological Sciences T. A. Korol'kova for much valuable advice and for their consultation.

APPENDIX

Calculation of Potential of

a Spherical Electrode

Let a spherical conducting electrode of radius R, with a specific resistance τ_1 (Fig. 2) be placed inside an infinite conducting medium with specific resistance τ_2. We can calculate the distribution of potential inside and outside this inclusion created by a point source of current I, located outside the inclusion at a distance a from its center. The origin of the spherical coordinates (r, θ, φ) is placed at the center of the inclusion, and the z axis is carried through the source.

It has been shown (Smythe, 1954) that the distribution of potential created within a homogeneous medium by a point source is determined by the Laplacian equation

$$\nabla^2 V(r,\theta) = 0. \tag{1}$$

The z axis is the axis of symmetry, and solution is independent of the angle φ. The solutions of this equation (Smythe, 1954) have the following form:

$$A r^n P_n(\cos\theta) + B \frac{P_n(\cos\theta)}{r^{n+1}}, \tag{2}$$

where n represents a whole number, A and B are constants, and $P_n(\cos\theta)$ is a Legendre polynomial of the n-th degree.

The solutions of equation (1) in the region r < R are obtained in the form

$$V_1 = \sum_{n=0}^{\infty} A_n r^n P_n(\cos\theta) + \frac{\tau_2 I}{4\pi} \sum_{n=0}^{\infty} r^n P_n(\cos\theta). \tag{3}$$

So that the potential does not increase indefinitely as $r \to 0$, the coefficients for terms containing $1/r^{n+1}$ are assumed to be equal to zero. The second sum is the field of a point source created in the region r < R when $\tau_1 = \tau_2$.

In the field R < r < a the solution is obtained in the form

$$V_2 = \sum_{n=0}^{\infty} B_n r^n P_n(\cos\theta) + \sum_{n=0}^{\infty} C_n \frac{P_n(\cos\theta)}{r^{n+1}} \tag{4}$$

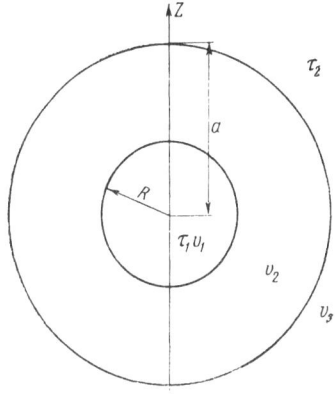

Fig. 2. Areas of integration of the Laplacian equation.

and in the region $R > a$,

$$V_3 = \sum_{n=0}^{\infty} D_n \frac{P_n (\cos \theta)}{r^{n+1}}. \tag{5}$$

In the expression for V_3 the coefficients for terms containing r^n are assumed to be equal to zero, so that when $r \to \infty$, the potential V_3 tends toward zero.

The boundary conditions of equality of the potentials are given by

$$V_1 (R) = V_2 (R), \tag{6}$$

$$V_2 (a) = V_3 (a). \tag{7}$$

The boundary conditions for equality of normal components of the current:

$$\frac{1}{\tau_1} \frac{\partial V_1}{\partial r} (R) = \frac{1}{\tau_2} \frac{\partial V_2}{\partial r} (R), \tag{8}$$

$$\frac{\partial V_2}{\partial r} (a) = \frac{\partial V_3}{\partial r} (a), \tag{9}$$

Let condition (8) be integrated over the surface Ω of a sphere $r = R$. Remembering that the integral for the potential of a point source located outside the sphere is

$$\int_{\Omega} \frac{\tau_2 I}{4\pi} \sum_{n=0}^{\infty} r^n P_n (\cos \theta), \tag{10}$$

in accordance with (3) and (4) we obtain

$$\frac{n}{\tau_1} A_n R^{n-1} = \frac{nB_n}{\tau_2} R^{n-1} - \frac{(n+1)C_n}{\tau_2 R^{n+2}}. \tag{11}$$

Taking into consideration this condition, and also conditions (6), (7), and (9), we obtain the following system of equations:

$$\left.\begin{array}{l} n\tau_2 R^{2n+1} A_n - n\tau_1 R^{2n+1} B_n + (n+1)\tau_1 C_n = 0, \\[4pt] R^{2n+1} A_n - R^{2n+1} B_n - C_n = -\frac{\tau_2 I}{4\pi} R^{2n+1}, \\[4pt] na^{2n+1} B_n - (n+1) C_n + (n+1) D_n = 0, \\[4pt] a^{2n+1} B_n + C_n - D_n = 0. \end{array}\right\} \tag{12}$$

Solving this system relative to A_n, B_n, C_n, and D_n, and substituting the solutions in equations (3), (4), and (5), we obtain

$$V_1 = \frac{\tau_1 \tau_2 I}{4\pi a} \sum_{n=0}^{\infty} \frac{2n+1}{(\tau_1 + \tau_2) n + \tau_1} \left(\frac{r}{a}\right)^n P_n(\cos\theta) \tag{13}$$

$$V_2 = \frac{\tau_2 I}{4\pi a} \sum_{n=0}^{\infty} \left(\frac{r}{a}\right)^n P_n(\cos\theta) - \frac{\tau_2 I}{4\pi} \sum_{n=0}^{\infty} \frac{nR^n(\tau_2 - \tau_1)}{(\tau_1 + \tau_2) n + \tau_1} \left(\frac{R}{ar}\right)^{n+1} P_n(\cos\theta) \tag{14}$$

$$V_3 = \frac{\tau_2 I}{4\pi} \sum_{n=0}^{\infty} \frac{1}{r} \left(\frac{a}{r}\right)^n P_n(\cos\theta) - \frac{\tau_2 I}{4\pi} \sum_{n=1}^{\infty} \frac{nR^n(\tau_2 - \tau_1)}{(\tau_1 + \tau_2) n + \tau_1} \left(\frac{R}{ar}\right)^{n+1} P_n(\cos\theta). \tag{15}$$

Hence it is clear that the potential in the center of the inclusion

$$V_1(r = 0) = \frac{\tau_2 I}{4\pi a} \tag{16}$$

is equal to the potential of the medium which it would have at the point r = 0 if $\tau_1 = \tau_2$.

The potential of a spherical electrode for which $\tau_1 = 0$, when

$$V_1(r,\theta) \equiv \frac{\tau_2 I}{4\pi a} \tag{17}$$

has the value given.

This conclusion is valid for any source of current which is a superposition of point sources, for example, a dipole, quadripole, etc.

Hence, if the potential in the center of a conducting spherical inclusion with any specific resistance is considered, the specific resistance of this inclusion may be considered to be equal to the specific resistance of the surrounding medium.

Bibliography

Abuladze, K. S. Zh. Vyssh. Nervn. Deyat., 1:647 (1951).

Adey, W.R. Internat. Rev. Neurobiol., 1:1 (1959).

Adey, W.R. Electroenceph. Clin. Neurophysiol., (Suppl. 20):41 (1961a).

Adey, W.R. Feder. Proc., 20:617 (1961b).

Adey, W.R. In: R. Stacy and B. Waxman (eds). Computers in Biomedical Research, Vol. 1, New York, Academic Press (1965), p. 223.

Adey, W.R. Progress in Physiological Psychology, Vol. 1 (1966). In press.

Adey, W.R., Segundo, J.P., and Livingston, R.B. J. Neurophysiol., 20:1 (1957).

Adey, W.R., Dunlop, C.W., and Hendrix, C.E. A.M.A. Arch. Neurol., 3:74 (1960).

Adey, W.R., Walter, D.O., and Hendrix, C.D. Exper. Neurol., 3:501 (1961).

Adey, W.R., Bell, F.R., and Dennis, B.J. Neurology, 12:591 (1962a).

Adey, W.R., Kado, R.T., and Didio, J. Exper. Neurol., 5:47 (1962b).

Adey, W.R., Walter, D.O., and Lindsley, D.F. A.M.A. Arch. Neurol., 6:194 (1962c).

Adey, W.R., Kado, R.T., Didio, J., and Schindler, W.J. Exper. Neurol., 7:282 (1963).

Adey, W.R., and Walter, D.O. Exper. Neurol., 7:282 (1963).

Adey, W.R., Kado, R.T., and Walter, D.O. Exper. Neurol., 11:190 (1965).

Adey, W.R., Kado, R.T., McIlwain, J.T., and Walter, D.O. Exper. Neurol., 15:490 (1966).

Adrian, E.D. J. Physiol., 47:460 (1914).

Adrian, E.D. (1935). The Mechanism of Nervous Action, University of Pennsylvania Press (1932).

Adrian, E.D., and Bronk, D.W. J. Physiol., 67:119 (1929).

Adrian, E.D., and Matthews, B.H.C. Brain, 57:355 (1934).

Adrian, E.D., and Yamagiwa, K. Brain, 58:323 (1935).

Airikyan, E.A., and Gaske, O.D. Fiziol. Zh. SSSR, 51:105 (1965).

Airikyan, E.A., and Gaske, O.D. Fiziol. Zh. SSSR. In press.

Alajouanine, T. (ed). Les Grandes Activités du Rhinencéphale. II. Physiologie et Pathologie du Rhinencéphale, Paris, Masson (1961).

Albe-Fessard, D., and Gillett, E. Electroenceph. Clin. Neurophysiol., 13:257 (1961).

Amassian, V.E. J. Neurophysiol., 14:445 (1951).

Amassian, V., Berlin, L., Macy, J., Hardress, J., and Waller, H. Trans. N.Y. Acad. Sci., Ser. II, 21(5):395 (1959).

Andersen, P. Acta Physiol. Scand., 47(1):3 (1959).

Andersen, P., and Eccles, J.C. Nature (Lond.), 196:645 (1962).

Andersen, P., Eccles, J.C., and Løyning, Y. Nature, 198:540 (1963a).

Andersen, P., Eccles, J.C., and Voorhoeve, P.E. Nature, 199:655 (1963b).

Andersen, P., Brooks, C., Eccles, J.C., and Sears, T.A. J. Physiol., 174(3):348 (1964a).

Andersen, P., Eccles, J.C., and Løyning, Y. J. Neurophysiol., 27:592 (1964b).

Andersen, P., Eccles, J.C., and Sears, T.A.   J. Physiol., 174(3):370 (1964c).

Andersen, P., Eccles, J.C., and Voorhoeve, P.E.   J. Neurophysiol., 27(6):1138(1964d).

Andersen, P., and Lφmo, T.   Exper. Brain Res., 2(3):247 (1966)

Andersen, P., Blackstad, T.W., and Lφmo, T.   Exper. Brain Res., 1(3):236 (1966).

Angel, A., Magni, F., and Strata, P.   Nature, 208:495 (1965).

Angyan, L., and Grastyán, E.   Acta Physiol. Acad. Sci. Hung., 25:297 (1963).

Angyán, L., Grastyán, E., and Sakhiulina, G.T.   Zh. Vyssh. Nervn. Deyat., 13:228
     (1963).

Anokhin, P.K.   In: Proceedings of a Conference on Philosophical Problems in Higher
     Nervous Activity and Psychology [in Russian], Moscow (1962).

Arduini, A., and Hirao, T.   Arch. Ital. Biol., 97:140 (1959).

Arvanitaki, A.   Arch. Internat. Physiol., 52:381 (1942).

Arvanitaki, A., and Chalazonitis, N.   Arch. Sci. Physiol., 9:115 (1955).

Arvanitaki, A., and Chalazonitis, N.   In: Nervous Inhibition (1961), p. 194.

Arvanitaki, A., and Chalazonitis, N.   In: Olfaction and Taste (1963), p. 377.

Arvanitaki-Chalazonitis, A.   C.R. Acad. Sci., 252:193 (1961).

Asafov, B.D., Zimkina, A.M., and Stepanov, A.I.   Fiziol. Zh. SSSR, 41:314 (1955).

Asratyán, E.A.   Fiziol. Zh. SSSR, 30:13 (1941).

Asratyán, E.A.   Proc. Internat. Union Physiol. Sci., XXIII Internat. Cong. Physiol.
     Sci. (1965), p. 629.

Baldwin, M., and Bailey, P.   Temporal Lobe Epilepsy, Thomas, Springfield, Mass.
     (1959).

Barker, S.A., Bayyuk, S.H., and Stacey, M.   Nature, 196:64 (1962).

Barlow, J., Brazier, M., and Rosenblith, W.   In: Quastler and Morowitz (eds). Proc.
     First Nat. Biophys. Conf., Yale University Press, New Haven, Conn. (1959),
     p. 622.

Barlow, J., and Freeman, B.   Electroenceph. Clin. Neurophysiol., 2:618 (1959).

Barrnett, R.J.   Trans. Am. Neurol. Assoc., 88:123 (1963).

Bartley, S.H.   J. Cell Comp. Physiol., 8:41 (1936).

Bartley, S.H., O'Leary, J., and Bishop, G.H.   Am. J. Physiol., 120(3):604 (1937).

Baru, A.V.   Abstracts of Proceedings of the 10th All-Union Congress of the I.P. Pavlov
     Physiological Society [in Russian], Vol. 2, No. 1 (1964), p. 79.

Baru, A.V., Gershuni, G.V., and Tonkonogii, I.M.   Zh. Nevropat. i Psikhiatr. im.
     Korsakova, 64:481 (1964).

Batini, C., Palestini, M., Rossi, G.F., and Zanchetti, A.   Arch. Ital. Biol., 97:26
     (1959).

Batuev, A.S.   Vestnik Leningrad. Gos. Univ., 15:98 (1963).

Baumgarten, R., and Schaefer, K.   Electroenceph. Clin. Neurophysiol., 9:953 (1957).

Baumgartner, G.   Pflüg. Arch. Ges. Physiol., 261(5):456 (1955).

Baumgartner, G.   Electroenceph. Clin. Neurophysiol., 10:195 (1958).

Bekhtereva, N.P., Smirnov, V.M., and Genkin, A.A.   Gagrskie Besedy (1965).

Bekkering, D.H., Kamp, A., and Storm van Leeuwen, W.   Electroenceph. Clin.
     Neurophysiol., 10:555 (1958).

Belenkov, N.Yu., and Kalinina, T.E.   Zh. Vyssh. Nervn. Deyat., 15:285 (1965).

Belenkov, N.Yu., and Chirkov, V.D.   Zh. Vyssh. Nervn. Deyat., 11:512 (1961).

Belenkov, N.Yu., and Chirkov, V.D.   Zh. Vyssh. Nervn. Deyat., 14:68 (1964).

Belenkov, N. Yu., and Chirkov, V. D.   Zh. Vyssh. Nervn. Deyat., 15:128 (1965).

Bennett, H.S.   J. Histochem. Cytochem., 11:14 (1963).

Bennett, M., Aljure, E., Nakajiaya, Y., and Pappas, H.   Science, 141:264 (1963).

Benoit, I.   C. R. Acad. Sci. (Paris), 199:1671 (1934).

Benoit, I.   Ann. N. Y. Acad. Sci., 117(1):23 (1964).

Berger, H.   Arch. f. Psych., 33(2):521 (1900).

Berger, H.   Arch. f. Psych., 103:445 (1935).

Bergman, P.S.   Arch. Neurol. Psychiatr., 78(6):568 (1957).

Beritov, I.S.   Trudy Inst. Fiziol. Akad. Nauk Gruz. SSR, 5:193 (1943).

Beritov, I.S.   Nervous Mechanisms of Behavior of Higher Vertebrate Animals [in
        Russian], Moscow (1961).

Berry, C.A., and Hance, A.J.   Electroenceph. Clin. Neurophysiol., 18(2):124 (1965).

Beurle, R.L.   Phil. Trans. Roy. Soc. Series B, 240:55 (1956).

Biederman, M.A.   Comp. Biochem. Physiol., 12:311 (1964).

Bishop, G.H.   Am. J. Physiol., 103(1):213 (1933).

Bishop, G.H.   Physiol. Rev., 36:376 (1956).

Bishop, P.O., and Davis, R.   J. Physiol., 154(3):514 (1960).

Bissonette, T.H.   Proc. Roy. Soc. B., 110:322 (1932).

Blinkov, S.M., and Rusinov, V.S.   Vopr. Neirokhir., 13:38 (1949).

Boldyreva, G.N.   In: Mathematical Analysis of Electrical Phenomena in the Brain
        [in Russian], Moscow (1965), p. 29.

Bol'shev, L.I., and Smirnov, I.V.   Tables of Mathematical Statistics [in Russian],
        Moscow (1965).

Bonnet, V.   C. R. Soc. Biol., 127:798 (1938).

Bowsher, D.   J. Comp. Neurol., 11 :213 (1961).

Brandt, P.W.   Circulation, 26:1075 (1962).

Bratus', N.V.   Fiziol. Zh. SSSR, 46:179 (1960).

Brazier, M.A.B.   Proc. 1st Internat. Cong. Cybernetics, Gauthier-Villars, Paris (1956),
        p. 856.

Brazier, M.A.B.   J. Nerv. Ment. Dis., 126:303 (1958).

Brazier, M.A.B.   Exper. Neurol., 2:123 (1960).

Brazier, M.A.B., and Casby, J.   Electroenceph. Clin. Neurophysiol., 4:201 (1952).

Brazier, M., and Barlow, J.   Electroenceph. Clin. Neurophysiol., 8:325 (1956).

Bremer, F.   C. R. Soc. Biol. (Paris), 118:1235 (1935).

Bremer, F.   Arch. Internat. Physiol., 51:51 (1941).

Bremer, F., and Terzuolo, C.   Arch. Internat. Physiol., 60:228 (1952).

Bremer, F., and Terzuolo, C.   Arch. Internat. Physiol., 61:86 (1953a).

Bremer, F., and Terzuolo, C.   J. Physiol. (Paris), 45:56 (1953b).

Bremer, F., and Terzuolo, C.   Arch. Internat. Physiol., 62:157 (1954).

Brodal, A.   Anatomical Aspects and Functional Correlations, Oliver and Boyd, Edin-
        burgh (1957).

Brooks, C., Koizumi, K., and Siebens, A.   Am. J. Physiol., 184:497 (1956).

Buchwald, N.A., Wyers, E.J., Okuma, T., and Heuser, G.   Electroenceph. Clin. Neuro-
        physiol., 13:509 (1961).

Buchwald, J.S., Halas, E.S., and Schram, S.   Abstr. 23rd Internat. Cong. Physiol. Sci.,
        Tokyo (1965), p. 466.

Bullock, T.H.   Behavior, 17:48 (1961).

Bullock, T.H., and Terzuolo, C.A.   J. Physiol., 138:341 (1957).

Bulygin, I.A.   Investigation of the Principles and Mechanisms of Interoceptive Reflexes [in Russian], Minsk (1959).

Bureš, J., and Burešová, O.   Proc. 23rd Internat. Cong. Physiol. Sci., Tokyo (1965), p. 359.

Bureš, J., Petran, M., and Zachar, J.   Electrophysiological Methods of Investigation [Russian translation], Foreign Lit. Press, Moscow (1962).

Burešová, O., and Bureš, J.   Acta Physiol., 24(1-2):53 (1965).

Burmistrov, Yu.M.   Biofizika, 10:90 (1965).

Burmistrov, Yu.M., and Shuranova, Zh.P.   Dokl. Akad. Nauk SSSR, 165:1450 (1965).

Burns, B.D.   The Mammalian Cerebral Cortex, Boston (1958).

Burns, B.D., Grafstein, B., and Olszewski, J.   J. Neurophysiol., 20(2):200 (1957).

Bychkov, M.S.   In: The Biological Action of the Very High Frequency Electric Field. Trans. S.M. Kirov Military Med. Acad., Vol. 7 [in Russian], Leningrad (1957), p. 58.

Cajal, S.R.   Histologie du Système Nerveux de l'Homme et des Vertébrés, Vol. 1, Maloine, Paris (1909).

Chalazonitis, N.   Ann. N. Y. Acad. Sci., 109:451 (1963).

Chang, H.T.   J. Neurophysiol., 15(1):5 (1952).

Chang, H.T.   In: Electrophysiological Investigations of Higher Nervous Activity [in Russian], Moscow (1962), p. 67.

Chapman, B.L.M.   Kybernetica, 2:152 (1959).

Chatrian, G.E., Petersen, C., and Lazarte, J.A.   Electroenceph. Clin. Neurophysiol., 11(3):497 (1959).

Chertok, L., and Fontaine, M.   J. Psychosom. Res., 7:229 (1963).

Chizhenkova, R.A.   Investigation of the Role of Specific and Nonspecific Structures in Electrical Responses of the Rabbit Brain to UHF and VHF Fields and to a Steady Magnetic Field. Candidate dissertation, Moscow (1966).

Cholokoshvili, E.S.   Trudy Inst. Fiziol. Akad. Nauk Gruz. SSR, 11:207 (1958).

Chorayan, O.G.   Fiziol. Zh. SSSR, 49:1026 (1963).

Chorayan, O.G.   Fiziol. Zh. SSSR, 51:1050 (1965a).

Chorayan, O.G.   Fiziol. Zh. SSSR, 51:463 (1965b).

Ciganek, L.   Ann. N. Y. Acad. Sci., 112(1):241 (1964).

Claes, E.   Arch. Internat. Physiol., 48:181 (1939).

Clare, M.   Das Nervensystem des Menschen, Leipzig (1941).

Clare, M.H., and Bishop, G.H.   Electroenceph. Clin. Neurophysiol., 7(1):85 (1955).

Clark, W.E. LeGros Trans. Ophthalm. Soc. U.K., 79:455 (1959).

Clark, W. E. Legros and Mejer, M. Brit. Med. Bull., 6:341 (1950).

Cohen, J., Boshes, L.D., and Snider, R.S. Electroenceph. Clin. Neurophysiol., 13:914 (1961).

Cohen, M.J., Landgren, S., Ström, L., and Zottermann, Y.   Acta Physiol. Scand., 40(Suppl. 135):1 (1957).

Cooper, R., and Mundy-Castle, A.   Electroenceph. Clin. Neurophysiol., 12:153(1960).

Cooper, R., Shipton, H., Shipton, J., Walter, V., and Walter, G.   Electroenceph. Clin. Neurophysiol., 9:375 (1957).

Cramer, H.   The Elements of Probability Theory, New York (1955).

Crapper, D.R., and Noell, W.K.   Physiologist, 3(3):42 (1960).

Creutzfeldt, O., and Meisch, J.J.   Electroenceph. Clin. Neurophysiol., (Suppl. 24): (1963).

Creutzfeldt, O., and Struck, S.   Arch. f. Psychiatr. u. Z. ges. Neurol., 203:708 (1962).

Creutzfeldt, O., Baumgartner, G., and Schoen, L.   Arch. f. Psychiatr. u. Z. ges. Neurol., 194(46):597 (1956).

Creutzfeldt, O., Fuster, J.M., Lux, H.D., and Nacimiento, A.   Naturwissenschaft, 51:166 (1964).

Cross, B.A., and Green, J.D.   J. Physiol., 148:554 (1959).

Czermak, Y.N.   Pflüg. Arch. Physiol., 7:107 (1873).

Danilewski, V.I.   Congr. Internat. de Psychol. Physiol. C.R. Bureau des Revues, Paris (1890), p. 79.

Danilov, I.V.   In: Problems in Electrophysiology and Encephalography [in Russian], Moscow (1960), p. 246.

Danilova, N.N.   In: Typological Features of Human Higher Nervous Activity [in Russian], Moscow (1963), p. 25.

Deglin, V.L.   Zh. Nevropat. i. Psikhiatr., 60(11):1494 (1960).

Dell, P., and Olson, R.,   C. R. Soc. Biol., 145:1084 (1951).

Denisov P.S., and Kupalov, P.S.   Arkh. Biol. Nauk, 33:5 (1933).

Dikareva, I.P.   Proceedings of the 15th Conference of Physiologists, Biochemists, and Pharmacologists of the South of the RSFSR [in Russsian], Makhach-Kala (1965), p. 105.

Dodge, F.A. and Cooley, J.W.   Abstr. 9th Ann. Biophys. Soc. Meetings (1965), p. 13.

Donchin, E., Wicke, J.D., and Lindsley, J.B.   Science, 141(3584):1285 (1963).

Doty, R.W.   J. Neurophysiol., 28:623 (1965a).

Doty, R.W.   In: Cybernetics of Neural Processes (1965b).

Doty, R.W., and Giurgea, C.   In: Brain Mechanisms and Learning, Oxford University Press, Oxford (1961), p. 133.

Doty, R.W., and Rutledge, L.T.   J. Neurophysiol., 22:428 (1959).

Doty, R.W., Rutledge, L.T., and Larsen, R.M.   J. Neurophysiol., 19:401 (1956).

Doty, R.W., Beck, E.C., and Kooi, K.A.   Exper. Neurol., 1:360 (1959).

Doty, R.W., Kimura, D.S., and Mogenson, G.J.   Exper. Neurol., 10:19 (1964).

Downman, C.B.B.   In: The Spinal Cord (1953), p. 92.

Downman, C.B.B., and Evans, M.H.   J. Physiol., 137:66 (1957).

Downman, C.B.B., and Hussain, A.   J. Physiol., 141:489 (1958).

Dubner, H.H., and Gerard, R.W.   J. Neurophysiol., 2(3):142 (1939).

DuBois-Reymond, E.   Untersuchungen über thierische Elektricität, Reimer, Berlin, Vol. 1 (1848); Vol. 2 (1849).

Duda, P.   Csl. Fisiol., 9:225 (1960).

Dumenko, V.N.   Electrophysiological Investigation of Interrelations of Nuclei of Analyzers and Their Reactivity to Combined Stimuli.   Candidate's dissertation, Moscow (1953a).

Dumenko, V.N.   Abstracts of Proceedings of the 16th Conference on Problems in Higher Nervous Activity [in Russian] (1953b), p. 81.

Dumenko, V.N.   Trudy Inst. Vyssh. Nervn. Deyat. Akad. Nauk SSSR, 1:335 (1955).

Dumenko, V.N.   Byull. Eksperim. Biol. i Med., 49(3):8 (1960).

Dumenko, V.N.   Zh. Vyssh. Nervn. Deyat., 12(2):281 (1961).

Dumenko, V.N.   Zh. Vyssh. Nervn. Deyat.,   15(5):769 (1965).

Dunlop, C.W.   Electroenceph. Clin. Neurophysiol.,   10:297 (1958).

Durinyan, R.A.   Dokl. Akad. Nauk SSSR, 137:739 (1961).

Dzidzishvili, N.N.   Gagrskie Besedy (Talks at Gagra), Vol. 4 [in Russian], Acad. Sci. Georgian SSR Press, Tbilisi (1963), p. 203.

Dzugaeva, S.B.   In: Problems in Morphology of the Nervous System [in Russian], Moscow (1960), p. 40

Eccles, J.C.   The physiology of Nerve Cells, Johns Hopkins Press, Baltimore (1957).

Eccles, J.C. (1959). See: Eccles, J.C. (1957). Russian translation.

Eccles, J.C.   In: Structure and Function of the Cerebral Cortex, Elsevier, Amsterdam (1960), p. 192.

Eccles, J.C.   In: G. Moruzzi, A. Fessard, and H.H. Jasper (eds). Progress in Brain Research. Brain Mechanisms, Vol. 1, Elsevier, Amsterdam (1963), p. 1.

Eccles, J.C.   The Physiology of Synapses, Springer Verlag, Berlin (1964).

Eccles, J.C.   Abstr. Papers 23rd Internat. Cong. Physiol. Sci., Vols. 1-9, Tokyo (1965a), p. 5.

Eccles, J.C.   In: Proc. 23rd Internat. Cong. Internat. Un. Physiol. Sci., Vol. 4, Tokyo (1965b), p. 84.

Eccles, J.C. (1966). See: Eccles, J.C. (1964). Russian translation.

Eccles, J.C., and Meulders, M.   Rev. Med. Louvain, 83(11):389 (1964).

Eccles, J.C., Kostyuk, P.G., and Schmidt, R.F.   J. Physiol., 161:237 (1962).

Eccles, J.C., Llinás, R., and Sasaki, K.   Exper. Brain Res., 1:161 (1966).

Eccles, R.M., and Lundberg, A.   J. Physiol., 43:267 (1936).

Eckert, R.   J. Gen. Physiol., 46:573 (1963).

Ectors, L.   Arch. Internat. Physiol., 43:26 (1936).

Elul, R.   Exper. Neurol., 6:285 (1962).

Elul, R.   In: First Conference on Information and Control Processes in Living Systems. N.Y. Acad. Sci. (1965a).

Elul, R.   Physiologist, 8:159 (1965b).

Elul, R.   In: Symposium on Slow Electrical Phenomena in the Central Nervous System.   Neurosciences Research Program Bulletin, Massachusetts Institute of Technology (1966). In press.

Elul, R., and Adey, W.R.   Phsyiologist, 8:98 (1965).

Engel, G.L., Ferris, E.B., and Romano, J.   Am. J. Med. Sci., 209:650 (1945).

Ermolaeva, V.Yu., and Chernigovskii, V.N.   Dokl. Akad. Nauk SSSR, 159(3):686 (1964).

Euler, C., Green, J., and Ricci, G.   Acta Physiol. Scand., 42:87 (1958).

Evans, M.H., and McPherson, A.   J. Physiol., 140:201 (1958).

Evarts, E.V.   J. Neurophysiol., 28(2):216 (1965).

Fatt, P., and Katz, B.   J. Physiol., 117:109 (1952).

Fessard, A.   Trudy 1-go Moskovsk. Med. Inst., 11:239 (1961).

Fielden, A., and Hughes, G.M.   J. Exp. Biol., 39:31 (1962).

Foerster, O.   In: Handbuch der Neurologie, Vol. 6, Springer, Berlin (1936), p. 358.

Foss, G., and Blackstad, T.W.   J. Ultrastruct. Res., 14:413 (1966).

Fray, A.H.   Psychol. Bull., 63(5):322 (1965).

French, J.D.   In: Reticular Formation of the Brain, Boston (1958), p. 491.

French, J.D., Amerongen, F.K., and Magoun, H. Arch. Neurol. Psychiatr. (Chicago), 68:577 (1952).

French, J.D., Hernandez-Peon, R., and Livingston, R.B. J. Neurophysiol., 18:74 (1955).

Frey, E. Schweiz. Arch. Neurol. Psychiatr., 66:67 (1950).

Freygang, W.H., and Landau, W.M. J. Cell. Comp. Physiol., 45(3):377 (1955).

Fromm, G.H., and Bond, W. Electroenceph. Clin. Neurophysiol., 18(5):520 (1965).

Fujita, Y., and Sato, T. J. Neurophysiol., 27:1011 (1964).

Fuortes, M.G.F., Frank, K., and Becker, M.C. J. Gen. Physiol., 40:735 (1957).

Furshpan, E., and Potter, D. J. Physiol., 145:289 (1959).

Garoutte, B., and Aird, R.B. Electroenceph. Clin. Neurophysiol., 10:259 (1958).

Gastaut, H. Rev. Neurol., 87(2):176 (1952).

Gastaut, H., and Hunter, J. Electroenceph. Clin. Neurophysiol., 2:263 (1950).

Gastaut, H., and Lammers, H.J. In: Les Grandes Activités du Rhinencéphale, Vol. 1, Paris (1961).

Gastaut, H., Gastaut, Y., Roget, J., Carriol, J., and Naquet, R. Electroenceph. Clin. Neurophysiol., 3:401 (1951).

Gastaut, Y. Rev. Neurol., 89(5):382 (1953).

Gavrilova, N.A. Zh. Nevropat. i Psikhiatr., 65(12):1855 (1965).

Gavrilova, N.A., and Aslanov, A.S. In: Mathematical Analysis of Electrical Phenomena of the Brain [in Russian], Moscow (1965), p. 57.

Gavrilova, N.A., Aslanov, A.S., and Kaganova, Z.I. In: Electrophysiology of the Nervous System [in Russian], Rostov-on-Don (1963), p. 90.

Gavrilova, N.A., Aslanov, A.S., Dzugaeva, S.B., and Kaganova, Z.I. Zh. Vyssh. Nervn. Deyat., 14(1):3 (1964a).

Gavrilova, N.A., Il'in, E.A., and Aslanov, A.S. In: Abstracts of Proceedings of the 10th Congress of the I.P. Pavlov All-Union Physiological Society [in Russian], Vol. 2, No. 1 (1964b), p. 182.

Genkin, A.A. Dokl. Akad. Pedagog. Nauk RSFSR, 4:99 (1962).

Gerasimov, V.D., Kostyuk, P.G., and Maiskii, V.A. In: Protoplasmic Membranes and Their Functional Role [in Russian], Kiev (1965), p. 24.

Gerebtzoff, M.A. Arch. Internat. Physiol., 51:365 (1941).

Gernandt, B.E., and Thulin, S.-A. J. Neurophysiol., 18:113 (1955).

Gershuni, G.V. Fiziol. Zh. SSSR, 48:241 (1962).

Gershuni, G.V. Zh. Vyssh. Nervn. Deyat., 13:887 (1963).

Gershuni, G.V. (Gerschuni, G.W.). Neuropsychology. Pergamon, Oxford (1965).

Gershuni, G.V., and Tonkikh, A.V. Trudy Inst. Fiziol. im. I.P. Pavlova, 3:11 (1949).

Gershuni, G.V., Gasanov, U.G., Zaboeva, N.V., and Lebedinskii, M.M. Biofizika, 9:597 (1964).

Gerstein, G.L., and Clark, N.A. Science, 143:3612 (1964).

Glivenko, E.V., Korol'kova, T.A., and Kuznetsova, G.D. Fiziol. Zh. SSSR, 48(4):384 (1962a).

Glivenko, E.V., Korol'kova, T.A., and Kuznetsova, G.D. Fiziol. Zh. SSSR, 48(9):1026 (1962b).

Glivenko, E.V., Korol'kova, T.A., Kuznetsova, G.D., Luchkova, T.I., and Trubnikova, R.S. Fiziol. Zh. SSSR, 51(8):943 (1965).

Gloor, P., Sperti, L., and Vera, C.L.   Electroenceph. Clin. Neurophysiol., 15(3):379 (1963).

Granit, R.   Electrophysiological Investigation of Reception [Russian translation], Moscow (1957).

Granit, R., Leksell, L., and Skoglund, C.R.   Brain, 67:125 (1944).

Granit, R., Kernell, D., and Shortness, G.K.   J. Physiol., 169:743 (1963a).

Granit, R., Kernell, D., and Smith, T.S.   J. Physiol., 168:890 (1963b).

Granit, R., Kellerth, J.O., and Williams, T.D.   J. Physiol., 174:435 (1964a).

Granit, R., Kellerth, J.O., and Williams, T.D.   J. Physiol., 174:453 (1964b).

Grastyán, E.   In: The Central Nervous System and Behavior, New York (1959), p. 119.

Grastyán, E., Lissak, K., and Szabo, J.   Acta Physiol. Acad. Sci. Hung., 7:187 (1955).

Grastyán, E., Lissak, K., Madarasz, I., and Donhoffer, H.   Electroenceph. Clin. Neurophysiol., 11:409 (1959).

Gray, J., Lissman, H.W., and Pumphrey, R.J.   J. Exper. Biol., 15:408 (1938).

Green, J., and Shimamoto, T.   Arch. Neurol. Psychiatr., 70:687 (1953).

Green, J.D.   In: Structure and Function of the Cerebral Cortex, Elsevier, Amsterdam (1959), p. 266.

Green, J.D., and Arduini, A.   J. Neurophysiol., 17:533 (1954).

Green, J.D., and Machne, X.   Am. J. Physiol., 181:219 (1955).

Green, J.D., and Adey, W.R.   Electroenceph. Clin. Neurophysiol., 8:245 (1956).

Green, J.D., Maxwell, D.S., Schindler, W., and Stumpf, C.   J. Neurophysiol., 23:403 (1960).

Green, J.D., Maxwell, D.S., and Petsche, H.   Electroenceph. Clin. Neurophysiol., 13:854 (1961).

Greenbaum, L.J., and Merlis, J.K.   Electroenceph. Clin. Neurophysiol., 18:109 (1965).

Griffith, J.S., and Horn, G.   Nature, 199:4896 (1963).

Grigor'ev, Yu.G.   Radiation Injuries and Compensation of Disturbed Functions [in Russian], Moscow (1963).

Grindel', O.M.   Zh. Vyssh. Nervn. Deyat., 13:577 (1963).

Grindel', O.M.   Fiziol. Zh. SSSR, 50:3 (1964).

Grindel', O.M.   In: Mathematical Analysis of Electrical Phenomena of the Brain [in Russian], Moscow (1965), p. 15.

Grindel', O.M.   Zh. Vyssh. Nervn. Deyat., 16:458 (1966).

Grindel', O.M., Boldyreva, G.N., Burashnikov, E.N., and Andreevskii, V.M.   Zh. Vyssh. Nervn. Deyat., 14(5):745 (1964).

Grüsser, O.-J., and Grützner, A.   Graefes Arch. f. Ophthalm., 160:565 (1958).

Grüsser, O.-J., and Kapp, H.   Pflüg. Arch. Ges. Physiol., 266:111 (1958).

Grüsser, O.-J., and Rabelo, C.   Pflüg. Arch. Ges. Physiol., 265:501 (1958).

Grützner, A., Grüsser, O.-J., and Baumgartner, G.   Arch. Psychiatr. Z. Ges. Neurol., 197:377 (1958).

Gurgenidze, R.V.   Proceedings of the 2nd Scientific Conference of Ophthalmologists [in Russian], Ordzhonikidze (1958), p. 44.

Gustson, P.P.   Dokl. Akad. Nauk SSSR, 150:945 (1963).

Hagiwara, S., and Bullock, T.H.   J. Cell Comp. Physiol., 50:25 (1957).

Hagiwara, S., and Morita, H.   J. Neurophysiol., 25:721 (1962).

Hagiwara, S., Watanabe, A., and Saito, N.   J. Neurophysiol., 22:554 (1959).

van Harreveld, A. Proc. Soc. Exper. Biol. (N.Y.), 34:428 (1936).

van Harreveld, A., Crowell, J., and Malhotra, S.K. J. Cell Biol., 25:117 (1965).

Harris, T.R., and Weaver, R. In: E. Herz (ed). San Diego Symposium for Biomedical Engineering (1965), p. 61.

Hartline, H.K., Wagner, H.G., and Ratliff, F. J. Gen. Physiol., 39:651 (1956).

Hassler, R., and Hess, W.R. Arch. Psychiatr., 192:488 (1954).

Hebb, D. The Organization of Behavior, J. Wiley, New York (1949).

Heinbecker, P., and Bartley, H. J. Neurophysiol., 3:219 (1940).

Hild, W., and Tasaki, I. J. Neurophysiol., 25:277 (1962).

Hodgkin, A.L. Proc. Roy. Soc., B, 148:1 (1958).

Hodgkin, A.L. The Conduction of the Nervous Impulse, Thomas, Springfield (1964), p. 108.

Hodgkin, A.L., and Huxley, A.F. J. Physiol., 117:500 (1952).

Hoffman, B.F., and Cranefield, P.F. Electrophysiology of the Heart [Russian translation], Moscow (1962).

Hollwich, F. Ann. N.Y. Acad. Sci., 117:105 (1964).

Holmgren, B., and Frenk, S. Nature, 192:1294 (1961).

Holmqvist, B., and Lundberg, A. Arch. Ital. Biol., 97:340 (1959).

Holmqvist, B., Lundberg, A., and Oscarson, O. Arch. Ital. Biol., 98:60 (1960).

Hori, Y., and Yoshii, N. Psychol. Rep., 16:241 (1965).

Horridge, G.A. In: Nervous Inhibition (1961), p. 395.

Howland, B., Lettvin, I.Y., McCulloch, W.S., Pitts, W., and Wall, P.D. J. Neurophysiol., 18:1 (1955).

Hubel, D.H., and Wiesel, T.N. J. Physiol., 160:106 (1962).

Hugelin, A., and Bonvallet, M. J. Physiol. (Paris), 49:212 (1957).

Hugelin, A., Bonvallet, M., and Dell, P. Rev. Neurol., 89:419 (1953).

Hunter, J., and Ingvar, D.H. Electroenceph. Clin. Neurophysiol., 7:39 (1955).

Hunter, J., and Jasper, H.H. Electroenceph. Clin. Neurophysiol., 1:113 (1949).

Iwama, K. Tohoku J. Exper. Med., 52:53 (1950).

Iwama, K., and Yamamoto, C. Jap. J. Physiol., 11:169 (1961).

Iwase, J., Uchida, T., and Ochi, J. Jap. J. Physiol., 11:13 (1961).

Izosimov, G.V. Radiobiologiya, 1(4):535 (1961).

Jarcho, L.W. J. Neurophysiol., 12:447 (1949).

Jasper, H.H. In: Brain Mechanisms and Consciousness, Blackwell, Oxford (1954), p. 374.

Jasper, H.H. In: W. Penfield and H. Jasper. Epilepsy and the Functional Anatomy of the Human Brain [Russian translation], Moscow (1958), p. 194.

Jasper, H.H. In: Inhibition in the Nervous System and Gamma-Aminobutyric Acid, Pergamon, Oxford (1960), p. 544.

Jasper, H.H., and Stefanis, C. Electroenceph. Clin. Neurophysiol., 18:541 (1965).

Jasper, H.H., Ajmone-Marsan, C., and Stoll, J. Arch. Neurol. Psychiatr. (Chicago), 67:155 (1952).

Jasper, H.H., Ricci, G., and Doane, B. Electroenceph. Clin. Neurophysiol., (Suppl. 13): 137 (1960).

Jasper, H.H., Ricci, G., and Doane, B. In: Electroencephalographic Investigations of Higher Nervous Activity [in Russian], Acad. Sci. USSR Press (1962), p. 194.

Jendralski, H.J. Klin. Monatsblätter f. Augenheilk., 118:318 (1951).

John, E.R. Ann. Rev. Physiol., 23:451 (1961).

John, E.R. In: W.S. Fields and W. Abbott. Information Storage and Neural Control, Springfield, Ill. (1963).

John, E.R., and Killam, K.F. J. Pharmacol. Exper. Ther., 125:252 (1959).

John, E.R., and Killam, K.F., J. Nerv. Ment. Dis., 131:183 (1960).

John, E.R., Ruchkin, D.S., and Villegas, J. Science, 141:429 (1963).

John, E.R., Ruchkin, D.S., Leiman, A., Sachs, E., and Ahn, H. Proc. 23rd Internat. Cong. Physiol. Sci. Internat. Cong. Ser. No. 87, Excerpta Medica, Amsterdam (1965).

Josephson, R.K. Comp. Biochem. Physiol., 5:45 (1962).

Jouvet, M. In: Brain Mechanisms and Learning, Oxford (1961), p. 445.

Jouvet, M., and Michel, F. C. R. Soc. Biol. (Paris), 152:1167 (1958).

Jung, R. Electroenceph. Clin. Neurophysiol., (Suppl. 4):57 (1953).

Jung, R., and Baumgartner, G. Pflüg. Arch. Ges. Physiol., 261:434 (1955).

Jung, R., Kornhuber, H., and Fonseca, J.S. In: Progress in Brain Research, Vol. 1 (1963), p. 231.

Kaada, B.K., and Jasper, H.H. Arch. Neurol. Psychiatr., 68:609 (1952).

Kadzhaya, D.V., and Narikashvili, S.P. Soobshch. Akad. Nauk Gruz. SSR, 28:745 (1962).

Kadzhaya, D.V., and Narikashvili, S.P. Soobshch. Akad. Nauk Gruz. SSR, 37:709 (1965).

Kamikawa, K., McIlwain, J.T., and Adey, W.R. Electroenceph. Clin. Neurophysiol., 17:485 (1964).

Kamp, A. Electroenceph. Clin. Neurophysiol., 15:164 (1963).

Kamp, A., Storm van Leeuwen, W., and Tielen, A.M. Electroenceph. Clin. Neurophysiol., 19:91 (1965).

Karapetyan, S.K. Role of Light in Physiological Stimulation of the Animal Organism [in Russian], Erevan (1961).

Karpenko, L.D. Fiziol. Zh. SSSR, 51:1192 (1965).

Katchalsky, A. In: Symposium "Connective Tissue: Intercellular Macromolecules." N.Y. Heart Assoc., Little Brown, Boston (1964), p. 9.

Kats, K. Nonspecific Response in Human EEG under Normal Conditions and in Brain Lesions. Dissertation, Moscow (1958).

Katz, B., and Miledi, R. J. Physiol., 168:389 (1963).

Keder-Stepanova, I.A., Ponomarev, V.A., and Chetaev, A.N. Biofizika, 11(1):123 (1966).

Kennedy, D. Am. J. Ophthalm., Ser. 3, 46:19 (1958).

Kennedy, D. J. Gen. Physiol., 46:551 (1963).

Kennedy, D., and De Mellon, F. Comp. Biochem. Physiol., 13:275 (1964).

Kerkut, G.A., and Taylor, B.J.R. Nature, 178:426 (1956).

Khananashvili, M.M. Abstracts of Proceedings of the 10th All-Union Congress of the I.P. Pavlov Physiological Society [in Russian], Vol. 2, No. 2 (1964), p. 371.

Kholodov, Yu.A. Effect of Electromagnetic Fields and Magnetic Fields on the Central Nervous System [in Russian], Moscow (1966).

Kholodov, Yu.A., and Zenina, I.N. In: Biological Action of Electromagnetic Fields
of Radiofrequencies [in Russian], Moscow (1964), p. 33.

Killam, K.F., and Hance, A.J. Abstracts 23rd Internat. Cong. Physiol. Sci., Tokyo
(1965), p. 469.

King, E.E., Naquet, R., and Magoun, H.W. J. Pharmacol. Exper. Ther., 119:48 (1957).

Klass, D.W., and Bickford, R.G. Electroenceph. Clin. Neurophysiol., 9:570 (1957).

Klee, M.R. In: D.P. Purpura and M.D. Yahr (eds). The Thalamus, Columbia University Press (1966), p. 287.

Klepach, G.S. Proceedings of the 15th Conference of Physiologists, Biochemists, and
Pharmacologists of the South of the RSFSR [in Russian], Makhachkala (1965),
p. 148.

Klosovskii, B.N., and Kosmarskaya, E.N. Active and Inhibitory States of the Brain
[in Russian], Moscow (1961).

Klüver, H., and Bucy, P.C. Arch. Neurol. Psychiatr., 42:979 (1939).

Klüver, H., and Barrera, E. J. Neuropath. Exper. Neurol., 12:400 (1953).

Knipst, I.N. Trudy Inst. Vyssh. Nervn. Deyat. Akad. Nauk SSSR. Ser Fiziol., 1:159
(1955).

Knipst, I.N. Trudy Inst. Vyssh. Nervn. Deyat. Akad. Nauk SSSR. Ser. Fiziol., 3:59 (1958).

Knipst, I.N. Trudy Inst. Vyssh. Nervn. Deyat. i Neirofiziol. Akad. Nauk SSSR, Ser.
Fiziol., 5:3 (1960).

Knipst, I.N. Trudy Inst. Vyssh. Nervn. Deyat. Akad. Nauk SSSR. Ser. Fiziol., 6:267
(1961).

Knipst, I.N., and Shul'gina, G.I. In: Symposium on Central—Peripheral Mechanisms
of Motor Responses [in Russian], Moscow (1966), p. 317.

Knipst, I.N., Korol'kova, T.A., and Livanov, M.N. Proceedings of the 9th Congress of
the All-Union Society of Physiologists, Biochemists, and Pharmacologists [in
Russian], Vol. 1 (1959), p. 238.

Kogan, A.B. Fiziol. Zh. SSSR, 44(9):810 (1958).

Kogan, A.B. In: Problems in Neurocybernetics. Abstracts of Proceedings of an Inter-
Institute Conference [in Russian], Rostov-on-Don (1962), p. 16.

Kogan, A.B. In: Reflexes of the Brain. International Conference to Commemorate
the Centenary of Publication of I.M. Sechenov's Book of This Name [in Russian],
Moscow (1965), p. 103.

Kogan, A.B. Fiziol. Zh. SSSR, 52:195 (1966).

Kogan, A.B., and Chorayan, O.G. Zh. Vyssh. Nervn. Deyat., 15(2):229 (1965).

Kogan, A.B., and Nikolaeva, N.I. In: Conference on Electrophysiology of the Nervous
System [in Russian], Moscow (1958), p. 65.

Kolmogorov, A.I. In: The Possible and Impossible in Cybernetics [in Russian], Moscow
(1963), p. 10.

Kolodnaya, A.Ya. Byull. Éksper. Biol. i Med., 17(3):31 (1944).

Kondrat'eva, I.N. Radiobiologiya, 2(4):569 (1962).

Kondrat'eva, I.N. Zh. Vyssh. Nervn. Deyat., 14:1069 (1964).

Kondrat'eva, I.N. In: Problems in Neurocybernetics. Abstracts of Proceedings of the
2nd Inter-Institute Conference on Neurocybernetics [in Russian], Rostov-on-Don
(1965), p. 52.

Kondrat'eva, I.N. In: Investigation of Sensory Processes at the Neuronal and Behavioral Level. Symposium, 18th Internat. Psychol. Cong., Vol. 15, Moscow (1966), p. 60.

Kondrat'eva, I.N., and Bakaeva, S.A. In: Electrophysiology of the Central Nervous System. Proceedings of the 5th All-Union Conference [in Russian], Tbilisi (1966), p. 156.

Kondrat'eva, I.N., and Volodin, B.I. Zh. Vyssh. Nervn. Deyat., 16(5):874 (1966).

Kononova, E.P. Anatomy and Physiology of the Occipital Lobes on the Basis of Clinical, Pathologoanatomical, and Experimental Data [in Russian], Moscow (1926).

Konorski, J., and Miller, S. In: Transactions of Academician I.P. Pavlov's Physiological Laboratory [in Russian], Vol. 6 (1936), p. 119.

Korol'kova, T.A., Luchkova, T.I., Trubinkova, R.S., and Ts'ai T'i-tao. Zh. Vyssh. Nervn. Deyat., 16(3):489 (1966).

Koshtoyants, Kh.S. Fundamentals of Comparative Physiology [in Russian], Vol. 2, Moscow (1957).

Kostyuk, P.G. In: Current Problems in Physiology and Pathology of the Nervous System [in Russian], Moscow (1965), p. 28.

Kozodoi, N.S., and Farber, D.A. Proceedings of the 7th Scientific Conference on Problems in Age Morphology, Physiology, and Biochemistry [in Russian], Acad. Pedagog. Sci. Press, Moscow (1965), p. 344.

Krnjević, K., Randić, M., and Straughan, D.W. Nature, 201:1294 (1964).

Kuffler, S.W., and Potter, D.D. J. Neurophysiol., 27:290 (1964).

Kuiper, J.W., and Leutscher-Hazelhoff, J.T. Nature, 207:1158 (1965).

Kukuev, L.A. Zh. Vyssh. Nervn. Deyat., 3:765 (1953).

Kulagin, V.K. Fiziol. Zh. SSSR, 16(6):754 (1955).

Kullanda, K.M. Dokl. Akad. Nauk SSSR, 124:1367 (1959).

Kumazawa, T. Electroenceph. Clin. Neurophysiol., 15:660 (1963).

Kupalov, P.S. Transactions of Academician I.P. Pavlov's Physiological Laboratory [in Russian], Vol. 5 (1933), p. 383.

Kupalov, P.S. Zh. Vyssh. Nervn. Deyat., 2:457 (1952).

Kupalov, P.S. In: Electroencephalographic Investigation of Higher Nervous Activity [in Russian] (1962), p. 9.

Landgren, S., Phillips, C.G., and Porter, R. J. Physiol., 161:91 (1962).

Lat, J. In: Central and Peripheral Mechanisms of the Motor Activity of Animals [in Russian] (1960), p. 70.

Lebedev, A.N. Participation of Specific and Nonspecific Pathways in the Early Bioelectrical Response of the Cerebral Cortex to Whole-Body Exposure to Ionizing Radiation. Candidate's dissertation, Moscow (1963).

Lelord, G. Electroenceph. Clin. Neurophysiol., 9:561 (1957).

Li, C.L. J. Physiol., 144:40 (1956).

Li, C.L. Science, 129:783 (1959).

Li, C.L., McLennan, H., and Jasper, H. Science, 116:656 (1952).

Li, C.L., Cullen, C., and Jasper, H. J. Neurophysiol., 19:111 (1956).

Li, C.L., Armando, O.-G., Shelley, N.C., and Saxton, Y.H. J. Neurophysiol., 23:592 (1960).

Liberson, W.T., Smith, R.W., and Stern, A. J. Neurophysiol., 3:28 (1961).

Libet, B., and Gerard, R. Proc. Soc. Exper. Biol. (N.Y.) 38:886 (1938).

Licklider, J.C.R. In: R.D. Luce (ed). Developments in Mathematical Physiology, Free Press of Glencoe, Chicago (1960), p. 169.

Lievens, P. Acta Neurol. Psychiatr. Belg., 7:638 (1960).

Lilly, J.C. "A 25-channel potential field recorder." 2nd Annual Joint CRE—AIEE Conference on Electronic Instrumentation in Nucelonics and Medicine, New York (1949).

Livanov, M.N. Zh. Sovrem. Nevropat., Psikhiatr. i Psikhogig., 3(11-12):98 (1934).

Livanov, M.N. Transactions of the Brain Institute [in Russian], Vol. 3, Moscow (1938).

Livanov, M.N. Fiziol. Zh. SSSR, 28:172 (1940).

Livanov, M.N. Zh. Obshch. Biol., 5:9 (1944a).

Livanov, M.N. Probl. Fiziol. Optiki, 2:106 (1944b).

Livanov, M.N. In: 50 Years of I.P. Pavlov's Theory of the Conditioned Reflex [in Russian], (1952), p. 248.

Livanov, M.N. Some Problems in the Action of Ionizing Radiation on the Nervous System [in Russian], Moscow (1962a).

Livanov, M.N. In: Biological Aspects of Cybernetics [in Russian], Moscow (1962b), p. 112.

Livanov, M.N. Zh. Vyssh. Nervn. Deyat., 12(3):399 (1962c).

Livanov, M.N. In: Electroencephalographic Investigation of Higher Nervous Activity [in Russian], Moscow (1962d), p. 174.

Livanov, M.N. In: Abstracts of Proceedings of an International Conference to commemorate the Centenary of Publication of I.M. Sechenov's Book "Reflexes of the Brain," Moscow (1963), p. 14.

Livanov, M.N. Proc. 23rd Internat. Cong. Internat. Union Physiol. Sci., Vol. 4 (1965), p. 600.

Livanov, M.N. Med. Electron. Biol. Eng., 3:137 (1965a).

Livanov, M.N. In: Problems in Contemporary Neurophysiology [in Russian], Nauka, Moscow (1965b), p. 36.

Livanov, M.N. In: Reflexes of the Brain. International Conference to Commemorate the Centenary of Publication of I.M. Sechenov's Book of This Name [in Russian], Moscow (1965c), p. 64.

Livanov, M.N. Proceedings of the 22nd Internat. Cong. Physiol. Sci., Vol. 1/2. Internat. Cong. Ser. No. 47, Excerpta Medica, Amsterdam (1966), p. 899.

Livanov, M.N., and Anan'ev, V.M. Electroencephaloscopy [in Russian], Moscow (1959).

Livanov, M.N., and Frenkel', G.M. Abstracts of Proceedings of the 14th Conference on Problems in Higher Nervous Activity [in Russian], Moscow (1951), p. 30.

Livanov, M.N., and Polyakov, K.L. Izvest. Akad. Nauk SSSR, Ser. Biol., No. 3, 286 (1945).

Livanov, M.N., and Rabinovskaya, A.M. Fiziol. Zh. SSSR, 33:523 (1947).

Livanov, M.N., and Korol'kova, T.A. In: 13th Conference on Physiological Problems, in Memory of I.P. Pavlov [in Russian], Leningrad (1948), p. 63.

Livanov, M.N., Korol'kova, T.A., and Frenkel', G.M. Zh. Vyssh. Nervn. Deyat., 1:521 (1951).

Livanov, M.N., Tsypin, A.B., Grigor'ev, Yu.G., Khrushchev, V.G., Stepanov, S.M., and Anan'ev, V. M. Byull. Éksperim. Biol. i Med., 49(5):63 (1960).

Livanov, M.N., Gavrilova, N.A., and Aslanov, A.S. Zh. Vyssh. Nervn. Deyat., 14(2):185 (1964).

Livanov, M.N., Gavrilova, N.A., Efremova, T.M., Korol'kova, T.A., and Aslanov, A.S. Med. Electron. Biol. Eng., 3:137 (1965).

Llinás, R., and Terzuolo, C. J. Neurophysiol., 27:579 (1964).

Llinás, R., and Terzuolo, C. J. Neurophysiol., 28:413 (1965).

Lloyd, D.P.C. Proc. Nat. Acad. Sci. USA, 45:586 (1959).

Lobanova, L.V. Dokl. Akad. Nauk SSSR, 5:1073 (1954).

Loève, M. Probability Theory, Princeton (1963).

Lømo, T., and Mollica, A. Electroenceph. Clin. Neurophysiol., 12:936 (1960).

Loomis, A.L., Harvey, E.N., and Hobart, G. Science, 83:239 (1936).

van der Loos, H. Abst. Proc. 22nd Internat. Cong. Physiol. Sci., Leiden (1962), p. 110.

Lorente de Nó, R. A Study of Nerve Physiology. Studies from the Rockefeller Institute for Medical Research, Vol. 131-132, New York (1947).

Luchkova, T.I. In: Structure and Function of the Nervous System. Proc. 2nd Conference of Junior Research Workers, Institute of the Brain and Institute of Higher Nervous Activity [in Russian], Moscow (1965), p. 68.

Luk'yanova, S.N. In: Problems in Hematology, Radiobiology, and Biological Action of Magnetic Fields [in Russian], Tomsk (1965), p. 368.

Lur'e, R.N., and Trofimov, L.G. Fiziol. Zh. SSSR, 42:348 (1956).

Lux, H.D., and Klee, M.R. Arch. Psychiatr. Nervenkr., 203:648 (1962).

Lux, H.D., and Nacimiento, A.C. Pflüg. Arch. Ges. Physiol., 278:66 (1963).

Lyubimov, N.N., and Trofimov, L.G. Zh. Vyssh. Nervn. Deyat., 8:617 (1958).

MacLean, P.D. Arch. Neurol. Psychiatr., 73:130 (1955).

MacLean, P.D. In: The Central Nervous System and Behavior, New York (1959), p. 31.

MacNichol, E.F. In: Information Processing in Sight Sensory Systems (1965). In press.

Maekawa, K., and Purpura, D.P. (1966). In preparation.

Magoun, H.W. Physiol. Rev., 30:459 (1950).

Magoun, H.W. Res. Publ. Ass. Nerv. Ment. Dis., 30:480 (1952).

Magoun, H.W. The Waking Brain [Russian translation], Moscow (1965).

Magoun, H.W., and Rhines, R. J. Neurophysiol., 9:165 (1946).

Majkowski, J. Acta Physiol. Pol., 9:565 (1958).

Mangold, E. Hypnose und Katalepsie bei Tieren zur menschlichen Hypnose, Fischer, Jena (1914), p. 82.

Markelov, G.I. Uchenye Zapiski Odessk. Psikho-Nevrol. Inst., 1:5 (1949).

Marshall, W.H. J. Neurophysiol., 4:25 (1941).

Marshall, W.H., Woolsey, C.N., and Bard, P. J. Neurophysiol., 4:1 (1941).

Martin, A., and Branch, C. J. Neurophysiol., 21:368 (1958).

Maruseva, A.M. Abstracts of Proceedings of the 18th International Psychological Congress [in Russian], Moscow (1966), p. 387.

McCulloch, W.S. In: Mechanization of Thought Processes. National Physics Laboratory Symposium (1959), p. 611.

McCulloch, W.S., and Pitts, W. Bull. Math. Biophys., 5:115 (1943).

McCulloch, W.S., Graf, C., and Magoun, H.W. J. Neurophysiol., 9:123 (1946).

Melekhova, A.M., and Shul'gina, G.I. In: Electrophysiology of the Nervous System (Proc. 4th All-Union Electrophysiol. Conf.) [in Russian], Rostov-on-Don (1963), p. 256.

Melik-Pashayan, M.A. One Electroencephalographic Correlate of Consciousness. Author's abstract of Candidate's dissertation, Erevan (1965).

Mering, T.A. Zh. Vyssh. Nervn. Deyat., 2:894 (1952).

Merkulova, O.S. Problems in the Physiology of Interoception [in Russian], Vol. 1 (1952), p. 323.

Merlis, J.K. Electroenceph. Clin. Neurophysiol., 18:118 (1965).

Meshcherskii, R.M. In: Reflexes of the Brain. International Conference to Commemorate the Centenary of Publication of I.M. Sechenov's Book of This Name [in Russian], Moscow (1963), p. 275.

Meshcherskii, R.M., and Gustson, P.P. Physiol. Bohemoslov., 13:236 (1964).

Meshcherskii, R.M., and Okudzhava, V.M. Dokl. Akad. Nauk SSSR, 153:974 (1963).

Mettler, F.A. J. Comp. Neurol., 61:221 and 509; 62:263; 63:25 (1935).

Mill, P.J. Comp. Biochem. Physiol., 8:83 (1963).

Minsky, M. Discussion in: Information Processing in the Nervous System, Leiden Symposium, 1962, Excerpta Medica, Amsterdam (1964), p. 303.

Moiseeva, N.I., and Orlov, V.A. In: Role of Deep Structures of the Human Brain in Mechanisms of Pathological Reactions [in Russian], Leningrad (1965).

Monakhov, K.K. Byull. Éksperim. Biol. i Med., 50(10):23 (1960).

Monakhov, K.K. Zh. Nevropat. i Psikhiatr., 63(12):1835 (1963).

Morin, G., Gastaut, H., Naquet, R., and Roger, A. J. Physiol. (Paris), 43:820 (1951).

Morison, R.S., and Dempsey, E.W. Am. J. Physiol., 135:281 (1942).

Morison, R.S., and Dempsey, E.W. Am. J. Physiol., 138:297 (1943).

Morrell, F. Electroenceph. Clin. Neurophysiol., (Suppl. 13):65 (1960).

Morrell, F., and Jasper, H. Electroenceph. Clin. Neurophysiol., 8:201 (1956).

Moruzzi, G., and Magoun, H.W. Electroenceph. Clin. Neurophysiol., 1:455 (1949).

Mountcastle, V.B. J. Neurophysiol., 20:408 (1957).

Mountcastle, V.B., Davies, P.W., and Berman, A.L. J. Neurophysiol., 20:374 (1957).

Mundy-Castle, A. Electroenceph. Clin. Neurophysiol., 9:221 (1957).

Murata, K., Lehmann, D., and Bac-y-Rita, P. Electroenceph. Clin. Neurophysiol., 20:100 (1966).

Murphy, J.P., and Gellhorn, E. J. Neurophysiol., 8:341 (1945).

Myers, R.E. J. Comp. Neurol., 118:1 (1962).

Myers, R.E. In: E.G. Ettlinger, A.V. Reuck, and R. Porter (eds). Functions of the Corpus Callosum. Ciba Foundn. Study Group No. 20, Vol. 1, Little, Brown and Co., Boston (1965), p. 133.

Myslobodskii, M.S. Zh. Vyssh. Nervn. Deyat., 16(2):303 (1966).

Nacimiento, A.C., Lux, H.D., and Creutzfeldt, O.D. Pflüg. Arch. Ges. Physiol., 281:152 (1964).

Narikashvili, S.P. Uspekhi Sovr. Biol., 52:257 (1961).

Narikashvili, S.P. Thalamo–Cortical Interrelationships. Textbook of Physiology [in Russian], Moscow (1966).

Narikashvili, S.P., and Kadzhaya, D.V. Fiziol. Zh. SSSR, 49:281 (1963).

Narikashvili, S.P., and Kadzhaya, D.V. In: Reflexes of the Brain. International Conference to Commemorate the Centenary of Publication of I.M. Sechenov's Book of This Name [in Russian], Moscow (1965), p. 227.

Narikashvili, S.P., Butkhuzi, S.M., and Moniava, É.S. Fiziol. Zh. SSSR, 46:653 (1960a).

Narikashvili, S.P., Moniava, É.S., and Kadzhaya, D.V. Dokl. Akad. Nauk SSSR, 134:229 (1960b).

Narikashvili, S.P., Moniava, É.S., and Butkhuzi, S.M. Fiziol. Zh. SSSR, 47:863 (1961).

Narikashvili, S.P., Moniava, É.S., and Kadzhaya, D.V. In: Electrophysiology of the Nervous System. Proceedings of the 4th All-Union Electrophysiological Conference [in Russian], Rostov-on-Don (1963), p. 278.

Naumova, T.S. Zh. Vyssh. Nervn. Deyat., 12:118 (1962).

Naumova, T.S. In: Electrophysiology of the Nervous System. Proceedings of the 4th All-Union Electrophysiological Conference [in Russian], Rostov-on-Don (1963), p. 276.

Naumova, T.S. Zh. Vyssh. Nervn. Deyat., 15:11 (1965a).

Naumova, T.S. On Interaction between Specific and Nonspecific Afferents during Formation of a Conditioned Defensive Reflex. Doctoral dissertation, Moscow (1965b).

Nebylitsyn, V.D. In: Typological Peculiarities of Human Higher Nervous Activity [in Russian], Moscow (1963), p. 47.

Neff, W.D. In: Sensory Communication, MIT Press (1961), p. 259.

Nelson, P.G., and Frank, K. J. Neurophysiol., 27:928 (1964).

Neubert, P.G. Naturwissenschaften, 47:526 (1960).

Nicholls, J.G., and Kuffler, S.W. J. Neurophysiol., 27:645 (1964).

Nicholson, P.W. Exper. Neurol., 13:386 (1966).

Niemer, W.T., and Jimenez-Castellanos, J. J. Comp. Neurol., 93:101 (1950).

Novikova, L.A. Electrical Activity of the Brain in Disturbances of Distance Receptors. Doctoral dissertation, Moscow (1965).

Novokhatskii, A.S. Anatomical Connections of the Optic Tract with the Hypothalamus. Dissertation, Khar'kov (1956).

Nudel', M.B. Abstracts of Proceedings of the 2nd Scientific Conference of the Society of Neuropathology and Psychiatry of the Kaunas Zone [in Russian], Kaunas (1962), p. 87.

Ochs, S. Feder. Proc., 21:642 (1962).

Ochs, S., and Booker, H. Exper. Neurol., 4:70 (1961).

Ogden, T.E. Electroenceph. Clin. Neurophysiol., 12:621 (1960).

Okudzhava, V.M. Activity of Apical Dendrites in the Cerebral Cortex [in Russian], Tbilisi (1963).

Olds, J. Proc. 23rd Internat. Cong. Physiol. Sci., Tokyo (1965), p. 372.

Olds, J., and Olds, M. In: G.F. Delafresnaye (ed). Brain Mechanisms and Learning, Blackwell, Oxford (1961), p. 153.

Palestini, M., Pisano, M., Rosadini, G., and Rossi, G.F. Electroenceph. Clin. Neurophysiol., 19:276 (1965).

Pampiglione, G. Electroenceph. Clin. Neurophysiol., 4:228 (1952).

Pangelova, T.K., and Lyudkovskaya. In: Abstracts of Proceedings of an All-Union Conference of Junior Research Workers on Molecular Biophysics [in Russian], Pushchino-on-Oka (1966), p. 88.

Passano, L.M. Proc. Nat. Acad. Sci. USA, 50:306 (1963).

Patton, H., and Amassian, V. In: Handbook Physiol., Sect. I. Neurophysiology, Washington (1960), p. 837.

Pavlov, I.P. Conditioned Reflexes. An Investigation Of the Physiological Activity of the Cerebral Cortex. Oxford University Press, London (1927), p. 430.

Pavlov, I.P. Complete Collected Works [in Russian], Vol. IV, Acad. Sci. USSR Press, Leningrad (1947), p. 56.

Pavlov, I.P. Complete Collected Works [in Russian], Vol. III, Moscow (1949).

Pearlman, A.L. Electroenceph. Clin. Neurophysiol., 15:426 (1963).

Pease, D.C. J. Ultrastruct. Res., 14:356 (1966).

Pecci-Saavedra, J., Wilson, P.D., and Doty, R.W. Nature, 210:740 (1966).

Penfield, W. In: Brain Mechanisms and Consciousness, Oxford (1954), p. 284.

Penfield, W. Res. Publ. Ass. Nerv. Ment. Dis., 36:210 (1958).

Penfield, W., and Jasper, H. Epilepsy and the Human Brain, Boston (1954).

Perkel, D.H., Schulman, J.H., Bullock, T.H., Moore, G.P., and Segundo, J.P. Science, 145:61 (1964).

Perl, E., and Casby, J. J. Neurophysiol., 17:429 (1954).

Peters, A., and Palay, S.L. Proc. Anat. Soc. Gt. Brit. Irel., 17 (1964).

Petrovich, V. K. Zh. Vyssh. Nervn. Deyat., 15(5):846 (1965).

Petsche, H., and Marko, A. Arch. Psychiatr. Nervenkr., 193:177 (1955).

Phillips, C.G. Quart. J. Exper. Physiol., 44:1 (1959).

Pigareva, Z.D., and Shilyagina, N.N. Proceedings of the 4th Scientific Conference on Problems in Age Morphology, Physiology, and Biochemistry [in Russian], Acad. Pedagog. Sci. RSFSR Press, Moscow (1955).

Pines, L.Ya., and Prigonnikov, I.E. Problems in Morphology of the Cerebral Cortex [in Russian], Vol. 51 (1936), p. 97.

Plonskaya, E.I. Zh. Vyssh. Nervn. Deyat., 9:593 (1959).

Poggio, G.F., and Mountcastle, V.B. J. Neurophysiol., 26:775 (1963).

Pollen, D.A. Electroenceph. Clin. Neurophysiol., 17:398 (1964).

Pollen, D.A., and Sie, P.G. Electroenceph. Clin. Neurophysiol., 17:154 (1964).

Pollen, D.A., Reid, K.H., and Perot, P. Electroenceph. Clin. Neurophysiol., 17:57 (1964).

Polyakov, G.I. The Problem of Origin of Reflex Mechanisms [in Russian], Moscow (1964).

Polyanskii, V.B. Evoked Potentials of the Rabbit Visual Cortex and Their Connection with Spike Activity of Neurons. Candidate dissertation, Moscow (1965a).

Polyanskii, V.B. Zh. Vyssh. Nervn. Deyat., 15:903 (1965b).

Pompeiano, O., and Swett, G.E. Arch. Ital. Biol., 100:311 (1962).

Porter, R., Adey, W.R., and Kado, R.T. Neurology, 14:1002 (1964).

Presman, A.S. Uspekhi Fizicheskikh Nauk, 86:263 (1965).

Preston, J.B., and Kennedy, D. J. Gen. Physiol., 45:821 (1962).

Preyer, W. Die Kataplexie und der thierische Hypnotismus, Fischer, Jena (1878), p. 100.

Prosser, C.L. J. Cell Comp. Physiol., 4:185 (1934a).

Prosser, C.L. J. Cell Comp. Physiol., 4:363 (1934b).

Prosser, C.L., and Buehl, C.C. J. Cell Comp. Physiol., 14:287 (1939).

Protopopov, V.P. Combined Motor Responses to Acoustic Stimulation. Dissertation, St. Petersburg (1909).

Puchinskaya, L.M. Local Changes to Light in the Human Electroencephalogram. Candidate dissertation, Moscow (1963).

Purpura, D.P. In: C.C. Pfeiffer and J.R. Smythies (eds). Internat. Rev. Neurobiol., Vol. 1, Acad. Press, New York (1959), p. 47.

Purpura, D.P. Ann. N.Y. Acad. Sci., 94:604 (1961).

Purpura, D.P. In: Mechanisms of the Intact Brain [Russian translation], Moscow (1963), p. 9.

Purpura, D.P. In: Reflexes of the Brain. International Conference to Commemorate the Centenary of Publication of I.M. Sechenov's Book of This Name [in Russian], Moscow (1965), p. 38.

Purpura, D.P. Nature, 211:1317 (1966).

Purpura, D.P., and Cohen, B. J. Neurophysiol., 25:621 (1962).

Purpura, D.P., and Grundfest, H. J. Neurophysiol., 19:573 (1956).

Purpura, D.P., and Malliani, A. Brain Research (1966). In press.

Purpura, D.P., and McMurtry, J.G. J. Neurophysiol., 28:166 (1965).

Purpura, D.P., and Shofer, R.J. J. Neurophysiol., 26:494 (1963).

Purpura, D.P., and Shofer, R.J. J. Neurophysiol., 27:117 (1964).

Purpura, D.P., Shofer, R.J., Housepian, E.M., and Noback, C.R. In: D.P. Purpura and J.P. Schade (eds). Progress in Brain Research. Growth and Maturation of the Brain, Vol. IV, Elsevier, Amsterdam (1964a), p. 176.

Purpura, D.P., Shofer, R.J., and Musgrave, F.S. J. Neurophysiol., 27:133 (1964b).

Purpura, D.P., Scarff, T., and McMurtry, J.G. J. Neurophysiol., 28:487 (1965a).

Purpura, D.P., Shofer, R.J., and Scarff, T. J. Neurophysiol., 28:925 (1965b).

Purpura, D.P., Frigyesi, T.L., McMurtry, J.G., and Scarff, T. In: D.P. Purpura and M.D. Yahr (eds). The Thalamus, Columbia University Press, New York (1966a), p. 153.

Purpura, D.P., McMurtry, J.G., and Leonard, C.F. Brain Research, 1:109 (1966b).

Purpura, D.P., McMurtry, J.G., Leonard, C.F., and Malliani, A. (1966c). In press.

Purpura, D.P., McMurtry, J.G., and Maekawa, K. Brain Research, 1:63 (1966d).

Rabin, A.G. Dokl. Akad. Nauk SSSR, 156:466 (1964a).

Rabin, A.G. Dokl. Akad. Nauk SSSR, 156:478 (1964b).

Rabin, A.G. Fiziol. Zh. SSSR, 51:159 (1965).

Rabinovich, M.Ya., and Trofimov, L.G. Byull. Éksper. Biol. i Med., 43(2):3 (1957).

Radionova, E.A. Abstracts of Proceedings of the 18th International Psychological Congress [in Russian], Moscow (1966), p. 385.

Radionova, E.A., and Popova, A.V. Fiziol. Zh. SSSR, 51:441 (1965).

Radulovacki, M., and Adey, W.R. Exper. Neurol., 12:68 (1965).

Rall, W.G., Shepherd, G.M., Reese, T.S., and Brightman, M.W. (1966). In press.

Rapoport, A. Bull. Math. Biophys., 12:109, 187, 317, and 327 (1950).

Reed, D.J., Woodbury, D.M., and Holtzer, R.L. A.M.A. Arch. Neurol., 10:604 (1964).

van Reeth, C. J. Physiol., 55:354 (1963).

Rhines, R., and Magoun, H.W. J. Neurophysiol., 9:219 (1946).

Roeder, K.D., and Roeder, S. J. Cell Comp. Physiol., 14:1 (1939).

Roger, A., Rossi, G.F., and Zirondoli, A. Electroenceph. Clin. Neurophysiol., 8:1 (1956).

Roger, A., Voronin, L.G., and Sokolov, E.N. Zh. Vyssh. Nervn. Deyat., 8(1):3 (1957).

Roitbak, A.I. Biolelectrical Phenomena in the Cerebral Cortex [in Russian], Tbilisi (1955).

Roitbak, A.I. Electroenceph. Clin. Neurophysiol., (Suppl. 13):91 (1960).

Roitbak, A.I. In: Electrophysiology of the Nervous System. Proceedings of the 4th All-Union Electrophysiological Conference [in Russian], Rostov-on-Don (1963), p. 322.

Roitbak, A.I. In: Current Problems in Electrophysiological Investigations of the Nervous System [in Russian], Moscow (1964), p. 64.

Rose, M. J. Psychol. Neurol., 43:353 (1931).

Rosenzweig, M.K., and Rosenblith, W.A. Psychol. Monogr. Gen. and Appl., 67(13):363 (1953).

Rossi, G.F., and Brodal, A. J. Anat. (Lond.) 90:42 (1956).

Rossi, G.F., and Zanchetti, A. Arch. Ital. Biol., 95:199 (1957).

Rowan, W. Biol. Rev., 13:374 (1938).

Rozental', I.E. Arkh. Biol. Nauk SSSR, 58(4):61 (1940).

Ruch, T.C., and Fulton, J.F. Medical Physiology and Biophysics, Oxford (1961).

Ruckebush, Y. Rev. Med. Veter., 115:793 (1964).

Rusinov, V.S. In: Electroencephalographic Investigation of Higher Nervous Activity [in Russian], Moscow (1962), p. 288.

Rusinov, V.S., and Rabinovich, M.Ya. Electroenceph. Clin. Neurophysiol., Suppl. 8, (1958).

Rutledge, L.T., Jr. Science, 148:1246 (1965).

Rutledge, L.T., Jr. and Kennedy, T.T. Exper. Neurol., 4:470 (1961).

Sakharov, D.A. Nauchn. Dokl. Vyssh. Shkoly. Biol. Nauki, 3:60 (1960).

Sakhiulina, G.T. In: Electroencephalographic Investigation of Higher Nervous Activity [in Russian], Moscow (1962), p. 199.

Sarkisov, S.A. Zh. Nevropat. i Psikhiatr., 8(2-3):11 (1938).

Sawyer, C.H., Everett, J.W., and Green, J.D. J. Comp. Neurol., 101:801 (1954).

Scharer, E. Schweiz. Med. Wochschr., 44:1521 (1937).

Scheibel, M., and Scheibel, A. In: D.P. Purpura and M.D. Yahr (eds). The Thalamus, Columbia University Press, New York (1966), p. 13.

Scheibel, M., Scheibel, A., Mollica, A., and Moruzzi, G. J. Neurophysiol., 18:309 (1955).

Schlag, J.D., and Balvin, R. J. Neurophysiol., 27:334 (1964).

Schlag, J.D., and Chaillet, F. Electroenceph. Clin. Neurophysiol., 15:39 (1963).

Schoolman, A., and Evarts, E.V. J. Neurophysiol., 22:112 (1959).

Schwartz, M., and Shagass, C. Electroenceph. Clin. Neurophysiol., 14:11 (1962).

Schwarz, B.E., and Bickford, R.G. J. Nerv. Ment. Dis., 124:433 (1956).

Sechenov, I.M. In: First Russian Investigations in Electroencephalography [in Russian], State Med. Lit. Press, Moscow (1949).

Sefton, A.J., and Burke, W. Nature, 205:1325 (1965).

Segundo, J.P., Naquet, R., and Buser, P. J. Neurophysiol., 18:236 (1955).

Serkov, F.N., Makul'kin, R.F., and Russev, V.V. Fiziol. Zh. SSSR, 46:408 (1960).

Serkov, F.N., and Makul'kin, R.F. Fiziol. Zh. SSSR, 52:6 (1966).

Shevchenko, D.G. Fiziol. Zh. SSSR, 52:329 (1966).

Shkol'nik-Yarros, E.G. Zh. Vyssh. Nervn. Deyat., 8:123 (1958).

Shmel'kin, D.G. Transactions of the 15th Session of the Ukrainian Psychoneurological Institute [in Russian], Vol. 24 (1949), p. 53.

Sholl, D.A. The Organization of the Cerebral Cortex, London (1956).

Shul'gina, G.I. Transactions of the Institute of Higher Nervous Activity, Acad. Sci. USSR, Vol. 5 (1960), p. 231.

Shul'gina, G.I. Investigation of Local High-Amplitude Waves of Potential in the Cerebral Cortex of Rabbits during Conditioned-Reflex Activity. Candidate's dissertation, Moscow (1962).

Shumilina, A.I. Fiziol. Zh. SSSR, 45:1176 (1959).

Shuranova, Zh.P. Investigation of Slow Electrical Potentials in the Rabbit Brain. Candidate's dissertation, Moscow (1964).

Shvets, T.B. Investigation of the Level of the Steady Potential of the Rabbit Cerebral Cortex during Creation of a Dynamic Focus and Formation of a Conditioned Defensive Reflex. Candidate's dissertation, Moscow (1963).

Silva, E.E., Estable, C., and Segundo, J.P. Arch. Ital. Biol., 97:167 (1959).

Simonoff, L.N. Arch. Anat. Physiol. (Lpz.), 33:545 (1866).

Simonov, P.V. Three Phases in Responses of the Organism to an Increasing Stimulus [in Russian], Moscow (1962).

Skrebitskii, V.B., and Voronin, L.L. Dokl. Akad. Nauk SSSR, 160:972 (1965).

Skrebitskii, V.B., and Voronin, L.L. Zh. Vyssh. Nervn. Deyat., 16:864 (1966).

Smythe, W.R. Static and Dynamic Electricity [Russian translation], For. Lit. Press, Moscow (1954). [Original: 2nd ed., New York (1950)].

Sokolov, E.N. Perception and the Conditioned Reflex [in Russian], Moscow (1958).

Sokolov, E.N. In: Problems in Electrophysiology and Encephalography [in Russian], Moscow (1960), p. 80.

Sokolov, E.N. Perception and the Conditioned Reflex, Pergamon, Oxford (1963).

Sokolov, E.N. Zh. Vyssh. Nervn. Deyat., 15:249 (1965).

Sokolov, E.N., Arakelov, G.G., and Levinson, L.B. Tsitologiya, 8:567 (1966a).

Sokolov, E.N., Arakelov, G.G., and Levinson, L.B. Évolyuts. Biokhim. i Fiziol. (1966b). In press.

Sokolov, E.N., Arakelov, G.G., and Levinson, L.B. Évolyuts. Biokhim. i Fiziol. (1966c). In press.

Sokolova, A.A. Zh. Vyssh. Nervn. Deyat., 8:593 (1958).

Sologub, E.B. Fiziol. Zh. SSSR, 50:681 (1964).

Spencer, W.A., and Kandel, E.R. J. Neurophysiol., 24:272 (1961a).

Spencer, W.A., and Kandel, E.R. Exper. Neurol., 4:149 (1961b).

Sprague, J., and Chambers, W.W. Am. J. Physiol., 176:52 (1954).

Stefanis, C., and Jasper, H. J. Neurophysiol., 27:828 (1964a).

Stefanis, C., and Jasper, H. J. Neurophysiol., 27:855 (1964b).

Stepanov, I.A. Abstracts of Proceedings of a Conference of Junior Researchers of Leningrad Research Institute of Medical Boarding and Resettlement of the Disabled [in Russian], Leningrad (1957), p. 15.

Sukhov, A.G. In: Problems in Neurocybernetics. Abstracts of Proceedings of the 2nd Inter-Institute Conference on Neurocybernetics [in Russian], Rostov-on-Don (1965), p. 92.

Suzuki, H., and Tukahara, Y. Jap. J. Physiol., 13:386 (1963).

Svetozarov, E., and Shtraikh, G. Uspekhi Sovrem. Biol., 12:25 (1940).

Sviderskii, V.L. Dokl. Akad. Nauk SSSR, 165:1204 (1965).

Svorad, D. Paroxysmalny utlm. (Paroxysmal inhibition), Slovak Acad. Sci., Bratislava (1956), p. 231.

Svorad, D. Arch. Neurol. Psychiatr. (Chicago), 77:533 (1957).

Szabo, I. Acta Morphol. Acad. Sci. Hung., 13:251 (1965).

Takagi, K., Nakayama, T., and Nagasaka, T. Electroenceph. Clin. Neurophysiol., (Suppl. 18):3 (1959).

Tasaki, I., Polley, E., and Orrego, F. J. Neurophysiol., 17:454 (1954).

Tauc, L. J. Physiol. (Paris), 47:769 (1955).

Tauc, L. C. R. Acad. Sci. (Paris), 248:1857 (1959).

Thompson, L., and Obrist, W. Electroenceph. Clin. Neurophysiol., 15:159 (1963).

Thorpe, W.H. In: J.A. Moore (ed). Ideas in Modern Biology. Proc. 16th Internat. Cong. Zool., New York (1965), p. 485.

Tkachenko, N.N. In: Electrophysiology of the Nervous System. Proceedings of the 4th All-Union Electrophysiological Conference [in Russian], Rostov-on-Don (1963), p. 382.

Tolkunov, B.F. Fiziol. Zh. SSSR, 51:285 (1965).

Tron, E.D., and Pressman, Ya.M. Problemy Fiziol. Optiki, 11:201 (1955).

Tsvilineva, V.A. Material on Evolution of the Thoracic Nerve Chain of Arthropods. Doctoral dissertation, Dushanbe (1964).

Tunturi, A.R. Am. J. Physiol., 196:1175 (1959).

Tunturi, A.R. Am. J. Physiol., 204:51 (1963).

Ukhtomskii, A.A. Collected Works [in Russian], Vol. 4, Leningrad Univ. Press, Leningrad (1945).

Urbakh, V.Yu. Mathematical Statistics for Biologists and Medical Scientists [in Russian], Moscow (1963).

Uttley, A.M. Proc. Roy. Soc. Ser. B., 144:229 (1955).

Vardapetyan, G.A. Abstracts of Proceedings of the 18th International Psychological Congress [in Russian], Moscow (1966), p. 390.

Vasilevskii, N.N. Zh. Vyssh. Nervn. Deyat., 15:529 (1965).

Vastola, E.F. J. Neurophysiol., 22:258 (1959).

Veprintsev, B.N., Krasts, I.V., and Sakharov, D.A. Biofizika, 9:327 (1964).

Verworn, M. Pflüg. Arch. Ges. Physiol., 65:63 (1897).

Verzeano, M. Science, 124:366 (1956).

Verzeano, M. Electroenceph. Clin. Neurophysiol., 14:419 (1962).

Verzeano, M., and Calma, I. J. Neurophysiol., 17:417 (1954).

Verzeano, M., and Negishi, K. J. Gen. Physiol., 43:177 (1960).

Verzeano, M., Laufer, M., Spear, P., and McDonald, S. Actual. Neurophysiol., 6^e Serie (1965), p. 223.

Verzilova, O.V. Zh. Vyssh. Nervn. Deyat., 8:437 (1958).

Vinogradova, O.S. The Orienting Reflex and Its Neurophysiological Mechanisms [in Russian], Moscow (1961).

Voeller, K., Pappas, G.D., and Purpura, D.P. Exper. Neurol., 7:107 (1963).

Voitkevich, A.A. Priroda, 2:34 (1944).

Volokhov, A.A., and Davydov, N.N. Proceedings of a Scientific Conference on Age Morphology and Physiology [in Russian], Moscow (1954), p. 49.

Von Neumann, J. In: C.E. Shannon and J. McCarthy (eds). Automata Studies, Princeton University Press, Princeton, N.J. (1956), p. 43.

Von Neumann, J. The Computer and the Brain, Yale University Press (1958), p. 82.

Voronin, L.G., and Kotlyar, B.I. Zh. Vyssh. Nervn. Deyat., 13:917 (1963).

Voronin, L.G., Kalyuzhnyi, L.V., and Zakharova, I.N. Zh. Vyssh. Nervn. Deyat., 15:364 (1965).

Vvedenskaya, I.V., Grachev, K.V., Dubikaitis, V.V., Dubikaitis, Yu.V., and Stepanov, T.S. Epilepsy [in Russian], Vol. 2, Moscow (1964), p. 108.

van der Waerden, B.L. Mathematical Statistics [Russian translation], Moscow (1960).

Walter, D.O. Exper. Neurol., 8:155 (1963).

Walter, D.O., and Adey, W.R. Exper. Neurol., 7:481 (1963).

Walter, D.O., Rhodes, J.M., and Adey, W.R. Electroenceph. Clin. Neurophysiol., 22:22 (1965).

Walter, D.O., Rhodes, J.M., Kado, R.T., and Adey, W.R. Electroenceph. Clin. Neurophysiol., 23:87 (1966).

Walter, W.G. In: Brain Mechanisms and Consciousness. A Symposium, Oxford (1954).

Walter, W.G. In: Reflexes of the Brain. International Conference to Commemorate the Centenary of Publication of I.M. Sechenov's Book of This Name [in Russian], Moscow (1965), p. 365.

Walter, W.G., and Shipton, H.W. Electroenceph. Clin. Neurophysiol., 3:281 (1951).

Walter, W.G., and Storm van Leeuwen, W. Excerpta Med., 37(98):113 (1961).

Ward, A.A., McCulloch, W.S., and Magoun, H.W. J. Neurophysiol., 11:317 (1948).

Wassner, L. Med. Monatschr., 8:530 (1954).

Watanabe, A., and Bullock, T.H. J. Gen. Physiol., 43:1031 (1960).

Watanabe, A., and Takeda, K. J. Gen. Physiol., 46:773 (1963).

Weiss, P. Proc. Nat. Acad. Sci. (Wash.), 52:1024 (1964).

Westrum, L.E., and Blackstad, T.W. J. Comp. Neurol., 119:281 (1962).

Whitteridge, D. In: E.G. Ettlinger, A.V.S. Reuck, and R. Porter (eds). Functions of the Corpus Callosum. Ciba Found. Study Group No. 20, Little, Brown and Co., Boston (1965), p. 115.

Widen, L., and Ajmone-Marsan, C. Exper. Neurol., 2:468 (1960).

Wiesel, T.N., and Hubel, D.N. J. Neurophysiol., 28:1029 (1965).

Wilson, D.M., and Davis, W.J. J. Exper. Biol., 43:193 (1965).

Wilson, P., Pecci-Saavedra, J., and Doty, R.W. Feder. Proc., 24:206 (1965).

Wolbarsht, M.L., Wagner, H.D., and MacNichol, E.F. "Receptive fields of retinal ganglion cells: extent and spectral sensitivity." In: R. Jung and H. Kornhuber (eds). The Visual System: Neurophysiology and Psychophysics, Springer, Berlin (1961) p. 170.

Yoshii, N. In: 22nd Internat. Cong. Physiol. Sci., Leiden. Abstracts, Vol. 2, (1962), p. 1088.

Yoshii, N., and Hockaday, W.J. Electroenceph. Clin. Neurophysiol., 10:487 (1958).

Yoshii, N., and Ogura, H. Med. J. Osaka Univ., 11:1 (1960).

Yoshii, N., Shimokochi, M., and Yamaguchi, Y. Med. J. Osaka Univ., 10:375 (1960).

Zenina, I.N. In: Biological Action of Electromagnetic Fields of Radiofrequencies [in Russian], Moscow (1964), p. 26.

Zhang Gin-ru. Acta Physiol. Sinica, 26:171 (1963).

Zimkina, A.M. Fiziol. Zh. SSSR, 42:372 (1956).

Zurabashvili, A.D., and Naneishvili, B.R. Simpozionul Lanturi Sicircuite Neuronale, Bucharest (1961), p. 103.

SUPPLEMENTARY BIBLIOGRAPHY

Translations of Russian articles cited, or readily available translations of closely related work by the same authors

Blinkov, S. M. and Karaseva, T. A. "Effect of unilateral lesions of auditory cortex on reaction time to auditory stimuli of varying duration," Neurosci. Trans. 1:248 (1968).

Chernigovskii, V. N. "Morphophysiologic architecture of subcortical and cortical projections of afferent vagus nerve fibers," Fed. Proc. Trans. Suppl. 24:T668 (1965).

Gerasimov, V. D., Kostyuk, P. G., and Maiskii, V. A. "Action potential production in giant neurons of mollusks," Fed. Proc. Trans. Suppl. 24 : T763 (1965).

Gerasimov, V. D., Kostyuk, P. G., and Maiskii, V. A. "Excitability of giant nerve cells of various pulmonate mollusks in sodium-free solutions," Fed. Proc. Trans. Suppl. 24 : T676 (1965).

Gerasimov, V. D., Kostyuk, P. G., and Maiskii, V. A. "The prolonged action potential of giant nerve cell," Fed. Proc. Trans. Suppl. 25 : T729 (1966).

Gerasimov, V. D., Kostyuk, P. G., and Maiskii, V. A. "Reactions of giant neurons to break of hyperpolarizing current," Fed. Proc. Trans. Suppl. 25 : T438 (1966).

Gershuni, G. V., Baru, A. V., and Karaseva, T. A. "Role of auditory cortex in discrimination of acoustic stimuli," Neurosci. Trans. 1 : 370 (1968).

Gershuni, G. V. and Zaboeva, N. V. Evaluation of functional significance of auditory system responses to exponentially increasing wideband noises and tones," Fed. Proc. Trans. Suppl. 23 : T275 (1964).

Glivenko, E. V., Korol'kova, T. A., and Kuznetsova, G. D. "Spatial correlation between cortical potentials of a rabbit during formation of a conditioned defensive reflex." Fed. Proc. Trans. Suppl. 22 : T411 (1963).

Grindel', O. M., Boldyreva, G. N., Burashnikov, E. N., and Andreevskii, V. M. "Possible application of correlation analysis to human electroencephalogram," Fed. Proc. Trans. Suppl. 24 : T753 (1965).

Gustson, P. P. "Effect of removal of visual cortex on responses in lateral geniculate body of rabbits," Fed. Proc. Trans. Suppl. 24 : T655 (1965).

Karpenko, L. D. "Correlation of structure and function in neurons of the subpharyngeal ganglion in the vine snail," Fed. Proc. Trans. Suppl. 25 : T743 (1966).

Kogan, A. B. and Zaguskin, S. L. "Correlated changes in ribonucleic acid and electrical activity for crayfish single stretch receptor neurons during excitation and inhibition," Fed. Proc. Trans. Suppl. 25 : T191 (1966).

Merkulova, O. S. and Popova, T. V. "Changes in electrical activity of the medulla during stimulation of receptors in urinary bladder," Neurosci. Trans. 1 : 81 (1967).

Meshcherskii, R. M., Fedorov, V. M., and Smirnov, G. D. "Efferent effects from visual cortex on lateral geniculate nucleus in rabbit," Fed. Proc. Trans. Suppl. 23 : T1111 (1964).

Monakhov, K. K. "Mechanism of spatial synchronization of cortical electrical activity. Fed," Proc. Trans. Suppl. 23 : T1145 (1964).

Narikashvili, S. P., Arutyunov, V. S., and Moniava, E. S. "Effect of somatosensory cortex on spontaneous activity in association cortex," Neurosci. Trans. 1:397 (1968).

Narikashvili, S. P. and Kadzhaya, D. V. "Cortical regulation of evoked responses in thalamic relay nuclei," Fed. Proc. Trans. Suppl. 23 : T252 (1964).

Narikashvili, S. P. and Moniava, E. S. "Effect of reticular stimulation on cortical responses to simultaneous peripheral stimulation of three afferent systems," Fed. Proc. Trans. Suppl. 24 : T150 (1965).

Narikashvili, S. P., Moniava, E. S., and Arutyunov, V. S, "Mechanism of the influence of the reticular formation on cortical neurons," Neurosci. Trans. 1 : 160 (1968).

Narikashvili, S. P., Timchenko, A. S., and Khadartseva, N. A. "Activity evoked in cortical association areas of cats," Neurosci. Trans. 1 : 277 (1968).

Polyanskii, V. B. "Connection between spike discharges and evoked potentials in rabbit visual cortex," Fed. Proc. Trans. Suppl. 25 : T753 (1966).

Rabin, A. G. "Selective cortical control of evoked activity in the reticular structures of the brain," Fed. Proc. Trans. Suppl. 25 : T1 (1966).

Radionova, E. A. "Electrophysiological characteristics of auditory activity in the frog medulla," Fed. Proc. Trans. Suppl. 23 : T227 (1964).

Radionova, E. A. "Differentiation of a useful signal from noise in neuronal activity of the cochlear nucleus of the cat," Fed. Proc. Trans. Suppl. 25 : T546 (1966).

Radionova, E. A. "Reactions of neurons in cochlear nucleus to acoustic signals of varying duration," Fed. Proc. Trans. Suppl. 25 : T389 (1966).

Radionova, E. A. and Popov, A. V. "Electrophysiological examination of neurons in cochlear nucleus of the cat," Fed. Proc. Trans. Suppl. 25 : T231 (1966).

Roitbak, A. I. and Oniani, T. N. "Effect of calcium and magnesium on dendritic potentials," Neurosci. Trans. 1 : 221 (1968).

Shkol'nik-Yarros, E. G. "Structure of the visual analyzer in connection with the problem of color vision," Fed. Proc. Trans. Suppl. 22 : T389 (1963).

Shuranova, Zh. P. "Slow shifts of cortical potential on stimulation of medial thalamic nuclei," Fed. Proc. Trans. Suppl. 25 : T577 (1966).

Sviderskii, V. L. "Nervous regulation of insect wing function," Fed. Proc. Trans. Suppl. 23 : T213 (1964).

Vardapetyan, G. A. "Classification of single unit responses in the auditory cortex of cats," Neurosci. Trans. 1 : 1 (1967).

Vasilevskii, N. N. "Properties of individual neurons of somatosensory cortex of adult rabbits and cats," Fed. Proc. Trans. Suppl. 25 : T391 (1966).

Vasilevskii, N. N., Aleksanyan, Z. A., and Soroko, S. I. "Activity evoked in cortical neurons and primary cutaneous afferent fibers by vibratory stimulation," Neurosci. Trans. 1 : 259 (1968).

Voronin, L. G. and Gusel'nikov, V. I. "Phylogenesis of internal mechanisms of analysis and integration in the brain," Fed. Proc. Trans. Suppl. 23 : T105 (1964).

Voronin, L. G., Kalyuzhnyi, L. V., and Zakharova, I. N. "Electroencephalographic observations on role of lateral and ventromedial nuclei of hypothalamus in formation of food conditioned reflexes," Fed. Proc. Trans. Suppl. 25 : T253 (1966).

Index